# INSTITUTIONS, SUSTAINABILITY, AND NATURAL RESOURCES

# Institutions, Sustainability, and Natural Resources

## Institutions for Sustainable Forest Management

*Edited by*

SHASHI KANT

*Faculty of Forestry, University of Toronto, Canada*

and

R. ALBERT BERRY

*Munk Centre for International Studies, University of Toronto, Canada*

 Springer

A C.I.P. Catalogue record for this book is available from the Library of Congress.

ISBN-10  1-4020-3479-2 (HB)
ISBN-13  978-1-4020-3479-4 (HB)
ISBN-10  1-4020-3519-5 (e-book)
ISBN-13  978-1-4020-3519-7 (e-book)

Published by Springer,
P.O. Box 17, 3300 AA Dordrecht, The Netherlands.

*www.springeronline.com*

*Printed on acid-free paper*

Dedicated to Hoshwati Yadav, Rachel Carson, and
Maurizio Merlo

**Companion volume:**
*Economics, Sustainability, and Natural Resources: Economics of
Sustainable Forest Management*

# CONTENTS

# FIGURES AND TABLES

**FIGURES**

**TABLES**

# ABOUT THE CONTRIBUTORS

**William Bentley** is the principal of Salmon Brook Associates in North Granby CT. He recently retired as Professor of Forest Policy and Management at the SUNY College of Environmental Science and Forestry, Syracuse NY, where he was Chair, Faculty of Forest and Natural Resources Management. He taught forest and natural resource economics at Michigan, Yale and other universities, managed Forestry Research for the Crown Zellerbach Corporation in the 1970s, worked for the Ford Foundation in New Delhi, India, and later was a senior program officer with Winrock International. He served as chair and facilitator of the Blue Ribbon Panel on America's Forest Research Policy. His paper with Richard Guildin draws on recent work by the US Forest Service to estimate US performance on the Montréal Process criteria and indicators.

**R Albert Berry** is Professor Emeritus of Economics and Research Director of the program on Latin America and the Caribbean at the University of Toronto's Center for International Studies. He holds a Ph. D. from Princeton University. His main research interests, with focus on Latin America, are labor markets and income distribution, agrarian structure, the economics of small and medium enterprise, and the economics of forests and of sustainability. Apart from his academic positions at Yale University, the University of Western Ontario and the University of Toronto, he has worked with the Ford Foundation, the Colombian Planning Commission and the World Bank. He is currently directing a research program on an employment strategy for Paraguay.

**Clark S. Binkley**, Managing Director and Chief Investment Officer, Hancock Natural Resource Group, is responsible for raising equity capital, client account management, and investment strategy and research functions. Prior to joining HNRG, Clark was Dean of the Faculty of Forestry at the University of British Columbia. He has served on the boards of directors of several publicly traded forest products companies and private timberland ventures and has consulted for numerous forest companies, governmental agencies and conservation groups. He has written more than 100 books and articles on forest economics, and is known worldwide for his research on timberland investment. Clark holds degrees in Applied Mathematics and Engineering from Harvard University and in Forestry and Environmental Studies from Yale University. From 1978-90, he was a member of the faculty at Yale University, both in the School of Forestry and in the School of Management, and in 1990 was named the Frederick K. Weyerhaeuser Professor of Forest Resource Management.

**Cherukat Chandrasekharan** holds a Ph.D. degree in Natural Resources Economics from the Michigan State University, East Lansing. He has    over 50 years of experience in tropical forestry. Before joining the FAO as a forestry expert in 1975, Chandrasekharan had served  in various capacities in India. During his 20 years of service in FAO, he has held several positions - Regional Forestry Economist of the

Asia Pacific Region at the FAO Regional Office in Bangkok, Forestry Institutions Specialist in Bangladesh, Team Leader of the FAO-World Bank Forestry Studies Project in Indonesia, Senior Forestry Policy and Planning Officer, Chief of the Non-wood Forest Products and Energy Branch. After retirement from FAO in 1995, Chandrasekharan has been actively engaged as a Consultant for ITTO, FAO, UNDP, ADB, CIFOR, Ford Foundation and IDRC. He is also a member of the Board of Governors of the Indian Institute of Forest Management, Bhopal. He has authored some 120 papers and publications.

**Graciela Chichilnisky** holds two Ph.D. degrees, Mathematics and Economics, from the University of California, Berkeley. She holds the UNESCO Chair of Mathematics and Economics and is Professor of Statistics at Columbia University. She is Director of Columbia's Center for Risk Management (CCRM) and created Columbia's Program on Information and Resources (PIR). She is a Senior Advisor to the University of Arizona Science Center. She is a recipient of the 1995 Lief Johansen Award from the University of Oslo, and has been the 1994-95 Salinbemi Professor at the University of Siena. Currently, she is Chairperson of Cross-Border Exchange, and from 1985-90, she was Chairperson and the Chief Executive Officer of FITEL She has served as advisor to the World Bank, the United Nations Economics Commission for Latin America and the Caribbean, the Organization of Economic Cooperation and Development, the United Nations and the Organization of Petroleum Exporting Countries (OPEC). She is the author of twelve books and some 190 scientific articles. She has been the editor or the member of the editorial boards of leading professional journals, including Advances in Applied Mathematics, Review of Economic Studies, Economic Letters, Journal of International Trade and Economic Development and Journal of Development Economics.

**Chimere M. Diaw** is a senior anthropologist and rural sociologist at the Centre for International Forestry Research in Yaoundé Cameroon. During his 24-year research career, he also worked for the International Institute of Tropical Agriculture, the Dakar-based Center of Oceanographic Research, and in several international research programs on Adaptive Collaborative Management of forests, Environmental Services and Rural Livelihoods, Environmental Decentralizations, Criteria and Indicators of sustainable forest management, West African fisheries and International fishing migrations. His research works include the mathematical modeling, and most comprehensive analysis-to-date, of the share system in coastal fisheries, the economic anthropology of non-wage systems and kinship, economic theory and non-conventional economic systems, community forestry, indigenous tenure systems and common property institutions, community conservation and co-management, participatory action research and interactive social methodologies. He currently coordinates CIFOR's Governance program activities in the Congo Basin and is a member of the World Commission on Protected Areas, the IUCN Commission on Environmental, Economic, and Social Policies, and the IUCN Collaborative Management Working Group.

**Alison J. Eagle** is a research associate in the Resource and Environmental Economics and Policy Analysis (REPA) research group in the Department of Economics at the University of Victoria. Her research focuses on biological and economic issues related to agriculture and forestry.

**Richard W. Guildin** is Director, Science Policy, Planning, Inventory, and Information, with the USDA Forest Service in Washington DC. He is responsible for the Forest Inventory and Analysis (FIA) program--America's forest census. He led the team that recently completed the National Report on Sustainable Forests-2003. This report was the first to describe the condition, health, and productivity of U.S. forests in terms of the 7 criteria and 67 indicators endorsed by the 13 Montreal Process countries for reporting on the sustainability of temperate and boreal forests. Among his other responsibilities are seeking new ways to bring the latest science to policy makers to better inform decision-making, strategic planning and budgeting for the agency's research and development program, and developing information management policies that safeguard long-term research data bases and assure that research results are based on high-quality information.

**William F. Hyde** is a forest economist and policy analyst who taught at Duke University and Virginia Tech and served as a senior researcher with Resources for the Future and the US government's Economic Research Service.  He has participated on the editorial boards of seven international economics and forestry research journals, including a term as editor of Forest Science, the leading international journal of forestry research. His research covers a range of forest policy and management issues: timber supply; public forest management and government regulation of private forestry; fuelwood, recreation, endangered species management; community forestry and community impacts of forestry; agroforestry; and investments in forest research.  His experience includes technical observations from, and research participation in, over thirty countries.  He has published more than 100 professional papers and books, including, most recently, *China's Forests: Global Lessons from Market Reforms* (co-authored and co-edited with Brian Belcher and Jintao Xu), Resources for the Future, 2003.

**John Hartwick** has taught at Queen's, Kingston, Ontario, since 1969. He has published in leading economics journals on urban structure and on exhaustible resources. In the 1970's he became interested in the idea of compensating the next generation for resource stock depletion by bequeathing them more produced capital, equal to the value of the current natural resource draw-down. A more general treatment of this issue is forthcoming in the Canadian Journal of Economics, with Kirk Hamilton. A recent pre-occupation has been the economic nature of sustained motion in classical mechanics. Capital theory turns out to be well-suited to this analysis. His work on deforestation has focused on the way a "new country" adjusts its land-use in the face of new external prices for its primary product exports.

**Shashi Kant** is Associate Professor of Forest Resource Economics and Management at the Faculty of Forestry, University of Toronto, Canada; he completed his Ph. D. in

Forest Resource Economics at the University of Toronto in 1996. He is a recipient of Premier's Research Excellence Award for his research on economics of sustainable forest management. His research interests include market as well as non-market signals of resource scarcity, institutional and evolutionary aspects of economics of sustainable forest management,.quantum theory and behavioural economics, game theoretic and agent-based models, social choice theory and forest management. He is an Associate Editor of Journal of Forest Economics, and Canadian Journal of Forest Research. He has published more than fifty refereed papers and has worked as a consultant to the Food and Agriculture Organization (FAO), United Nations Environment Programme (UNEP), Swedish International Development Agency (SIDA), International Network for Bamboo and Rattan (INBAR), and Ontario Ministry of Natural Resources (OMNR).

**M.K. (Marty) Luckert** is Professor of Forest and Natural Resources, Economics and Policy, in the Department of Rural Economy, University of Alberta. He has broad experience investigating property rights to natural resources in Canada and abroad. His research has assessed how property rights influence incentives for firms to manage natural resources, and whether these incentives further the pursuit of policy objectives. Much of his work has dealt with Canadian forest tenures and subsistence economies in southern Africa.

**Harry Nelson** is a Senior Research Associate at the Forest Economics and Policy Analysis Unit of the University of British Columbia. Dr. Nelson has written about and conducted research on the interaction between international trade and forest management, as well as other topics such as the nature of competition in wood product markets, the Canada-US softwood lumber dispute, and the economic impacts of pollution upon the Canadian forest sector. He has also prepared studies for forest product companies, First Nations, the Federal government, and provincial governments, and has degrees from the Kennedy School and the University of British Columbia.

**Roger A. Sedjo** is a Senior Fellow and the Director of the Forest Economics and Policy Program at Resources for the Future (RFF), and the President of the Environmental Literacy Council (ELC). He has authored or edited fourteen books related to forestry, natural resources and the environment. His work focuses on both commercial timber industry issues such as timber supply, forest plantation and biotechnology in forestry, as well as on the conservation and environmental aspects of forests. He edited a recent book, *A Vision for the US Forest Service* (2000), on Forest Service issues and future potentials. Other recent books include *Sustainability in Temperate Forests (1998)*, with A. Goetzl and S.O. Moffat, and the *Economics of Forestry.* Other important books on the forestry and timber supply include *The Long-Term Adequacy of World Timber Supply* (1990) and *The Comparative Economics of Plantation Forestry* (1983).

**G. Cornelis van Kooten** is a Professor of Economics and Canada Research Chair in Environmental Studies and Climate in the Department of Economics at the

University of Victoria in Canada. He also holds a courtesy appointment in the Agricultural and Rural Policy Group at Wageningen University in the Netherlands. He has published more than 100 refereed journal articles in agricultural, forest and natural resource economics, and eight books, including *Climate Change Economics* (2004), *Forest and Land Economics* (2004) with Henk Folmer, and *The Economics of Nature* (2000) with Erwin Bulte.

**Arild Vatn** is a Professor in Environmental Sciences at the Department of Economics and Resource management, Agricultural University of Norway (AUN). His research is focused on institutional aspects of environmental resource management. His theoretical work has centered on issues concerning human motivation, the formulation of policy measures, choice of resource regimes and environmental valuation. His empirical work concerns analyses of different resource problems ranging from agricultural pollution to biodiversity protection to various regime issues. He has been head of several interdisciplinary research programs both at AUN and the Norwegian Research Council. Presently he is leading the program 'Environment and food production' at AUN. He is also leading a research team working on biodiversity protection in forests. He is vice president of the European Association of Ecological Economics. He has authored approximately 100 papers, books and scientific reports. He has just finished a book titled *Institutions and the Environment, which will be published by Edward Elgar in 2005.*

**Ilan Vertinsky:** Dr. Ilan Vertinsky is the Director of the Forest Economics and Policy Analysis (FEPA) Research Unit. He is also the Director of the Center for International Business Studies and the W. Maurice Young Entrepreneurship and Venture Capital Research center at the University of British Columbia. His major research interests are: international trade and competitiveness issues, innovation and forest sector business strategies. Currently, in collaboration with a team of researchers from across Canada, he is investigating new designs of tenure systems and their potential impact on sustainable forest development. Ilan has published more than 230 journal articles, book chapters, monographs and books on diverse topics including economics, operation research and animal ecology.

# PREFACE AND ACKNOWLEDGMENTS

In the short-term human beliefs and values are heavily influenced by existing social, cultural, economic, and environmental conditions, while in the long-term these conditions are in turn influenced by human behavior. These continuous interactions underlie the dynamic nature of human beliefs and values, as well as the surrounding social, cultural, economic, and environmental conditions. The increasing support for and dedication to sustainable forest management (SFM) reflects an evolution in the human value system, which in turn reflects the social, cultural, economic, and environmental conditions of the late twentieth and early twenty-first century, conditions which are quite different from those of the nineteenth and early twentieth century. The economic principles, theory, and models of SFM need to reflect the realities of the twenty-first century.

The concept of SFM incorporates human preferences for timber and non-timber products, preferences for marketed as well as non-marketed products and services, the preferences of industrial as well non-industrial agents, including Aboriginal and other local people, and the preferences of future generations as well as the present one. It takes account of diversity of preferences across agents, communities, time, and generations, and incorporates preferences that are revealed through the market as well as through non-market mechanisms. Forests, in the context of SFM, are valuable for their contributions to ecosystem functioning as well as their physical outputs. However, the existing paradigm of forest economics, which is focused on sustained yield timber management and has its roots in the conventional neoclassical paradigm of economics, is based on the combination of utility maximizing rational agents and the 'invisible hand' leading to an efficient general equilibrium. In this framework, peoples' preferences are internally consistent, static and revealed through the market only; public inputs are selected on the basis of market signals; all systems, including ecosystems, can be commoditized, which converts them into functionally-disjointed and discrete units; and there are no commitments and moral judgments attached to the domains of forest values. It is evident that the basic premises of the existing paradigm of forest economics are in serious contradiction of the realities and expectations of SFM, and the economics of SFM will thus require an extension of the boundaries of forest economics.

Keeping the unique features of SFM and the need to extend the boundaries of forest economics in perspective, Shashi Kant published, "Extending the boundaries of forest economics" in Volume 5 (2003) of Forest Policy and Economics. Response to the publication of this article revealed that there were many other forest and resource economists who shared our vision of extending the boundaries of forest economics. We then planned an International Conference on the Economics of Sustainable Management, at the University of Toronto, on May 22-24, 2003, but due to the outbreak of Severe Acute Respiratory Syndrome (SARS) in Toronto, the conference had to be rescheduled to May 20-22, 2004. In fact, the SARS outbreak was a good example and a reminder to economists of natural uncertainties.

We are pleased to announce that this volume is the second of the new series "Sustainability, Economics and Natural Resources". The papers in this volume and its companion "Economics, Sustainability, and Natural Resources: Economics of

Sustainable Forest Management" were originally presented at the conference. (In addition, a special edition of Forest Policy and Economics, Volume 6, Issues 3-4, also includes papers from the conference.) The volume is not a mere re-printing of conference papers, however. The original selection of papers and the rewriting, and reworking of them after the conference have been designed to cover the institutional issues related to SFM in an integrated and reasonably comprehensive way. We are thankful to the authors for responding positively to our suggestions.

In this volume leading  institutional economists discuss appropriate institutions for sustainable forest management, markets for environmental services, deforestation and specialization, and some country experiences related to institutions for carbon emissions and sequestration (Kyoto Protocol), international trade, biodiversity conservation, and sustainable forest management in general.  The companion volume mentioned above focuses on selected key aspects of the economics of SFM, including complexity, ethical issues, consumer choice theory, intergenerational equity, non-convexities, and multiple equilibria.

The conference was organised by the Faculty of Forestry, University of Toronto in collaboration with the Groups 4.04.02 and 4.13.00 of the International Union of Forestry Research Organizations (IUFRO). We are thankful to the late Prof. Maurizio Merlo and to Prof. Hans A. Joebstl, Group Leaders of IUFRO Groups, for their support.

The conference was made possible through the financial and overall support of the Canadian Forest Service, Ford Foundation, Forest Products Association of Canada, International Paper, Living Legacy Trust, Ontario Ministry of Natural Resources, Sustainable Forest Management Network, and Weyerhaeuser Canada. Along with our thanks to these organisations we would like to specifically recognize the contributions of – Gordon Miller, Jeffrey Campbell, Michael L. Willick, Paul K. Perkins, Sharon G. Haines, Karan Aquino, and Mark Hubert – who supported us throughout the period of about two years. We also express our thanks to Prof. Robert J. Birgeneau, Ex-President, University of Toronto, Prof. Rorke Bryan, Dean, Faculty of Forestry, University of Toronto, and Brian Emmett, Assistant Deputy Minister, Canadian Forest Service for their support and participation in the conference.

Special thanks are due to Amalia Veneziano and Sushil Kumar who were instrumental in the organization of the conference, with the assistance of other students and staff members of the faculty.

Finally, we would like to thank Springer Publishers and their staff members, specifically Paul Ross, Henny Hoogervorst, and Esther Verdries for taking up this project and Sushil Kumar who did a great job turning the manuscript into camera ready copy.

Shashi Kant
R. Albert Berry

# CHAPTER 1

# SUSTAINABILITY, INSTITUTIONS, AND FOREST MANAGEMENT

SHASHI KANT

*Faculty of Forestry, University of Toronto*
*33 Willcocks Street, Toronto, Canada M5S 3B3*
*Email: shashi.kant@utoronto.ca*

R. ALBERT BERRY

*Munk Centre for International Studies, University of Toronto*
*1DevonshirePlace, Toronto, Canada M5S 3K7*
*Email: berry2@chass.utoronto.ca*

**Abstract.** This chapter provides an overview of the contents of this volume. To put the contents in perspective, first the developments related to the concept of sustainable development and sustainable forest management (SFM), institutions, institutional economics, and their importance to SFM are discussed. Next, the relevance of markets and other institutions to sustainable forest management is discussed. Finally, an overview of each chapter included in the five parts of this volume is provided.

## 1. INTRODUCTION

The word "sustainable" is not new to the forestry profession, including forest economists, but the dynamics of societal values, specifically those related to forest resources and environment, have added new dimensions to thinking about sustainability of forest resources and forest management. The recent concerns about sustainability, signaled by the publication of 'The Limits to Growth' by Meadows et al. (1972) and 'Our Common Future' by WECD (1987), are not limited to a specific product or resource but include all natural systems and human life. The roots of the concept of sustainable forest management (SFM) can be found in these two publications, but it derived impetus from several global events, including the 1992 United Nations Conference on Environment and Development (UNCED) held in Rio de Janeiro, Brazil, the Intergovernmental Panel on Forests (1995-97), the Intergovernmental Forum on Forests (1997-2000), the United Nations Forum on Forests that came into existence in 2001, and the Johannesburg Summit in 2002. In

*Kant and Berry (Eds.), Institutions, Sustainability, and Natural Resources: Institutions for Sustainable Forest Management, 1-20.*
© *2005 Springer. Printed in Netherlands.*

terms of human activities and outcomes, concerns with global warming, declining energy sources, environmental pollution, biodiversity loss, and deforestation and degradation of forest resources raised global concerns about the sustainability of natural and human systems. The role of forest resources and sustainable forest management in sustainable development (or sustainability of society) can be gauged from the fact that forest resources are a critical component of most of the sustainability related international agreements, such as the Biodiversity Convention, the Kyoto Protocol, and the Agenda 21.

The parameters of SFM, after the Rio Summit, gained further clarification through the criteria and indicators initiatives such as the Montreal Process and the Helsinki Process, as well as through various forest certification schemes. In general, SFM refers to the ways and processes of managing forest resources to meet society's varied (social, economic, and ecological) needs, today and tomorrow, without compromising the ecological capacity and the renewal potential of the forest resource base. In economic terms, the main distinguishing features of SFM are the recognition of diverse and dynamic preferences of local people and other stakeholders, the incorporation of multiple sources of value and utility from the forests (including non-market values), the incorporation of multiple products and services in the production process, public participation in management decisions through non-market mechanisms, inter-generational equity, and a systems approach to forest management. Hence, SFM involves a complex matrix of interactions between social, economic and natural systems, and the resulting outcomes. In short, the transformation of forest management from sustained yield timber management (SYTM) to sustainable forest management is equivalent to the change in natural resource management from a "reductionist-mechanistic" or Newtonian approach to a "holistic-evolutionary" or Post-Newtonian approach. .

In the context of the global recognition of the concept of sustainability and the global goal of sustainable development, the challenge to the current and future generations of economists is to build a new economic paradigm—based on a more organic, holistic, and integrative approach than the reductionist neo-classical paradigm. The concept of sustainability offers a challenge to economists to bring the profession closer to the real world. As Einstein once observed, problems cannot be solved at the same level of thinking that lead to their creation (Ikerd 1997). Hence, the economic theory of sustainability and the economics of sustainable forest management cannot be based on the reductionist approach of neo-classical economics that has contributed to many problems related to sustainability[1], and a new economic theory, rather than a new public policy based on old theory, will be needed to guide humanity toward sustainability or sustainable development.

In the companion to this volume—Economics, Sustainability, and Natural Resources: Economics of Sustainable Forest Management—leading economists from behavioral economics, complexity theory, forest resource economics, post-Keynesian economics, and social choice theory discuss selected specific aspects of the economics of SFM, such as complexity of economic systems, ethical issues, consumer choice theory, intergenerational equity, non-convexities, and multiple equilibria. Institutional aspects are, broadly speaking, another critical dimension of sustainability and SFM; some scholars, such as Spangenberg (2002), identify institutions as a fourth dimension of sustainability, along with the social, economic, and environmental sides, while others, such as Opschoor and van der Straaten (1993), have used an institutional approach to suggest a new non-neoclassical economic framework for sustainable development. In fact, such institutional economists as Veblen, Kapp, and Myrdal, now thought of as members of the "old institutional economics"[2] stream, have been among the strongest critics of neo-classical economic theory. This volume focuses on institutions, institutional economics and sustainable forest management.

Institutional economists have found the classificatory distinctions between economic and non-economic factors and between economic and social processes to be unacceptable (Kapp, 1976). For them economics is much more than the study of a particular form of behavior—that of the "rational" economic agent—and of the determination of general equilibria, as is evident from the definition of economics provided by Gruchy:

> Economics is concerned with "the study of structure and functioning of the evolving field of human relations which is concerned with the provision of material goods and services for the satisfaction of human wants [...] it is the study of the changing patterns of cultural relations which deals with the creation and disposal of scarce material goods and services by individuals and groups." (Gruchy, 1947, pp. 550-552)

The key differences between neoclassical economics and institutional economics can be summarized in the words of Söderbaum:

> "Thus the neoclassical economist tends to believe in very clear boundaries between economics and other disciplines and in the possibility of giving useful advice on the basis of highly specialized knowledge .... Institutional economists on the other hand emphasize a holistic or inclusionist (as opposed to exclusionist) approach to economics and policymaking. Specialization and division of labour is seen not only as a positive possibility, but also as a danger.. .. Equilibrium theory has been mentioned as an example of the mechanistic tendencies of neoclassical economics. Institutional economists in turn have a preference for evolutionary thinking. "Patterns modelling" (Wilber & Harrison, 1978) is a characterisation of this interest in how technology, institutions, habits, values and the economy at large evolve through time (cf. also Norgaard, 1985). Where neoclassical economists use models that are closed in a mathematical sense, institutionalists prefer models which in the same sense are open-ended or only partially closed (Myrdal, 1978). (Söderbaum, 1992, pp.131-132)

The main problem of neo-classical economics, at least with respect to sustainability, is the common aphorism "economists know the price of everything and the value of nothing"[3], and the direct evidence of this claim is the exclusion of non-priced natural resources, including environment benefits, from neo-classical production analysis. However, as mentioned earlier, sustainable forest management and sustainable development involve not only priced goods and services but also values far beyond the reach of market mechanisms. In such circumstances, the neoclassical weighing premise, based on prices, has to be replaced by a hierarchical approach based on values. In this hierarchy, there may be some ultimate values, ones which cannot be substituted and which therefore can neither be protected nor managed through market mechanisms; management decisions around such values should not be based on an aggregation of individual preferences for them. The starting point for sustainable forest management (and sustainability) has to be the value society places on forest (and all other natural) resources. In this context, markets and market prices can constitute one of the many institutional arrangements to support and strengthen a broader set of institutions for sustainable forest management. Hence, though the discussion of the institutional aspects of sustainable forest management has to go well beyond markets and prices, it does include them.

This volume's authors articulate many of the institutional aspects of SFM. This introductory chapter provides an overview of the contents, after a brief discussion of markets, institutions and sustainable forest management designed to provide some perspective.

## 2. SUSTAINABLE FOREST MANAGEMENT, MARKETS, AND INSTITUTIONS

Sustainable forest management, just as sustainability, must be defined in its broadest sense. People will always disagree as to exactly how much should be preserved for future generations, about the legacy of resources, technologies, and aesthetics each generation should leave for the next. But, broadly speaking, the idea of sustainability implies that the legacy passed on should be adequate or acceptable. Most people would accept that decisions bearing on the future will often involve tradeoffs. Saving some species may have high costs in terms of foregone human consumption; saving some may come at the expense of saving others, but most people would also want to avoid a future world without forests, clean air or water, irrespective of the cost this might impose on the present generation. One of the tasks of the physical and social sciences, especially of economics, is to distinguish the items/values which are beyond trade-offs —ultimate value or "merit-goods" (James, Janesn, & Opschoor, 1978; Hueting, 1980)—and to clarify the terms of trade-offs among the goods which can be substituted for each other. In this task "merit goods" are beyond the boundaries of markets, and other institutions must play the key role in the decisions involving these goods or attributes of natural resources. Among those goods which are substitutable, some may not be subject to market transactions, due to the non-existence of markets for them[4], in which case their valuation and hence decisions regarding their tradeoffs with marketable goods and services will also require the support of some other institutions. Markets are useful in establishing appropriate rates of trade-off among the goods, those which can be effectively traded in the market. Hence, it is important to have an understanding of how the relevant markets function and what instruments governments can bring to bear to make them work better.

Markets are a social construct and operate in the context of a set of institutions, which greatly affect how well the market mechanisms serve human needs. To work well, markets must be well designed; they do not simply and automatically appear in their optimal form. As has been commonly noted, although perfectly competitive markets can often be shown to be the most socially efficient variant, most of the participants in such markets would prefer a different variant, one in which they have some monopoly or monopsony power, and will when possible bend their efforts towards the achievement of that socially inferior variant. Adam Smith's "invisible hand" which directs the functioning of markets towards a good social outcome may be invisible when working well, but is by no means automatically present and is usually under attack from very visible hands. The role of the state and the institutions it supports or constrains is thus important to good market functioning.

Institutions and policies interact with markets in at least three important ways. "Good markets", in the sense of ones which are socially beneficial sometimes owe part of their effectiveness to conscious design, always owe some of it to surrounding social institutions[5], and often require regulation to prevent some form of malfunction (such as abuse of monopoly power). Thus there is limited meaning to the frequently discussed question of how well markets work, since the answer must by definition be couched in terms of what the surrounding institutional framework is like, how effective government regulation is, and so on. This is not to imply that comparison of markets vs. governments as resource allocators has no meaning, since markets can have a greater or smaller role in the process; but it does imply that in its simplest form (which disassociates markets from the surrounding institutions) the question is indeed meaningless.

Discussion of the role of markets in any resource allocation process, including those surrounding forests, can be conveniently divided into two parts. First, what sort of institutions, regulations and support from public policy can in principle raise the performance of markets in doing what they can do relatively well? Second, in what respects or circumstances can markets not be expected, even under the best of circumstances, to work well in the sense of promoting social welfare. Public policy in the first area is designed to make markets work better, in the sense of helping them to achieve the best that can be expected of them. For example, a set of regulatory actions may assure that a given market approximates the ideal of pure competition, by blocking actions which would make entry more expensive. In the second area policy is designed to replace, completely or partially the function of the market as resource allocator. For example, it is recognized that markets cannot be relied on for decisions on the level of output of public goods or for the allocation of "merit-goods". The dividing line between these two types of state involvement is somewhat fuzzy. In any case, the role of the state and other social institutions is critical in both the situations.

What is the nature of the choices that need to be made as to the optimal roles of markets and of governments or other resource-allocating devices in a country? In broad terms the alternatives involve any number of combinations of market roles or functions with government (or other collective) roles and functions. For allocation to occur effectively in a given domain, say that of agriculture, requires that whatever markets do perform allocative functions be adequately supported by informal or formal (often state) institutions. Thus a particular type of market, which only works well when provided with some specific type of support (say to avoid theft) will not work well unless that support is available. Broadly speaking, there are upper limits to how well both markets and collective action can contribute to the effective functioning of an economy. Where, say, the conditions which make markets work well are scarce (e.g. a high level of information by buyers as to what they are getting), the function which that market might in principle perform well were the information conditions satisfied may be better carried out directly through collective action (government being a prominent form of this). Alternatively where effective collective action is very scare (because people do not trust each other or are excessively corrupt or rent-seeking) then those market-collective action partnerships which require effective collective action may be impossible or may be best left to relatively unsupported markets, even though these function poorly (everything being relative). As economic theory makes clear, markets supported only by weak institutions cannot be expected to come even close to constituting a perfect management system for an economy.

For each of the various challenges which people have in mind when they use the term "sustainable forest management" one may ask the question, "What combination of institutions, other than markets, (including prominently those of collective action) and markets may best respond to the challenge? A useful starting point is a list of the major challenges and of the combinations of institutions and markets which may be relevant to each of them. At least six major challenges are currently receiving priority attention: (i) global warming and the role of forest resources in that process; (ii) loss of biodiversity and the role of forest resources in that process; (iii) the recognition of Aboriginal and other local groups' rights on forests and incorporation of their values into forest management systems; (iv) the threat of an energy scarcity and the possible contribution of forests in addressing it; (v) possible scarcity of wood and other forest products for non-energy uses; and (vi) poverty and the extent to which forest use or misuse contributes to it, especially for the half billion or so people who derive considerable shares of their income from forest products.

How well various market and other institutions work cannot be summarized in any way which is both brief and satisfactory. One useful starting point, familiar to many students of economics, focuses on the conditions under which markets can achieve a sort of social optimum; this invites consideration of how those conditions can best be satisfied. The remaining weaknesses or incapabilities which exist even when the markets are functioning as well as can possibly be expected must be dealt with in some other way. Traditional Western economic theory focuses heavily on the merits of the perfectly competitive market, which allocates goods efficiently between suppliers and demanders (producers and users). For the "perfectly efficient" outcome which theory describes to come to pass requires (i) many buyers and sellers (to assure that the market is competitive enough, i.e. not subject to monopoly distortions); (ii) perfect knowledge and foresight on the part of all buyers and sellers; (iii) no externalities whose effects cannot be "marketized"; and (iv) no public goods (i.e. goods which are non-rival and non-exclusive). In simple textbook theory, the first limitation calls for regulation to prevent or control monopoly behavior; the second calls for regulations on accuracy in advertising and attempts by the state to improve the quality of information about goods and services sold in markets; the third calls for taxes or subsidies to offset the effects of such externalities; the fourth requires the state to make the allocative decision about such public goods since theory implies that markets cannot do so effectively, even if supported by the best institutions imaginable.

But even after responsibility for public goods is given to the state and the best supports, controls and regulations which real world governments can design and implement to make markets work well are in place, several major weaknesses typically remain, creating difficult challenges and forcing difficult decisions.

## 2.1 Markets Support Resourceful and Powerful People

Markets which are efficient in the narrow neoclassical sense weight the welfare of each person in proportion to his/her purchasing power, and allocate goods and services to people roughly in accordance with their initial wealth. In any society in which initial assets are very unequally distributed, "perfectly functioning" markets will thus reproduce that inequality; the economic system will give 100 times more weight to the welfare of one person than to that of someone else with only 1/100th as much wealth. The morality of such a system is obviously open to question. In some societies the state undertakes considerable redistributive activity, while in others it does not. At the world level, where the richest decile of people has about 65 times as much purchasing power as the poorest decile (Berry & Serieux, 2004), only a minuscule amount of redistribution in favour of the poor occurs. Both within countries and between them, the better off are able to use many instruments of power to maintain or even enhance their relative position. The institutions which surround markets and the governments which have the responsibility of making markets work well and fairly are all vulnerable to the self-serving tactics of the rich, from the use of legal systems in unfair ways to the use of informal pressure, the taking advantage of superior information, and finally the use of power to access valuable assets.

## 2.2 Human Preferences, Competition, and Jealousy

Human preferences may be such as to make human satisfaction difficult to achieve. Whether this is due to deep cultural characteristics or to the deliberate manipulation of economic agents for their own benefit, it can be problematic. Thus, when each individual in a society can be satisfied only by abusing others, demonstrating superiority to others or by other actions which involve "zero-sum games", then the task of making everyone happy becomes impossible. Businesses often promote jealousy ("keeping up with the Joneses") in order to increase their sales. State manipulation of preferences is a tricky ethical subject, but all states inevitably engage in such activities up to a point (if only in trying to dissuade people from the view that it is appropriate to abuse others in certain ways). How far the state or the collective should go in discouraging preference creation which has a zero or negative sum feature to it is an important social question.

## 2.3 Market Power as a Source of Economic Inefficiency and Inequality

Market power remains an important source of both economic inefficiency and inequality in nearly all countries of the world. It is often commingled with political and social power related to income and wealth inequality. It often takes the form of large enterprises, owned by wealthy people, taking unfair advantage of smaller firms owned and operated by lower-income people.

*2.4 Incomplete and Asymmetric Information as a Source of Inefficiency and a Limitation to Sustainability*

Incomplete and asymmetrical information contributes to market inefficiency even in the absence of inequalities of wealth and political and social power; in the presence of those patterns its impact is the more perverse. One prominent sort of inefficiency occurs as businesses try to mislead potential buyers with respect to the qualities of their products.

Incomplete information, especially about the future, leads to dynamically unstable and unpredictable paths of economies. Instability and the related path dependency detracts from the capacity of economies to achieve optimal outcomes at each point in time, and calls for state action to stabilize, to steer the economy towards superior equilibria and away from inferior ones. Even when markets are relatively adequate in allocating resources effectively at a point of time, or over a short period of time, they are generally much less efficient in allocation across long periods of time. The particular inefficiency of futures markets constitutes a limitation to conservation and sustainability, both in general and in the forest domain. Disagreements on how future and present values should be compared show up in debates on the appropriate discount rate which should be applied to future production, a matter of special importance in the context of forestry, where the growing period is long.

The papers included in this volume discuss important aspects of the institutional dimension of SFM. The volume starts with various theoretical perspectives on institutions for sustainable forest management and closes with the integration of the thirteen chapters by highlighting the linkages between institutions and the basic principles of the economics of SFM. In between, three other major themes—markets and SFM, deforestation, specialization and SFM, and country-specific institutional experiences—appear in the volume. .

## 3. THEORETICAL PERSPECTIVES ON INSTITUTIONS FOR SUSTAINABLE FOREST MANAGEMENT

Within the category of sustainable forest management issues, some are essentially technical ones and permit reasonably general answers which in turn allow one to proceed to a consideration of the sort of institutions best able to implement a clearly identifiable strategy. Others are not technical in the same sense, and thus have no general answer identifiable in technical terms; where, for example, societal values matter, some institutions will tend to produce "better" decisions than others. It is thus important to think about appropriate institutional design, appropriate decision-making systems and the like. What such institutions may be and how well any of them are likely to function depends on the level of knowledge and understanding of the role of forests vis a vis the challenges noted above. In this section, four chapters examine the theoretical aspects of institutions for SFM.

In Chapter 2, Luckert focuses on difficulties associated with the definitions and descriptions of institutions and the complexities of linkages/interactions between institutions and behavior of economic agents, and highlights the difficulties in identifying or even defining optimal institutions for SFM. He argues that due to these complexities it is very difficult to pin down the links between institutions and economic processes, and it further complicates the understanding of such links as do the sometimes ambiguous social objectives of SFM. He argues that even if we stick to the simplistic version of institutions which focuses only on the rules related to property rights, the combination of rules and situations covers a dauntingly wide range. The author emphasizes that more effective linking of institutions to economic behavior in pursuit of social objectives will require: (i) refinements in the understanding and characterization of institutions; (ii) refinements in the understanding of non-institutional determinants of behavior (such as socio-economic characteristics of firms and their time and risk preferences); (iii) a wider recognition of potential co-dependence (as opposed to cause and effect relationships) between institutions and economic behavior; (iv) more explicit recognition of transactions costs and belief systems; and (v) clearer specifications about what we want sustainable forest management to achieve.

Luckert concludes that although the tradition of theoretical abstraction and mathematical expression has contributed to our understanding of the impact of incentives and institutions on behavior and outcomes, we need to weigh the benefits and costs of such reductionism. He argues for a cross-disciplinary approach and more holistic thinking, and warns economists that without this balance their analysis may give answers which are precisely wrong, or precisely irrelevant. Readers will find many similarities between Luckert's arguments about complexity, the need for a cross-disciplinary approach, and holistic thinking and the chapters by Colander (2005), Kant (2005), and Khan (2005) in the companion volume—Kant and Berry (2005).

In Chapter 3, Diaw elaborates on the complexity of institutions and the weaknesses of the neo-classical or Western view that the optimal tenure system everywhere will be built around individual rights (though of course such a system applies much less to forest land than to agricultural and urban land even in Western countries). Diaw argues that customary tenure (mingled with state law and occasional private titling) continues to predominate on African rural and forest lands, in spite of prediction of its demise by evolutionist theories and the destructive attempts by colonial and post-colonial policies. He develops an anthropological conceptualization of embedded tenures, with examples from Africa and various parts of the world, and highlights the factors that account for the flexibility, adaptability and resilience of this type of institutional (tenure) system. He argues that embedded tenure has been able to cope with economic stress and hostile policies because of the unique way in which it nests private entitlements into the commons, and into collective property and long-lasting social institutions.

Diaw emphasises that the reductionist economic (neo-classical) interpretation of non-market systems, including kinship, common property and non-wage systems, have contributed to a dismissive attitude towards customary tenure systems, which tend to be complex and embedded in other social institutions, and to relegate their most innovative aspects to the limbo of "imperfect markets." He also highlights the failure of the Common Property Regime (CPR) literature, which focuses on the crafting of institutions, to duly recognize the theoretical and policy implications of the nesting of appropriation regimes in embedded tenure systems. The author enumerates the policy mistakes that can derive from such reductionism and concludes that, to rescue African forest policies from the mistakes (with attendant social costs) of the past, it is necessary to take due account of the complexity and validity of embedded tenure institutions and their demonstrated ability to adapt to legal pluralism and commodity markets.

In Chapter 4, Kant and Berry extend the arguments of the Luckert and Diaw into the area of institutional dynamics, arguing that neither the distinction between private and state regimes nor the price-dependent dynamics of institutions can adequately explain institutional dynamics. They also argue that organizational factors, along with institutional factors, play an important role in how institutions evolve. They propose a framework for institutional analysis which takes account both of factors internal to the institutions and organizations and of the external setting – the social, environmental, economic (including markets) and international factors, and suggest adaptive efficiency as a more helpful indicator than allocative efficiency in the analysis of institutional changes that are path-dependent rather than simply price or market-dependent.

The authors apply this framework to analyse the evolution of Indian forest regimes, finding that institutional evolution have been incremental and path-dependent, with the exception of the sudden shift from the dominance of community regimes in the pre-British period to that of state regimes in the British period. The dominant causal factors in this pattern of incremental change have varied markedly over time: institutional inertia of informal institutions in pre-colonial India; "organisational energy" during colonial period; self-reinforcing mechanisms at the level of the Legislative Wing and "organisational inertia" of the Executive Wing of the government during the first thirty years after independence, and "organisational energy" of the Legislative Wing, the external setting, and "organisational surges" of the Executive Wing during the recent periods. Prices and market factors were not the dominant determinants of change in Indian forest regimes. The authors conclude that the concept of property rights, as applied in neoclassical economics, is not sufficiently subtle to explain the success or failure of forest regimes, and that prescriptions for sustainable forest management should address institutional and organizational aspects in an integrative manner. This chapter contributes another dimension—organizations and organizational inertia—to the broad institutional story.

In the last chapter of this section, Chapter 5, Vatn extends the discussion to a specific category of institutions—value articulating institutions, and discusses a set of issues involving the evaluation of biodiversity in the context of forest ecosystems. He argues that choosing evaluative instruments implies choosing between different perceptions both of the good and of the (potential) rationalities involved.

Vatn argues that cost-benefit analysis, with its focus on monetary evaluation/contingent valuation, fails to treat the issues involved in a way consistent with the characteristics of the good and the ethical concerns involved, and makes a case for the use of deliberative value articulating institutions for the valuation of biodiversity. He accepts that there are important differences across the range of deliberative institutions, but emphasizes that these institutions generally offer a better response to the problems involved, such as cognitive limitations (where the potential for communication between citizens and experts is pivotal) and normative issues (the process by which we develop an understanding of the ethical issues and dilemmas involved). The author supports the view that these institutions offer possibilities for learning about and handling competing or incommensurable perspectives, and ways to handle issues where radical uncertainty is involved, by providing the necessary opportunity to resolve the relevant cognitive and normative issues in a reasoned way. He concludes that only deliberative value-articulating institutions can offer biodiversity valuation that is context-consistent with the type of cognitive and normative issues involved.

## 4. MARKETS AND SUSTAINABLE FOREST MANAGEMENT

The chapters of Section Two focus on markets related to environmental services, specifically the carbon sequestration of forest ecosystems. The common message of the section is that the creation of markets for environmental services is necessary but not sufficient for sustainable forest management. It thus reinforces the point made in section 2 of this chapter, that markets are only one category of institutions and cannot work efficiently in the absence of other supporting institutions.

The opening chapter by Binkley emphasizes the need for environmental services markets, and observes that much work remains to be done in designing the details of those markets. He identifies four main features of the forestry sector: the capital intensity of forest management; the material value of environmental services from forests; an increasing national and international emphasis on using markets to secure the material value of these services for the society; and finally, the possibility of capitalizing the value of environmental services into investment decisions. He observes that the evidence seems to be consistent with the first three of these, but, regrettably, not the fourth.

Binkley argues that the enthusiasm for markets for environmental services has so far not been matched by the reality. Economists, in espousing markets for environmental services, commonly focus on the misallocations associated with the absence of markets, but forget the transaction costs associated with the creation and effective functioning of markets. He identifies three kinds of transaction costs associated with markets for environmental services: political cost (associated with the negative reaction of the losers from environmental regulations), measurement cost (the cost of measuring environmental services), and actual financial transactions costs associated with the designing and developing market instruments such as licenses to deal in financial derivatives, and trading mechanisms for carbon credits. The author observes that in the face of the obvious problems and the clear economic prescriptions for solving them, economists commonly imagine policy makers to be stupid or venal because they do not jump to adopt market-based mechanisms; in his experience though policy makers may be venal they are rarely stupid. The author concludes with a note of optimism for markets of environmental services citing as example the New South Wales Greenhouse Abatement Certificates (NSWGAC) system in Australia.

In Chapter 7, Chichilnisky adds a critical dimension—equity—to the markets for environmental services which are public goods. She argues that the origin of today's global environmental problems is a historic difference in property rights regimes between industrial and developing countries, the North and the South. In developing countries, ill-defined and weakly enforced property rights lead to the over-extraction of natural resources, and these resources are exported at low prices to the North that over-consumes them. The international market amplifies the tragedy of the weak property regimes, leading to inferior solutions for the world economy. However, in developing countries, the conversion of natural resources regimes from community or state property regimes to private property regimes faces formidable opposition due to heavy dependence of local and poor people on these resources. Chichilnisky argues that the weakness of property rights in *inputs* to production, such as timber and oil, could be compensated by assigning well defined and enforceable property rights to products or *outputs* such as environmental services (carbon sequestration or carbon emission).

The author identifies environmental services as privately produced public goods, and argues that the markets for these goods are naturally different from those for private goods. Market efficiency in the case of privately produced public goods requires an additional condition which alters fundamentally Coase's conclusion about initial property rights; this is the Lindahl, Bowen, and Samuelson condition whereby the marginal rate of transformation equals the sum of the marginal rates of substitution among the traders. This additional condition required for efficiency 'over-determine.' the market equilibrium. Therefore while market solutions exist, they are not efficient in general. Distributing properly the initial rights to emit allows one to reach solutions that clear the markets and are, simultaneously, efficient in the use of the global public good. Hence, markets that trade public goods require a measure of equity to ensure efficiency, a requirement different than the markets for private goods. The author cites the 1997 Kyoto Protocol as an example.

## 5. DEFORESTATION, SPECIALIZATION AND SUSTAINABLE FOREST MANAGEMENT

The rapid deforestation occurring in many parts of the World has been a matter of widespread concern during the last few decades; the more gradual loss of world forest cover has worried some people for much longer. The felling of forests has been part of the development process everywhere. Population growth creates pressure to shift land from forests to agricultural use. Where development is successful this process eventually abates and reverses itself as population growth slows or disappears and land productivity in agriculture reduces the demand for land for that purpose. By this logic, and given sufficient optimism that currently developing countries are following and will follow the path previously traversed by the now developed countries, one might presume that the pressure for deforestation would eventually cease. Pessimists express concern that many developing regions are still very far from the turning point at which that pressure begins to abate, that some may never reach it (witness Haiti) or—currently the bigger concern, that the turning point may coincide with far too little remaining forest for global needs on the climatic and biodiversity fronts. In fact, the main cause behind the evolution of the concept of sustainable forest management has been tropical deforestation. In this section, Hartwick and Hyde examine some economic aspects of deforestation, and Sedjo discusses the role of specialization in sustainable forest management.

In Chapter 8, Hartwick presents an interesting interpretation of the long-run mechanisms linking population and economic growth to deforestation. His conceptual framework provides the tools to analyse the interaction among population, forest use, and agricultural land. While noting the obvious (Malthusian) possibility that population growth may cause deforestation, he also highlights the possibility that deforestation may contribute to economic growth and thereby forestall the negative impact of population on subsistence. In the extreme, the felling of forests may provide the exports which buy the imports (e.g. machinery and equipment for industrialization) which accelerate growth, slow population pressure and eventually lead to afforestation as the need for agricultural land diminishes. The combination of being able to tap an existing store of wealth (the timber) and get access to land for agricultural production has been productive of economic growth in a number of historical cases. At the other end of the "growth-promotion" spectrum would be those cases in which the export of timber has simply produced revenues for a narrow elite which has transferred the funds to other countries and left land of little value for agriculture. Evidentially, the relationship between clearing of land, population, and economic growth can vary widely, and it is pivotal to make key distinctions according to the key mechanisms at work.

The author highlights the fact that, in a situation of geographic isolation and a small resource stock, deforestation may lead to crisis as forests shrink and livelihood is imperilled; under such conditions sustainable forestry may be possible, whereas population growth is not. The conditions most conducive to sustainability of population and forests in an Easter Island type of scenario are property rights, social order, and a not trivial cost of harvesting. In contrast to geographically isolated areas like Easter Island, nations in Europe and cities in China have benefited from trading networks and an extensive hinterland providing resources and migration possibilities.

Deforestation has abetted population growth for centuries. As for the effect of population on deforestation, increases in population mean an increased demand for food and fuel, but the demand for food and fuel is a function not only of market size (population) but of market value (per capita income). Many increases in deforestation associated with increased population are actually the result of independent causes that boost per capita income, changes such as improved weather conditions and improved technology. In sum, high consumption levels as well as high population growth rates threaten world forests.

One evidence of the damaging impact of population growth on forest resources would be a historical link between wood scarcity, timber prices and population. In the twentieth century up to 1950, timber prices actually declined while world population growth was at record high rates. Though timber prices did jump in the 1970's along with other primary resource prices, they have not displayed an overall upward trend since the 1950's. Hence, markets are not signaling a basic timber scarcity in spite of the aggressive deforestation of the past; this may be due to defects in markets, or perhaps dire scarcity of forests and forest products is a thing of the remote future. The best we can do is to think deeply and carefully about the past and its links to the present and future, and exhort prudence where aggressive timber harvesting practices continue.

In the second chapter of this section, Hyde reviews the lessons for sustainability from the observed pattern of forest development. He emphasizes that sustainability in its narrowest sense, a "permanent forest estate with unchanging boundaries," is a futile objective. A more reasonable approach is to first determine what to sustain—critical habitat, characteristics of global climate, perpetual options on the use of forest resources, or whatever—and then consider the feasible means for achieving each objective. In terms of either national or world goals for sustainable forest management, a considerable degree of flexibility must be maintained; often it is the total amount of forest that matters more than which pieces of land remain under forest. Thus an appropriate strategy from a world perspective involves "total or general conditions"—assuring enough forest to take care of carbon sequestration needs, biodiversity needs, etc., and also specific needs, related to the fact that certain forests are especially important for particular uses (e.g. if many poor people get livelihood from them), that for certain purposes forests cannot be sustained below a certain size, etc.

Hyde argues that sustainable forest management will not be achieved until we attain a higher state of general economic development than is common in substantial parts of the world today. Until the poorer countries do develop, the wealthier must provide necessary support, on a reliable long-term basis, to assist the institutions of the developing countries with the responsibility for managing their forest resources. Hyde's realistic assessment of SFM and its function puts an important and appropriate spotlight on overall economic development as the key ingredient in dealing with the current challenges related to forests and forest management and suggests several follow up questions.

In Chapter 10, Sedjo discusses the economics of specialization in the production of forest-based goods and services, its potential for increased productivity and lowered costs, and the associated intra regional and international trade. He notes that much of the modern environmental movement is opposed to such specialization and instead stresses the goal of individual forest sustainability in a spectrum of outputs, an approach which is the opposite of economic specialization. The author attempts to reconcile these conflicting approaches by emphasizing the substantial differences in the output mix generated by different forests.

Sedjo argues that forests have at least three distinct roles in contemporary society: to provide commodities such as wood; to provide a host of useful, indeed essential, local environmental goods and services such as watershed protection, and to provide global environmental goods, e.g., biodiversity. He suggests that sustainable forest management requires not a single model, but rather at least three complementary models, if not more. The first model is drawn from the industrial revolution and modern agriculture and focuses predominantly on timber production. This is the intensely managed cropping system, where the other outputs of the forest are of minimal interest and the forest can be located in many places. The second model focuses on non-timber and non-market outputs, with the focus of providing ecosystem services, largely to a particular location. The third model relates to maintaining habitat that is conducive to the provision and continuity of biological diversity, largely native biodiversity. These models, at one level, may appear to be largely independent, but in a broad global context, they are highly complementary. Such a system would allow for specialization among the various components in order to generate a sustainable overall global forest system.

## 6. COUNTRY-SPECIFIC INSTITUTIONAL EXPERIENCES

In this section of the volume, authors evaluate the different institutional interventions undertaken in a variety of developed and developing countries. In the first three chapters, the focus is on the international regimes/institutions of Canada and the United States while the fourth chapter focuses on national and local level institutions in India, Indonesia, Malaysia, and Papua New Guinea.

In Chapter 11 van Kooten and Eagle discuss the Kyoto Protocol, the role of forests in meeting the Kyoto targets of carbon emission reduction, and the economics of carbon sequestration through afforestation, reforestation, and other forest management activities, specifically in Canada. They first review the main relevant features of the Kyoto Protocol—carbon sinks in lieu of $CO_2$ emission reductions, the potential carbon sinks allowed in forestry, and the discounting of physical carbon and its impacts on estimates of the costs of carbon sequestration. They then investigate the costs and limitations of creating carbon credits in forest ecosystems through land use, land use change, and forestry (LULUCF) activities. Their conclusion is that while potentially a significant proportion of required $CO_2$ emission reductions could be addressed using carbon sinks, once the opportunity cost of land and the ephemeral nature of sinks are taken into account, the cost of such carbon uptake is likely to be substantial. Carbon uptake via forest activities varies substantially depending on location (tropical, Great Plains, etc.), activity (forest conservation, tree planting, management, etc.), and the assumptions and methods upon which the cost estimates are based. Once one eliminates forestry projects that should be pursued because of their biodiversity and other non-market benefits, or because of their commercial profitability, there remain few projects that can be justified purely on the grounds that they provide carbon uptake benefits.

Further, the authors note landowners' tend to show reticence to tree planting programs, which will increase carbon uptake costs, and that trading of carbon credits and conversion of temporary into permanent removal of carbon will not emerge automatically. They acknowledge that there has been some trading of carbon credits but these have been limited to only large industrial emitters (LIEs) in a limited geographic area, and trading has been focused on industrial emissions but has not included agricultural or forestry offsets. The authors raise some concerns, similar to those of Binkley in Chapter 6, about a market-based approach to carbon sinks, arguing that in well-functioning market for carbon offset credits requires legislation that delineates the rights of landowners, owners of trees and owners of carbon. Similarly, landowners need clear guidelines as to how their activities would qualify for carbon offsets and how credits are to be certified so that they have a well-defined 'commodity' to sell in the carbon market.

In Chapter 12 Nelson and Vertinsky explore the structure of the international regimes related to forest management, and their impacts on Canadian forests and forest management. They look at regional and international trade agreements and multilateral environmental agreements; international criteria and indicator processes; and international forest certification systems. The first two categories are largely government regulated while the third is mainly a private regulatory system enforced by market behavior. The authors examine the interactions between these three categories of institutions and their interactions with the domestic regulatory system within Canada to directly affect sustainable forest management (SFM).

The authors report that the international forest regime shapes and interacts with Canadian policy-making processes in complex ways, as it moves through multiple layers, filtered by national and provincial policy-making processes. Domestic legislation and policies have had to respond to a rapidly changing international regime in which trade, and increasingly environmental, issues play a greater role. Canada has made more significant changes in its forest management policies over the past two decades than the US in part because of its dependence on export markets. The main impact on Canadian industry to date has been the result either of US trade pressure or of increases in regulatory costs resulting from international market pressures to protect the environment. Reduced prices and higher regulatory costs have simply provided the industry with greater incentives to rationalize production further and become even more competitive, although this has come at a high cost in terms of forest communities' sustainability. The question of who decides and what weight should be given to public participation at different levels is a difficult one, still unresolved in Canada, and is made even more so by changing of norms over time. The ambiguity and uncertainty as to what constitutes SFM and what weight should be given to different "publics" make it hard for many countries to commit to specific obligations for many of the values embedded in SFM. The introduction of certification has opened up the policy process to a wider range of groups within Canada than have traditionally participated in forest policy planning by incorporating to varying degrees (depending upon the system) a role for public participation and the promotion of social and environmental values. Given the increasing attention paid to environmental issues, and the environmental scrutiny Canadian forests receive, Canadian forest policies have and will continue to incorporate a number of important ideas and values developed in international environmental agreements. These ideas and values will also be reinforced through certification systems.

In the third chapter of this section, Chapter 13, Bentley and Guldin extend the discussion in another important direction—linking Ciriacy-Wantrup's definition of conservation to the criteria and indicator approach to sustainable forest management, and examining the state of the American forests using the Montréal Process criteria and indicators of SFM. The authors acknowledge the limitations of economic theory when seeking long-term optima; risk, uncertainty, and ambiguity regarding the future make it difficult to be at all precise about future outcomes. The fundamental problem is that risk or probabilistic models compound with time, and over a few economic and ecological events, the models "explode," thus becoming useless for making predictions that can guide future decisions. The authors highlight that Ciriacy-Wantrup's definition of conservation, shifting resource use toward the future, is an interesting rule that leads from the short-term to the long-term. The authors argue that reinterpreting the Simulated National Forest (SNAFOR) "funnel" model of information and knowledge about the future and Fedkiw's pathway model of learning through time in light of Shackle's work on uncertainty could lead to a practical understanding of Ciriacy-Wantrup's definition applied to sustainable forest management.

When they use national forest inventory data to evaluate the state of American forests, the authors find that it is not yet possible to make a firm, defensible statement about whether forests in the United States are being managed on a sustainable basis. This ambiguity is not just the result of incomplete data availability. Sustainable forest management requires a joint consideration of forest conditions and trends together with the values that society places on the many different facets of the forest. The 67 indicators of the Montreal (Criteria and Indicator) Process represent these values. Since different people place different values on the various forest attributes, views on whether sustainable forest management being practiced will also differ even if all the detailed information of interest is available and accurate. The authors conclude that focusing on Ciriacy-Wantrup's conservation criteria keeps the arguments concrete and practical in a subject that tends to be elusive.

In the last chapter in this section, Chandrasekharan argues that sustainable forest management and forest certification have provided primacy for stakeholders, and they substantially influence the way forest resources are managed, through their claims for benefits and related tactics. Most often they are competitors and their interests are in conflict, as a result they tend to view each others with suspicion and get involved in power struggles. Synergies are, however, developed in situations where there are no 'better' alternatives to co-operative action or where clear policy incentives foster and nurture development of such synergies. The understanding and appreciation of SFM by stakeholders are conditioned by the extent to which their claims are satisfied; this determines the success or failure of SFM implementation.

The author uses four cases studies—the Out-grower Farms of Clonal Trees of ITC Paperboard and Specialty Paper Division in India, PT. Sari Bumi Kusuma in Indonesia, the Matang Mangrove Forest of Perak State, Malaysia, and Vanimo Forest Products Ltd in Papua New Guinea—to analyse different sets of institutional arrangements for SFM and their outcomes. He concludes from these cases and the respective country contexts that the major constraint to SFM is not a lack of technology, but institutional factors which militate against the application of the best available technology. These institutional factors take the form of short-term perceptions and time preferences of the investors and other stakeholders. Due to the long time horizon involved, technology-based models of SFM often face implementation problems, and plans are vitiated by intervening developments in political, economic and social arenas. To address such situations, a long-term, policy-based commitment of the stakeholder community is crucial.

## 7. EPILOGUE

The chapters of this volume include a considerable diversity of arguments and approaches. Some reader may note that some of the papers appear to have more in common with the neoclassical approach than with the institutional approach put forward in section 2 of this chapter. Such diversity reflects our view that the economic analysis of forestry requires contributions from a range of approaches. We believe that each school of thought within economics has both strengths and weaknesses, and we should take an inclusionist approach not only across different disciplines but also among different schools of economics.

Keeping this in view, the volume provides a spectrum of institutional approaches as well as institutional issues related to SFM. The approach of the four chapters in the first section is close to that of the "old institutional economics", while the two chapters in the second section are closer to the "new institutional economics", and the three chapters in the third section tend to overlap the boundary between new institutional economics and neo-classical economics. The chapters in the fourth section use many elements of all the three streams of economics, and provide the real life experiences from various countries about the outcomes of different institutional arrangements. At the same time, most of the chapters comment on or reflect the weaknesses and incompleteness of the neoclassical approach when applied to sustainable forest management. In addition, there are many common themes across the chapters in this volume and the chapters in its companion volume. In the last chapter of this volume, Kant integrates the contents of this volume into the common framework of his four principles of the economics of sustainable forest management—principles of existence, relativity, uncertainty, and complementarity (Kant, 2003), and provides linkages to the chapters in the companion volume.

## NOTES

---

[1] The following observations of Robert Sollow and Partha Dasgupta are good examples of the vision of neo-classical economists about sustainability:

> ...history tells us an important fact, namely, that goods and services can be substituted for one another. If you don't eat one species of fish, you can eat another species of fish. Resources are, to use a favorite word of economists, fungible in a certain sense. They can take the place of each other. That is extremely important because it suggests that we do not owe to the future any particular thing. There is no specific object that the goal of sustainability, the obligation of sustainability, requires us to leave untouched.... Sustainability doesn't require that any *particular* species of fish or any *particular* tract of

forest be preserved. (Solow, 1993, p.181)

..I show that the idea of "sustainable development", as it is typically thought of, is far too loose to be of any use...In short, it is foolishly conservative....Recent simulation models of global warming suggest that nothing substantial needs to be done in the near future about greenhouse emissions." (Dasgupta, 1994, p.35)

One of the main causes of current problems related to sustainability, such as greenhouse gases and loss of biodiversity, is this neoclassical approach that every thing can be substituted, and nothing should be left untouched. Neoclassical economists, as Colander (2005) observed, are concerned about their own and their professions' sustainability and not the sustainability of natural systems and society.

[2] A detailed discussion of the old and the new versions of institutional economics is available in Chapter - 4. In this section, the term "institutional economics" refers to the old institutional economics, also termed "evolutionary economics" by some economists, specifically members of the US-based Association of Evolutionary Economics.

[3] McFadden (1999) identified this aphorism as a characteristic of the scientific priorities of main stream economics.

[4] Some goods, specifically public goods and systems' goods, cannot be traded through market due to their physical characteristics, and therefore there are no markets for such goods. However, in many cases, such as various non-timber forest products specifically in developing countries but including some developed countries, there are no markets even for private goods.

[5] Thus markets are always less effective, because transactions costs are higher, when a significant share of the participants would steal rather than purchase if given the chance to do so without penalty. When social convention and pressure rules out the possibility of theft, the seller need not expend resources to prevent it and the transactions costs are automatically lowered.

## REFERENCES

Berry, A., & Serieux, J. (2004). All about the giants: Probing the influences on World growth and income inequality at the end of the 20th Century. *CESifo Economic Studies*, 50(1), 133-170.

Colander, D. (2005). Complexity, muddling through, and sustainable forest management In S. Kant, & R. A. Berry (Eds.), *Economics, sustainability, and natural resources: Economics of sustainable forest management* (pp.23-38). Amsterdam: Springer.

Dasgupta, P. (1994). Optimal versus sustainable development. In I. Serageldin., & A. Steer (Eds.) *Valuing the environment: Proceedings of the First Annual International Conference on Environmentally Sustainable Development* (pp.35 -46). Washington D.C. : The World Bank.

Gruchy, A.G. (1947). *Modern economic thought.* New York: Prentice Hall.

Hueting, R. (1980). *New scarcity and economic growth.* Amsterdam: North-Holland.

Ikerd, J.E. (1997). *Toward an economics of sustainability.* Available from /www.ssu.missouri.edu/faculty/jikerd/papers/econ-sus.htm Accessed on August 1, 2004.

James, D. E., Janesn, H.M.B., & Opschoor, J.B. (1978). *Economic approaches to environmental problems.* Amsterdam: Elsevier Scientific Publishing Company.

Kant, S. (2003). Extending the boundaries of forest economics. *Journal of Forest Policy and Economics*, 5, 39-58.

Kant, S. (2005). Post-Newtonian economics and sustainable forest management. In S. Kant, & R. A. Berry (Eds.), *Sustainability, economics, and natural resources: Economics of sustainable forest management* (pp.253-268 ). Amsterdam: Springer.

Kant, S., & Berry, R. A. (Eds.) (2005). *Sustainability, economics, and natural resources: Economics of sustainable forest management.* Amsterdam: Springer.

Kapp, K. W. (1976). The nature and significance of institutional economics. *KYKLOS*, 29, 209-232.

Khan, M. A. (2005). Inter-temporal ethics, modern capital theory and the economics of sustainable forest management. In S. Kant, & R. A. Berry (Eds.), *Sustainability, economics, and natural resources: Economics of sustainable forest management* (pp.39-66 ). Amsterdam: Springer.

McFadden, D. (1999). Rationality for economists? *Journal of Risk and Uncertainty*, 19(1-3), 73-105.

Meadows, D.H., Dennis L.M., Jorgen R. & William B. III. (1972). *The limits to growth*. New York: Universe Books.

Myrdal, G. (1978). Institutional economics. *Journal of Economics Issues*, 12, 771-783.

Norgaard, R.B. (1985). Environmental economics: an evolutionary critique and a plea for pluralism. *Journal of Environmental Economics and Management*, 12, 382-393.

Opschoor, H., & van der Straaten, J. (1993). Sustainable development: an institutional approach. *Ecological Economics*, 7, 203-222.

Söderbaum, P. (1992). Neoclassical and institutional approaches to development and environment. *Ecological Economics*, 5, 127-144.

Solow, R. (1993). Sustainability: An economist's perspective. In R. Dorfman, N. Dorfman (Eds.), *Economics of the environment* (pp.179-187). Selected readings. 3rd ed. New York: Norton.

Spangenberg, J.H. (2002). Institutional sustainability indicators: An analysis of the institutions in Agenda 21 and a draft set of indicators for monitoring their effectivity. *Sustainable Development*, 10, 103-115.

WCED (World Commission on Environment and Development). (1987). *Our Common Future*. Oxford, U.K.: Oxford University Press.

Wilber, C. K., and Harrison, R. S. (1978). The methodological basis of institutional economics: pattern model, story telling, and holism. *Journal of Economic Issues*, 12, 61-89.

# CHAPTER 2

# IN SEARCH OF OPTIMAL INSTITUTIONS FOR SUSTAINABLE FOREST MANAGEMENT: LESSONS FROM DEVELOPED AND DEVELOPING COUNTRIES

M.K. (MARTY) LUCKERT

*Department of Rural Economy, University of Alberta,*
*Edmonton, Alberta, Canada T6G 2H1*
*Email: marty.luckert@ualberta.ca*

**Abstract.** Despite optimism in the economics literature in the early 1970s regarding our future ability to design institutions, we have yet to progress to a state where we can contribute much towards the definition of optimal institutions for sustainable forest management in developed and/or developing countries. As we have progressed in our knowledge, the concept of specifying an "optimal institution" seems more, rather than less, elusive. Complexities with respect to defining and describing institutions and preferences of economic agents have left us with little progress on connecting institutions with economic behavior. Further complexities regarding the potential endogeneity of institutions within economic behavior have also proven difficult. All of these intricacies must be pursued in a context where the social objectives associated with sustainable forest management are variable and ambiguous. Connecting institutions to economic behavior in pursuit of social objectives may require further refinements in: our understanding and characterization of institutions; our understanding of non-institutional determinants of behavior (such as socio-economic characteristics of firms and their time and risk preferences); a wider recognition of a potential co-dependence (as opposed to cause and effect relationships) between institutions and economic behavior; more explicit recognition of transactions costs and belief systems; and clearer specifications about what we want sustainable forest management to achieve.

## 1. INTRODUCTION

The 1960s were exciting years for institutional economists. Economists were increasingly expanding new approaches to supplant classical marginalism. In their literature review of 1972, Furubotn and Pejovich conjecture forward with optimism about the future potential of using the concept of rational firms within alternative institutional environments to analyze economic behavior and welfare results. They conclude there review: "Substantial advances have already been achieved and the

*Kant and Berry (Eds.), Institutions, Sustainability, and Natural Resources: Institutions for*
*Sustainable Forest Management, 21-42.*
© 2005 *Springer. Printed in Netherlands.*

literature gives evidence of continuing vitality and promise of future accomplishments."

More than thirty years later, we find that many of the contributions by some of the pioneers in institutional economics have had profound influences on several economic fields. Indeed, a number of Nobel prizes have gone to some of the key people in the field of institutional economics (i.e. Myrdal and von Hayek, Buchanan, Coase, Fogel and North). Following on this historic performance, it's not too surprising that in our euphoria, economists frequently think about designing optimal institutions. Indeed among the "Key Issues for Discussion" of this conference is "optimal institutions".

Despite these advances, there seem to be significant roadblocks facing resource economists in their ability to contribute towards designing optimal institutions. For example, in a recent email exchange, Daniel Bromley[1] states:

> As for linking property rights to "economic behavior" I now believe that this will be a difficult if not impossible task. There are simply too many intervening variables--time preference rates, age of individual (somewhat related to time preference, income, etc. etc. etc). I wish you luck, but I am now skeptical that anything coherent will be found.

The two quotes above embody a substantial change regarding future prospects of our ability to design optimal economic institutions. As we started to head down the paths laid out by Furubotn and Pejovich in 1972, which were supported over time with contributions from several Nobel prize winners, we ran into levels of complexity that were unforeseen.

The purpose of this paper is to investigate these sources of complexity. As economists, we frequently think about issues in terms of optimality: optimal levels of production, optimal time paths, optimal rotations, etc. Given the levels of complexity that we witness, I question whether we are ready to apply the concept of optimality to issues regarding institutions for SFM.

In pursuing this purpose, this paper will investigate the pursuit of institutions for sustainable forest management (SFM) with examples drawn from developed and developing countries (Canadian and southern Africa examples, respectively). Furthermore, while my main line of approach will be from an economic perspective, I will entertain ideas from other disciplines.

The next section will begin by discussing some key terms. Next, I will describe a number of areas that have proven problematic regarding our ability to contribute towards the development of optimal institutions. Highlighting these areas will provide the basis to conjecture about what future paths may increase our ability to contribute towards the designs of optimal institutions for SFM.

## 2. KEY TERMS: SFM AND OPTIMAL INSTITUTIONS

Although there are innumerable definitions of SFM (e.g., Burkhardt, 2001), for my purposes, it is useful to start by defining SFM in terms of what it is not. SFM is not the way we have managed forests historically. As such, SFM is largely a concept associated with a panacea that is supposed to replace outdated paradigms. Thus, in general terms, SFM is not the history from whence we come. Rather, it is the direction that we are going.

Continuing from the "negative definition" above, a positive definition may be constructed by highlighting a number of specific ills of the past that SFM is meant to solve (e.g. Adamowicz and Veeman, 1998; Luckert, 1997). First, SFM explicitly considers values of a broader range of forest goods and services than have historically been considered, including social, cultural and economic values. Second, by considering more than one forest output, trade-offs between different types of competing forest goods, services, uses and managed forest states may be assessed. Third, SFM emphasizes key goals of maintaining the health, integrity and biodiversity of forest ecosystems at multiple scales. Finally, given the uncertainty in dealing with complex forest issues, SFM stresses the importance of managing adaptively such that options are not foreclosed and new information may be used as it becomes available.

In economies where timber has historically been the main income source, the task of SFM is largely to fix the problems associated with an outdated sustained yield paradigm that largely focused on timber values. In developed economies, societies have basically decided that more attention needs to be paid to non-timber forest resources to ensure the sustainability of the many different use and non-use values associated with forests.

In developing economies, the task of SFM is to halt degradation of forest and woodland resources so that continued contribution of these lands to livelihoods is ensured. Furthermore, for some of the more valuable timber stocks, there is a desire that they be used to develop local industry and serve as a source of badly needed foreign currency. The focus in these countries is frequently on use-values of forests, necessary for day to day living and for diversifying livelihoods and economies so that they may better withstand shocks such as drought or economic upheavals.

Economic concepts of institutions frequently evolve around some search for an optimum. However, to a sociologist, the concept of an optimal institution would likely be laughable. In reviewing concepts of institutions, Cortner *et al.* (1994) refer to institutions as, "expressions and mechanisms of collective experiences". Such breadth in the concept of institutions allows it to capture a great many different kinds of rules, processes and organizations. However, to consider a concept of optimality within this scope seems impossible. Accordingly, economists generally take a much narrower view of institutions.

In contrast to the sociological concepts of institutions, the economic definition of institutions largely falls out of the work by North (1990) who defines institutions as: "the humanly devised constraints that structure political, economic and social interaction". Contrary to sociological concepts of institutions, North (1993) distinguishes his definition clearly from the concept of organizations: "Institutions are the rules of the game....Organizations are the players...".

The "rules of game" can cover a vast array of circumstances. Some rules are formulated to govern policy-making processes, such as who is allowed to have a voice and under what conditions. Other rules may arise as policy results that serve to govern the daily actions of members of society beyond those engaged in the policy-making processes. Property rights arise as benefit streams, derived from assets, which may be captured within the context of such rules.[2] In practice, most economic analyses of institutions have not addressed the breadth of institutional concepts to which North (1990) refers. That is, economists have not traditionally addressed all types of rules that make up institutions. Rather, they have frequently concentrated on property right rules, such as exclusiveness and transferability, that specify conditions regarding who and how firms may use goods and services. Therefore, when most economists consider institutions, they are actually considering property right rules, which may be considered a subset within the broader category of economic concepts of institutions. In this limited context, the concept of optimal institutions may be tractable. In the next section, I begin, by discussing complications within this narrow concept of optimal institutions. Subsequently broader contexts of institutions are more fully addressed by including rules, other than property right structures, and organizations.

Considering the concepts of optimal institutions and SFM in concert, optimal institutions may be thought of as the means towards the social goal of SFM. However, in practice, we see that the distinction between these two concepts is frequently blurred. That is, SFM is sometimes considered to be an institution into itself. This phenomenon partially arises out of ongoing policy processes of defining criteria and indicators for SFM. Haener and Luckert (1998) describe how the initial intent of criteria and indicators was to add more clarity to defining the objectives of SFM. However, as this process has progressed, these criteria and indicators have turned into policy rules as part of forest certification schemes. That is, because policy makers have failed to consider alternative rule structures for pursuing SFM, the ends (i.e. criteria and indicators) have also become the means. This phenomenon is largely due to the interconnectedness of policy processes and results that will be discussed further below.

## 3. DIFFICULTIES IN PURSUING OPTIMAL INSTITUTIONS FOR SFM

Within the more narrow confines of economic concepts of institutions as property right rules, the basic problem associated with identifying an optimal institution seems quite simple. As shown in Figure 2.1, there are numerous institutional/property right rule possibilities (i.e. combinations of rules, $I_j$), each which can result in a given allocation of resources ($A_j$), each of which creates some level of utility ($U_j$).[3] If we take as our social objective to maximize the aggregate utility consistent with SFM,[4] then the optimum institution may be defined as that combination of rules which results in that allocation of resources that best meets our social objective.

Conceptually, this process of finding an optimum institution looks fairly straightforward. We could, conceivably, set up a constrained maximization problem where institutional constraints are imposed on a utility/production function and solve for first order conditions (FOCs) in order to maximize some measure of welfare.[5] So why have we had such a hard time in finding optimal institutions for SFM? First, each of the components of Figure 2.1 has proven more complex than we initially imagined. Second, we have found that there are many relevant concerns not captured in Figure 2.1.

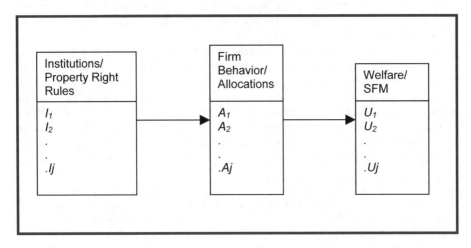

*Figure 2.1. A Common Economic Approach to Pursuing an Optimal Institution*

### 3.1 Institutions/Property Rights

Even if we stick to a simplification of the North (1990) definition and think of institutions as property right rules, the complexity we find with respect to the different combinations of rules governing forests is daunting. In many developed countries, such as Canada, there are complex institutional arrangements, frequently referred to as tenure policies, or concessions, which provide a framework that allows private companies to manage forests on public lands (Haley & Luckert, 1990). The numerous different conditions which govern the behavior of private firms are difficult to describe, much less analyze. Aspects of current tenure policies in Canada may be specified in: federal legislation that looks after some forest resources such as fisheries; provincial forest acts; forest regulations or guidelines issues by provincial forest management agencies; individual tenure agreements between the province and a specific company; and in policies that are not explicitly stated, but are known within the industry. In short, there is a hierarchy of many different institutional levels where rules are specified.

Although some forestry concessions in developing countries tend to mimic tenure arrangements in Canada,[6] many of the institutions governing forests in southern Africa primarily regulate household livelihood uses of forests, rather than industrial uses. Accordingly, the types of rules adopted look markedly different from developed country institutions. Despite this difference in purposes and rules for developing vs. developed country forestry, the complexity of rules throughout a hierarchical structure may be similar. For example, at the national level, Zimbabwe may be described as having three different kinds of property right systems: state land, communal land, and commercial land (Moyo *et al.* 1991). However, in addition to the national level regulations that define such areas, there are also, regional and district level rules that govern the use of natural resources (Mandondo, 1998). Furthermore, at yet a more local level (i.e. at the village level and below), there are customs, norms and courtesies between households that influence the use of natural resources (Mandondo, 1997).

In the face of such complexity regarding institutions governing forest resources, the property rights literature has frequently simplified descriptions of property right structures by relying on concepts such as "public" vs. "private forests" (e.g. Gamache, 1984). However, this distinction is not very enlightening for analytic purposes. As Alchian and Demsetz (1973) have explained long ago, that when sufficient control of an asset has been transferred from private to public control to constitute a difference in private vs. public property is ..."a moot point." Furthermore, even if we just consider "private forests", Randall (1987) explains:

> To say that for efficiency, property rights must be nonattenuated does not in itself tell us everything we need to know about property rights. One can conceive of many different sets of rights with respect to a particular kind of property object, all of which are exclusive, transferable, and enforced, but each of which is specified differently from the others.

Along these lines, Bromley (1991) states that there are as many different property rights structures as there are combinations of social structures.

In short, the complexity that we see with respect to the rules making up institutions has made it difficult for us to even describe and/or specify this part of the equation.

## 3.2 Firm Behavior/Allocations

The complexity of institutions (even using the simplified economic concept of institutions as property right rules) is one key reason why economists have had a hard time predicting behavior under alternative institutional structures. In the face of such complexity, the usual response has been to simplify the property rights specification by looking for significant differences in natural resource management between discrete institutional structures.[7] Although such studies have identified significant differences in performance between categories of institutions, explanations as to why resource management differs between these different sets of rules have been largely conjectural. Problems arise because it is not clear how the complex incentives created by combinations of rules, taken together, influence behavior. More recently, a few studies have started to tease out behavioral impacts of specific rules within institutions.[8]

A key challenge associated with trying to tease out impacts of individual rules on economic behavior involves trying to find sufficient variability in the desired rule, while not having too much variability to account for in other factors that contribute to behavioral variation. In many cases, economic behavioral consequences of institutions are assessed in local level case studies. However, variability in rules may be slight, as one institutional structure may govern the actions of many. In order to introduce more variability into studied institutions, multiple jurisdictions may be sampled. However, increased variability in the institutional variables invariably leads to increased variability in other potential determinants that must be accounted for.

Another fundamental question to address in trying to model economic behavior is the definition of the firm. In developed country settings, this question is generally pretty straight forward. We are frequently dealing with a defined corporate entity making decisions frequently tending towards profit maximization.[9] However, in developing country settings, there may be instances where decision-making may occur at the individual level (sometimes with gender and or age differentiation,) at the household level, or at the village level dealing with common property. Therefore, merely deciding how to define the firm may be a significant issue.

A further problem arises in trying to understand non-institutional determinants of economic behavior. Data on many of the potential non-institutional determinants, (e.g. socio-economic variables including wealth, marital status, age, and gender) may be readily available. However, there are at least two other potential determinants that represent major challenges in themselves – time preferences and risk preferences.

Given the long temporal nature of many forestry decisions, time preferences can have large impacts on economic behavior. Unfortunately, time preferences represent an area of research where the more we do, the more difficult the situation appears. In developed country settings, economic experiments have shown that a number of factors may influence time preferences including: socio-economic variables as described above (e.g. Harrison, Lau, & Williams, 2002); the length of period in question (i.e. duration dependent discounting; e.g. Cropper, Aydede, & Portney,. 1992); the type of good or service being considered (e.g. Sultan and Winer; 1993); and whether a utility stream involves gains first and losses second, or visa versa (e.g. Loewenstein, 1987).

In developing country settings, the same types of determinants effecting time preferences may be evident (e.g. Kundhlande, 2000a). However, we know less about these factors because experiments are rare.[10] Notable in these settings is a vast range of time preference estimates (ranging from negative values to over 100%). Given the sensitivity of welfare streams to the time preference rate being used, and the vast range of potential time preference rates, behavioral responses become difficult to predict.

Risk preferences may also have significant impacts on economic behavior, and, similar to time preferences, represents another challenging area. Although significant work regarding risk preferences has been done in agriculture[11], there is very little done in forestry. This, despite the fact that the long time horizons associated with some forestry activities allow more time for unexpected changes.

In pursuing behavioral models that reflect impacts of institutional structures, key differences become evident between developing and developed country settings. In addition to significant differences in the determinants discussed above, the general nature of the economies within which firms operate in each setting causes its own set of challenges. In developed economies we may have highly developed markets for commercial inputs and outputs from forests, where forces of supply and demand come together to create market prices. In contrast, many developing country settings have thin or non-existent markets for numbers of outputs and inputs. Lacking infrastructure, many subsistence economies are isolated and do not feed into national, much less international, markets. Accordingly, price information may be thin or lacking in modeling economic behavior.

We also find that these subsistence economies tend to be much more tied to the land than developed economies. Thus, the link between household production and the environment may be direct and visible. The closeness of economic activities to the land, and the localized nature of these economies may cause economic behavior to approach ecological behavior. For example, in the absence of opportunity costs of time, travel costs may be measured in calories (Hatton, Adamowicz, & Luckert, 2001), taking us close to optimal foraging models in grazing science. Closer ties to the land and living in poverty also make survival pressures evident in developing countries. In subsistence economies, incentives for efficiency and the discipline of the market may manifest itself as whether or not a household survives the next drought. In short, we find very lean systems in developing economies without much room for error, causing rationality to potentially become quite evident if we define firms' objectives and constraints correctly. However, these models may be difficult to identify and specify given the lack of robust prices and the presence of potentially multiple types of firms (e.g. individuals, households) making decisions with the context of complex institutions and constraints. Efficiency may also be hard to identify when characterized by investments in social capital and diversification, rather than specialization, in the presence of extreme resource limitations and risk.

## 3.3 Welfare/SFM

Even if we were able to describe/specify rules of alternative institutions in ways that allowed us to investigate behavior, and even if we were able to account for values that households receive and the many other potential factors that we believe influence behavior, we would still face the task of identifying some welfare measure(s) that correlate with the goals of SFM.

We could attempt to use the traditional concept of Pareto efficiency. The extension of this criterion to institutions, following Figure 2.1 above, would recognize that resource allocations are endogenous to rule structures (see footnote 3). Therefore, Pareto efficiency would entail finding that set of rules that leads to an allocation of resources such that no change in rules can lead to a reallocation of resources that can make one person better off, without making another person worse off. But such constructs, while valuable in theory, don't work when trying to derive optimal institutions in real settings. When rules and resource allocations change somebody is invariably worse off.

Unfortunately, we also find that seeking Potential Pareto Improvements becomes difficult to use as a criterion for optimality when seeking optimal institutions for SFM. Objectives associated with SFM, in developed and developing countries tend to have elements related to increasing or maintaining efficiency, equity, and environmental integrity. While the first of these objectives may be adequately expressed with the Potential Pareto Improvement criterion,[12] the last two frequently lie largely outside such measures.

In the absence of a generally defined measure of welfare, we must instead resort to measuring performance against the extent to which we achieve the panacea, SFM. Unfortunately, like all panaceas, utopia lies in the eyes of the beholders that understandably differs within and between developing and developed countries, and changes over time. Thus, a further dilemma is that we are lacking a criterion with which to define an optimum.

This problem in itself is enough to keep us from finding optimum institutions for SFM. However, as economists, we certainly don't let the absence of clearly defined social objectives keep us from looking for an optimum. Instead, we adopt one or a number of criteria that presumably are correlated with society's vision of the SFM panacea. So, for the sake of argument, let's assume that there are clear objectives that we can quantify such that we can call one set of results better than any other to establish an "optimum". Such criteria may evolve around some of the key concepts associated with SFM including looking after forest health, biodiversity, and practicing adaptive management, discussed above. We nonetheless have a number of other complications to face.

### 3.4 Direction of Causation between Institutions and Firm Behavior/Allocations

Historically, much of the economic thinking guiding links between institutions and behavior were concentrated on a single direction of causality, as indicated in Figure 2.1. That is, the basic question was, what incentives do rules provide for economic behavior? However, more recently, economists are increasingly considering how economic behavior may influence rules. That is, people are wondering whether behavior may have important implications with respect to the rules that are formed. Indeed I would like to posit two empirical observations regarding this phenomenon:

1.  Anytime a behavioral model with institutional variables is considered for publication, there is likely to be one or more reviewers that question the direction of causality of the institutional variable.

2.  The probability of this occurrence increases if one or more of the reviewers are not an economist.

The first observation I will support immediately with an example, the second is supported later in the discussion.

The case of the security of property rights in developing country settings is a case in point regarding increased importance being paid to the question: what really is endogenous here? Earlier studies, for example Feder and Onchan (1987), concentrated on the potential affects of insecurity on economic behavior. However, as time went on authors such as Besley (1995), Kundhlande (2000b) and Bledsoe (2003) began indicating that causation between security of property rights and behavior should also be considered in reverse. That is, people may be investing in land to improve security.

In developed country forestry settings, I am unaware of any who have contemplated the potential significance of this reversed causality involving security. Perhaps this omission has occurred because we consider institutions more developed and stable in developed countries than in developing countries. However, in cases where private firms operate on public lands, such as in Canada, there are concerns regarding the security of tenure, and their influence on investment decisions[13]. Furthermore, firms are able to potentially increase their tenure security through responsible forest management. So, we have motive and opportunity, suggesting that the allocative behavior of tenure holders is likely to be influenced by incentives to improve the security of property rights.

In addition to considering reversed causation with respect to economic behavior and tenure security, there are also those that have begun thinking more about how other aspects of institutions may evolve in response to economic behavior. For example, contrary to early work on non-exclusive property rights by Cheung (1970), Sethi and Somanathan (1996) investigate "The Evolution of Social Norms in Common Property Resource Use". That is, instead of allowing rents to dissipate, firms will have incentives to observe informal norms which regulate resource use and thereby prevent rent dissipation.

In the discussion thus far, we have restricted ourselves to thinking about how individual firms, making consumption and/or production decisions, may influence the evolution of institutions. As such, we have largely begun with a Coasean (1960) approach[14] to thinking about how property rights may evolve. However, when the Pandora's Box of evolving institutions is opened, we are quickly drawn into recognizing that rules may also evolve due to Pigouvian (1940) type regulatory actions,[15] which leads us to consider the role of organizations in influencing institutions and economic behavior.

### 3.5 Organizations

In addition to neglecting the potential for reverse causality between economic behavior and institutions, economists have also frequently neglected the role of organizations. As North (1993) states "Development economists have typically treated the state as either exogenous or as a benign actor in the development process." As this quote implies, it may well be important to think beyond how economic firms influence institutions, to thinking about how groups of economic agents, potentially organized as larger agents (i.e. organizations such as "the state" or NGOs), may influence the evolution of institutions. Likewise, it may be important to consider how rules regarding policy-making processes influence the strategies of these organizations.

The potential importance of organizations in developed and developing country forestry contexts is difficult to deny. In Canada, we find that while historically government and industry, sometimes with organized labor, have largely controlled the evolution of forest policy (Howlett and Rayner, 1995), the current trend is towards many more groups becoming influential. A key change, evident historically in the United States and now increasingly in Canada, is the importance of national and international groups which hold passive-use values in forests. These NGOs have not only been lobbying for changes in national, regional and local government rules, but they have also played an active role in pursuing criterion and indicator processes used in forest certification.

The role of organizations in certification in developing countries is also prominent. Some of the logic here is that certification is needed in developing countries, because these countries lack strong institutional structures. However, such viewpoints may fail to recognize the strong, highly developed organizations and sub-structures that are often in place at local levels that are designed and/or have evolved in response to important subsistence values of forests. Moreover, the overall potential for certification to influence SFM objectives is likely less in developing countries than in developed countries. Certification relies on market forces to influence forest manager/harvesters. However, in much of the developing world, many forest goods and services never reach those international markets where certification could matter.

Despite the potential importance of organizations in influencing institutions, they are frequently ignored by economic studies of forests.[16] Part of why we may have failed to pay more attention to organizations likely arises out of the definition of institutions that we have adopted. By excluding the concept of organizations in the economic definition of institutions it makes it easier to ignore the interplay between organizations and institutions. After all, we are trying to define optimum institutions, not optimal institutions and organizations. The omission of considerations of organizations is convenient. As discussed above, leaving out organizations allows us to kid ourselves into thinking that we may be able to specify "optimal institutions".

The great irony in this state of affairs is that our concepts of institutions have been heavily molded by North, who explicitly recognizes the important interactions among institutions and organizations (e.g. North, 1993). However, I believe that it is fair to say, that those disciplines that have explicitly included organizations in their concepts of forestry institutions (such as sociologists and political scientists) have likely paid more attention to the potential impact of organizations on rules.

Despite the criticisms that have been leveled at economists for not considering impacts of forestry organizations on rules, the converse of this criticism is frequently applicable for other disciplines. That is, disciplines such as political science and sociology, frequently fail to consider what impact rules have on incentives of households or individuals. The end result is that many policy studies, across disciplines, seem to consider separately policy processes and policy results. In economics, this division defines public choice (i.e. policy processes) and social choice (i.e. policy results) approaches to policy analysis.[17]

### 3.6 Public vs. Social Choice Approaches

The failure to consider policy processes and results simultaneously is part of a long standing dilemma with deep roots among economists. In his presidential address delivered to the American Economic Association, Sen (1995) compares advantages and disadvantages of social choice and public choice approaches. Regarding the proclivity of social choice approaches to ignore public choice complications, Sen states: "the practical reach of social choice theory, in its traditional form, is considerably reduced by its tendency to ignore value formation through social interactions." Along the lines of North (1993), Sen calls for a more integrated approach: ".many important lessons have emerged from the discipline of social choice theory as well as the public choice approach. In fact, we can get quite a bit more by *combining* these lessons."

The basic concept is that either approach in isolation of the other is likely to fail to find an optimum institution. That is, just as desired ends cannot justify all means, desired means cannot justify all ends. Utility is derived, not only from the social choice aspects of the policy, but also from the public choice aspects. For example, in the context of US forest policy, Leman (1984) states: "the political process is often messy, biased, and inconclusive, but the public revels in it and would not accept any outcome that did not run this gauntlet. "

In the case of forests in developing and developed countries, much of the analysis regarding national, regional and local policies falls neatly into the categories of social vs. public choice approaches. We frequently find economists trying to assess the social choice impacts of alternative policies, while sociologists, anthropologists and political scientists investigate the public choice processes. However, such divisions have sometimes been to our detriment. For example, Sobel (2002) describes how concepts of social capital, a potential indicator of a desirable policy from a social choice perspective, may not be very useful unless the concepts are considered within the context of public choice recognitions of evolving institutional environments. As Bourdieu (1986) states: "One can acquire social capital through purposeful actions and can transform social capital into conventional economic gains. The ability to do so, however, depends on the nature of the social obligations, connections, and networks available to you. "

The importance of considering both public and social choice aspects in forest policy is also evident as some Canadian provinces begin to experiment with "management by objective" policies[18] As an alternative to prescriptive command and control policies, management by objective policies seek to allow tenure holders more discretion in pursuing forestry objectives. For example, firms may be required to meet future yield projections, rather than specific requirements for regeneration. Thus, the commands become higher order objectives that are pursued within the presence of controls exerted by governments, and/or certification schemes. Within these types of forest policies, the line between policy processes and results becomes somewhat blurred. As firms seek alternative means of achieving higher order objectives, they are simultaneously dealing with policy processes and policy results.

### 3.7 The Broader Environment of Transactions Costs and Belief Systems

With the complication of considering public and social choice approaches in the context of reversed causality and organizations, two issues, heretofore in the background of Figure 2.1, become more prominent; transactions costs and held values.

With simplified approaches to economic institutions (i.e. Figure 2.1) transactions costs may enter in as administrative costs of establishing, monitoring and enforcing a given policy. However, when public choice questions are entertained, we must begin to consider costs of forming the new policy as well. For example, in management by objectives policies, public participation processes, designed to choose which of the many possible paths to take in pursuit of higher order objectives, may be costly. Therefore, transactions costs play a key role in influencing social and public choice questions, and interactions between the two.

Likewise, in the simplified picture of Figure 2.1, factors influencing economic behavior are not explicit. Economists undertaking social choice analysis typically rely heavily on influences of assigned values on economic behavior. However, belief systems (sometimes referred to as held values) influence assigned values and may be instrumental in dictating economic behavior (Brown, 1984). While such considerations are frequently not considered when adopting the model in Figure 2.1, it becomes much more difficult to ignore held values when complications of evolving institutions and organizations are included. Institutions and organizations must evolve as reflections of belief systems if they are to be supported by the society that is creating them (North, 1993). Therefore, if we are to better understand institutions and organizations, we must seek to better understand the belief systems that underlay these social structures. Assigned values are only the tip of the iceberg in understanding a broader view of economic behavior implied in combining social choice and public choice perspectives.

Within the context of forestry in developing and developed countries, we see clear indications that transactions costs and belief systems are influencing organizations, institutions, behavior and welfare. With respect to transactions costs, some have hypothesized that the absence of exclusive rights may occur because the costs of enforcing such rights and developing organizations are too high, relative to low values of resources. (e.g. Campbell *et al.* 2001). With respect to belief systems, we see that ethics and norms may be strong sources of institutional control (Hegan, Hauer, & Luckert, 2003). Along these lines, understanding trends towards communal forestry may necessitate understanding held values associated with extended family and/or community ties.

*3.8 Summary of Complications for Pursuing an Optimal Institution*

Figure 2.2, building on Figure 2.1, provides a summary of the additional complications that we face in assessing the optimality of institutions. The potential for reverse causality is indicated by the bi-directional arrow between firm behavior/allocations and institutional rules. Furthermore, the explicit presence of organizations is included. Much like individual behavior, organizational behavior (specified as strategies, Sj) interacts with institutions/rules. Figure 2.2 considers rules more generally than the property right rules specified in Figure 2.1, in order to be inclusive of rules regarding policy processes. Furthermore, utility is shown to be derived from organizational processes of policy formations, as it is from policy results. By including the potential for organizations and individuals to affect rules, and by showing that utility may be derived from these processes, we have added considerations of public choice to our previously social choice dominated model. Finally, the omnipresent environment of belief systems and transactions costs has been added.

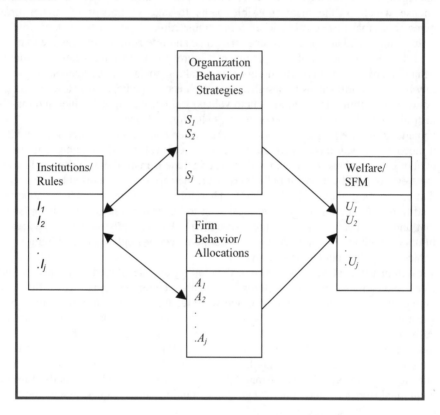

*Figure 2.2. Complications in Pursuing an Optimal Institution*

## 4. CAN WE MAKE PROGRESS IN THE FACE OF THESE DIFFICULTIES?

The complexities described above, involved with pursuing optimal institutions, represent a daunting array of potential roadblocks. Although each of the areas of difficulties described above suggests reasons why the concept of an optimal forest institution may not be tractable, they also provide potential means of making the economic analyses of forest institutions more relevant.

With respect to defining institutions, economists (and others) are beginning to look at institutions with a finer degree of resolution; recognizing the importance in specific combinations of rules. As economists, we have a long history of trying to key in on specific attributes of property rights that have been shown to be theoretically and empirically important in influencing economic behavior (e.g. exclusiveness, Cheung, 1970; and security, Feder and Onchan, 1987), and we are increasingly starting to describe property rights in terms of some of these key attributes to understand crucial differences in institutional structures. (e.g. Sjaastad and Bromley, 2000)

With respect to understanding economic behavior, we are slowly starting to find experimental situations where individual attributes of rule structures can be assessed *ceteris paribus* (e.g. footnote 8). That is, it is not always necessary for us to understand the impacts of all potential determinants simultaneously if we can find situations where some (or many) of those attributes do not vary. Although, such approaches do not supply us with complete behavioral models, they nonetheless allow us to chip away at unknowns regarding economic behavior, and increase our knowledge base incrementally.

With respect to defining welfare associated with SFM, it is unlikely that we are ever going to find an acceptable criterion against which to measure performance. However, this is not new territory for economists. Appropriate welfare criteria seem to always be elusive, but that does not keep us from adopting one or a number to add to the policy debate. Our historic experience in searching for policy relevant criteria will be put to the test as we are called on to participate in discussions to establish criteria and indicators for SFM. Furthermore, as economists, we will play a key role in distinguishing objectives from policy instruments and illustrating the importance of incentives in regulatory policy, so that alternatives to command and control are considered.

In searching for appropriate criteria for SFM, we face numerous problems that economists have grappled with for decades, such as the potential for irreversibilities among forest resources, social rates of discount, and distribution issues within and between generations. As we grapple with these issues within the context of SFM, we will also be doing so within institutional frameworks that confine the range of interests of some key players, therefore failing to provide them with incentives to consider criteria beyond their institutionally constrained range of interests. For example, in the context of Canadian forest tenures that are largely limited to granting exclusive harvesting rights, society expects private companies to establish criteria and indicators for SFM. What incentive is there for these forest companies, with such attenuated property rights, to collect information to consider how their harvesting operations affect recreational activities, biodiversity, or regional economic stability and/or growth?

With respect to the direction of causality among economic behavior and institutions, the "new institutional economics" has a long history of recognizing the importance of rule formation (e.g. North, 1990), and more economists are beginning to address the potential importance of firms' behavior on institutions (e.g. Sethi and Somanathan, 1996). Furthermore, as a discipline we are used to dealing with the co-determination of variables.[19] Nonetheless, we have been slow to consider the explicit importance of organizations on the evolution of rules, particularly in forest and developing country situations, and will likely benefit greatly from a wider disciplinary perspective in dealing with these potential interactions.

Within the economics discipline, we also have a strong history of considering public and social choice aspects of problems (e.g. Sen, 1995). However, curiously enough, we have not been horribly successful in merging these two approaches, even within the single discipline of economics. Therefore, trying to merge social and public choice approaches between disciplines is likely to be an ominous challenge.

Finally, as economists we have a long history of assessing the importance of transactions costs to economic organizations and market functions (e.g. Coase, 1960). However, we have only infrequently dealt with the underlying belief systems, or held values, that impact the assigned values that we as economists concentrate on (e.g. Brown, 1984). Until we better understand preference evolution and formulation, we will likely make little progress on sorting out the co-evolution of institutions and economic behavior.[20] Failure in this regard could have serious implications for our ability to design policy to further social welfare. Again, interdisciplinary approaches in dealing with these broader types of values will be crucial, but difficult.

## 5. CONCLUSIONS

In conclusion, we have made significant advances in understanding the importance of institutions to economic outcomes, and will hopefully continue to do so. Nonetheless, the state of our abilities relative to the complexities of the problems should influence how we respond in policy debates. I do not believe that there is any way to define an optimum forest institution for SFM in developing or developed countries using first order conditions (i.e. there is no FOCing way). Moreover, the concept of an optimum institution may undermine a basic economic premise. In the words of two Nobel Laureates, Stigler and Becker (1977): "De Gustibus non est Disputandum". Loosely translated, tastes are indisputable. Tastes are not right or wrong, they just are. If institutions are reflections of held values, as discussed above, is it possible to think of them in terms of optimality?

To speak of specifying an optimum institution may be at best arrogant, and at worst foolish. Instead, I would suggest that we portray our accomplishments and contributions as trying to understand processes and results that may lead to socially desirable SFM policies. However, this observation should not keep us from continuing our search for better forestry institutions. Starting with a healthy dose of humility in the face of complexity, we can then seek a more productive role in policy processes. Instead of seeing ourselves as designers of policies, our role may be to merely describe the impacts of alternative policies.[21] This is not to say that we should abandon thoughts about designing policy.[22] However, the "optimality" of these policies is not for us to determine. Rather, the policies and their impacts are for us to describe, while politicians and society will determine what is optimal by deciding whether they are improvements over what we currently have. [23]

Our ability to contribute will largely be dictated by how well our theoretical and empirical tools work in addressing issues. Accordingly, I suspect that we will have markedly different roles in developing and developed country settings. In the case of developed countries, we have a distinct advantage. To begin with, we generally have a long history that has created a research base for us to draw on. We also frequently have fairly developed markets around many forest goods and services, which give us price information. We often have fairly high values of forest resources that can potentially support complex regulatory structures, including democratic processes within institutions. Finally, the tools that we use to analyze these environments have largely been developed within the same belief systems that govern the forestry issues that we are attempting to address. With all of these factors in our favor, we are likely going to be able to take a more confident stance in debating forestry institutions in developed countries, proposing alternative options to current institutional structures.

In the case of developing countries, we are at a disadvantage relative to all of the advantages just discussed for developed countries. We generally have a smaller economics research base to draw on. We also frequently have thin or non-existent markets for forest product inputs and outputs in subsistence economies, leading to a dearth of price information. We sometimes see fairly low values of forest resources incapable of supporting complex regulatory structures and democratic processes. Finally, the tools that we use to analyze these environments have largely been developed outside of the belief systems that govern the forestry issues that we are attempting to address.

The importance of differences in belief systems between those trying to provide development and those trying to develop is difficult to over-emphasize. When economic tools are used to assess developed economies we are using ideas that evolved within the system that they are attempting to improve. When economic tools are used to assess issues in developing economies, the ideas being used are being transplanted to belief systems where they may not fit. Institutions are founded on belief systems, and it is not honors to change belief systems. Rather, our role is to try and propose options that may fit with belief systems that exist.

With all of the difficulties that we face in applying our economic tools in developing country settings, I believe that we are likely going to have to take a much less active stance in debating forestry institutions than we do in developed countries. Along these lines, I would suggest that instead of advocating whole scale changes, we will likely be giving advice on small changes that may work within the context of what already exists. The irony (and tragedy) of this conclusion is that the greatest demand for change, in terms of great needs and potential gains from improving forestry institutions, lies in developing countries where we are on less secure footing, and therefore likely have less to offer. Nonetheless, it will remain an important endeavor to attempt to assess how rules systems and belief systems co-evolve to provide insights to funding agencies and governments working in these arenas. Even small improvements in such dire conditions can have profound consequences.

As economists, we might also seek to mimic strategies that many other disciplines are adopting in pursuit of Sustainable Forest Management. That is, we may consider using concepts of adaptive management, commonly used in addressing natural science complexities of SFM (e.g. MacDonald, Arnup, & Jones, 1997), for addressing complexities of institutions. Although politicians and publics may be hesitant to accept institutional experiments, they have also been historically hesitant of landscape level natural science experiments that ecologists are beginning to conduct. Just as attitudes have begun to change regarding large natural experiments, they may also be changing regarding the need for institutional experiments. North (1993) sums it up well: "there is no greater challenge facing today's social scientist than the development of a dynamic theory of social change that will....give us an understanding of adaptive efficiency".

Much of the discussion above is structured around the complexities that economists have generally not yet directly faced in contemplating optimal institutions for SFM. Yet there are good reasons why we have neglected many of these complexities. As economists have grappled with concepts of optimal institutions, our tradition in theory and mathematics has caused us to abstract away many of the complexities that we face. These tools have allowed us to make unique and effective contributions in terms of understanding incentives and resulting behavior that institutions create. However, as we proceed, we must constantly weigh the benefits and costs of reductionist modeling relative to the problem we are trying to address. I would suggest, that in addition to pursuing our reductionist approaches, we expand our efforts cross-disciplinarily. In essence I am advocating that we expand our precise reductionism to include more holistic thinking. While as a group, economists tend to abhor the waffling that may be associated with such imprecision, without this balance, we may suffer from being precisely wrong, or precisely irrelevant.

**Acknowledgements:** Thanks to Vic Adamowicz, Bruce Campbell, Albert Berry and Shashi Kant for comments on an early draft. Also, thanks to the Centre for International Forestry Research and the Sustainable Forest Management network for supporting research from whence some of these ideas developed.

## NOTES

[1] Email dated July 28, 2001 from D. Bromley, Anderson-Bascom Professor of Applied Economics, University of Wisconsin.

[2] Indeed, the quotes in the introduction to this paper are made within the context of institutions as property rights. For a review of property right definitions, see Haley and Luckert (1990).

[3] Randall (1987) refers to this chain of events as "property rights and the nonuniqueness of Pareto-efficiency"

[4] The ambiguity regarding links between SFM and social welfare are discussed further below.

[5] Indeed this has frequently been done for individual rules. See for example, Boyd and Hyde (1989).

[6] For example, Mollo (2003) describes forest tenures in Cameroon modeled after Canadian policy.

[7]Studies relating discrete types of tenure to management performance include examples from agriculture and forestry. With regards to forest resources, see for example Zhang (1996), Zhang and Pearse (1996, 1997), Deacon (1994), and Luckert and Haley (1990). With regards to agricultural production see for example, Anderson and Lueck (1992), Feder and Onchan (1987), Feder, Onchan, and Chalamwong (1988), Migot-Adholla (1991), and Place and Hazel (1993).

[8] See for example; Place (1995), Hansen (1998), and Hegan, Hauer, and Luckert (2003) who have isolated effects of specific rules on economic behavior in Africa.

[9] There is of course the Non-Industrial Private Forest (NIPF) literature, where firms may be maximizing a combination of profits and utility (e.g. Hyberg and Holthausen 1989).

[10] Developing country studies of empirical rates of time preference include Cuesta, Carlson, and Lutz (1994), Kundhlande (2000a), Pender (1996), Street (1990) and Zuhair (1986)

[11] Again, there seems to be more literature on measuring risk preferences in developed countries than in

developing countries. For developed countries see for example Pope and Just (1991). For developing countries, see for example Pannell and Nordblom (1998).

[12] Efficiency may also be a difficult concept to handle with Pareto efficiency. For example, efficiency is frequently defined in economics as achieving maximum value of output from a given base of resources. This result only corresponds to maximizing social utility if the marginal utility of income is equal for everyone in society. I thank Albert Berry for providing this example.

[13] See for example Luckert (1991).

[14] Coase (1960) described how individual firms may have incentives to internalize externalities by establishing property rights through negotiations.

[15] Pigou (1940) discusses the need for governments to internalize externalities through taxes.

[16] There are many notable exceptions to this statement for other types of natural resource/environmental issues. For example, there are numerous studies that exemplify the public choice literature described further below (e.g. Magee, Brock, & Young, 1989). A significant part of this literature deals with the political economy of environmental regulations (e.g. Oates and Portney 2001; Stavins 2004) that disclose lessons likely applicable to forestry. Nonetheless, such studies are notably absent in forestry, and generally concentrate on behavior within the political economy, without extending analyses to resulting behavior of firms in the market economy. As North (1993) states: "...we still know all too little about the dynamics of institutional change and the interplay between economic and political markets."

[17] Social choice analyses ask questions such as, what is the best set of rules (e.g. property right policy) to create desired economic behavior? Public choice analyses ask questions such as, what are the best political processes (e.g. policy-making rules) for coming up with desirable policies?

[18] British Columbia and Alberta are notable examples of provinces considering such policies.

[19] For example, we have long struggled with the co-directional causality of prices and quantities.

[20] I am grateful to Vic Adamowicz for bringing to my attention the importance of this point.

[21] It is interesting that while many of us scorn natural scientists for stepping over the academic line into advocacy, as economists, we rarely consider ourselves in this vein.

[22] Historically, thinking about improved policy designs has caused economists to come up with policy instruments that have had enormous impacts. Dales (1968) and tradable emissions permits provides a case in point.

[23] Alston Chase developed this point well in his 1998 Forest Industry Lecture, University of Alberta, Edmonton.

# REFERENCES

Adamowicz, W., & Veeman, T. (1998). Forest policy and the environment: Changing paradigms. *Canadian Public Policy*. 24S, 51-S61.

Alchian, A.A., & Demsetz, H. (1973). The property rights paradigm. *Journal of Economic History*, 33(1), 16-27.

Anderson, T.L., & Lueck, D. (1992). Land productivity and agricultural productivity on Indian Reservations. *Journal of Law and Economics*, 24, 427-454.

Besley, T. (1995). Property rights and investment incentives: Theory and evidence from Ghana. *Journal of Political Economy*, 103(5), 903-937.

Bledsoe, D. (2003). Using land titling and registration to alleviate poverty. *Presentation to the World Bank.* January 29, 2003. Washington D.C.

Bourdieu, P. (1986). Forms of capital. In J.G. Richardson (Ed.), *Handbook of theory and research for the sociology of education (pp. 241-258)*. Westport CT: Greenwood.

Boyd, R.G., & Hyde, W.F. (1989). *Forest sector intervention: The impacts of regulation on social welfare.* Ames: Iowa State University Press.

Bromley, D. W. (1991). Testing for common versus private property: Comment. *Journal of Environmental Economics and Management*, 21, 92-96.

Brown, T. C. (1984). The concept of value in resource allocation. *Land Economics*, 60 (3), 231-246.

Burkhardt, R. (2001). New directions needed: perspectives on progress towards sustainable forest management in Canada. *The Forestry Chronicle*, 77(1), 75-76.

Campbell, B.M., de Jong, W., Luckert, M., Mandondo, A., Matose, F., Nemarundwe, N., & Sithole, B. (2001). Challenges to proponents of CPR systems - despairing voices from the socialforests of Zimbabwe. *World Development*,, 29(4), 589-600.

Cheung, S. N. S. (1970). The structure of a contract and the theory of a non-exclusive resource. *Journal of Law and Economics,* 13, 49-70.

Coase, R. H. (1960). The problem of social cost. *Journal of Law and Economics,* 3, 1-44.

Cortner, H.J., Shannon, M.A., Wallace, M.G., Burke, S., & Moore, M.A. (1994). Institutional barriers and incentives for ecosystem management: A problem analysis. *Issue Paper Number 16,* Water Resources Research Center, University of Arizona.

Cropper, M.L., Aydede, S.K., & Portney, P.R. (1992). Discounting human lives. *American Journal of Agricultural Economics,* 82(2), 469-472.

Cuesta, M., Carlson, G., & Lutz, E. (1994). *An empirical assessment of farmers' discount rate in Costa Rica and its implications for soil erosion conservation.* World Bank: Washington D.C.

Dales J.H. (1968). *Pollution property and prices.* Toronto: University of Toronto Press.

Deacon, R.T. (1994). Deforestation and the rule of law in a cross-section of countries. *Land Economics,* 70(4), 414-430.

Feder, G., & Onchan, T. (1987). Land ownership security and farm investment in Thailand. *American Journal of Agricultural Economics,* 69, 311-320.

Feder, G., Onchan, T., & Chalamwong, Y. (1988). Land policies and farm performance in Thailand's forest reserve areas. *Economic Development and Cultural Change,* 36(3), 483-501.

Furubotn, E., & Pejovich, S. (1972). Property rights and economic theory: A survey of recent literature. *Journal of Economic Literature,* 10, 1137-1162.

Gamache, A.E. (1984). (Ed.). *Selling the federal forests.* The University of Washington, College of Forest Resources. Seattle, WA.

Haener, M.K., & Luckert, M.K. (1998). Forest certification: Economic issues and welfare implications. *Canadian Public Policy,* XXIV, supplement 2, S83-S94.

Haley, D., & Luckert, M.K. (1990). *Forest tenures in Canada: A framework for policy analysis.* Information Report E-X-43 (English and French versions). Forestry Canada, Ottawa.

Hansen, J. (1998). *Tree planting under customary land and tree tenure systems in Malawi: An investigation into the importance of marriage and inheritance patterns.* M.Sc. thesis. Department of Rural Economy, University of Alberta, Edmonton.

Harrisson, G.W., Lau, I.M., & Williams, M.B. (2002). Estimating individual discount rates in Denmark: A field experiment. *American Economic Review,* 92(5), 1606-1617.

Hatton, M.D., Adamowicz, W.L., & Luckert, M.K. (2001). Fuelwood collection in Northeastern Zimbabwe: Valuation and caloric expenditures. *Journal of Forest Economics,* 7(1), 29-52.

Hegan. L., Hauer, G., & Luckert, M.K. (2003). Is the tragedy of the commons likely? Investigating factors preventing the dissipation of fuelwood rents in Zimbabwe. *Land Economics,* 79(2),181-197.

Howlett, M., & Rayner, J. (1995). Do ideas matter? Policy network configurations and resistance to policy change in the Canadian forest Sector. *Canadian Public Administration,* 38(3), 382-410.

Hyberg, B.T., & Holthausen, D.M. (1989). The behavior of nonindustrial private forest landowners. *Canadian Journal of Forest Research,* 19, 1014-1023.

Kundhlande, G. (2000a). Empirical measures and determinants of individual rates of time preferences: A case study of communal lands households in Zimbabwe. Pp. 10-41. *In Economic Behavior of Developing Country Farm-Households.* Ph.D. Dissertation, Department of Rural Economy, University of Alberta, Edmonton, Canada.

Kundhlande, G. (2000b). Risk and insurance in a household economy: The role of cattle as a buffer stock.. In *Economic Behavior of Developing Country Farm-Households* (pp. 42-74). Ph.D. Dissertation, Department of Rural Economy, University of Alberta, Edmonton, Canada.

Leman, C.K. (1984). The revolution of saints: The ideology of privatization and its consequences for the public lands. In A.E. Gamache (Ed.), *Selling the federal forests* (pp. 93-162). Seattle, WA: The University of Washington, College of Forest Resources.

Loewenstein, G. (1987). Anticipation and the valuation of delayed consumption. *Economic Journal,* 97, 666-684.

Luckert, M.K. (1991). The perceived security of institutional investment environments of some British Columbia forest tenures. *Canadian Journal of Forest Research,* 21, 318-325.

Luckert, M.K. (1997). Towards tenure policy framework for sustainable forest management. *Forestry Chronicle.* 73(2), 211-215.

Luckert, M.K., & Haley, D. (1990). The implications of various silvicultural funding arrangements for privately managed public forest land in Canada. *New Forests,* 4, 1-12.

MacDonald, G.B., Arnup, R., & Jones, R.K. (1997). *Adaptive forest management in Ontario: A literature review and strategic analysis*. Ontario: Queen's printer.

Magee, S.P., Brock, W.A., & Young, L. (1989). *Black hole tariffs and endogenous policy theory: Political economy in general equilibrium*. Cambridge University Press.

Mandondo, A. (1998). *Institutional framework for community natural resource management in Zimbabwe*. WWF Southern Africa Regional Programme, Office, Harare.

Mandondo, A. (1997). Trees and spaces as emotion and norm-laden components of local ecosystems in Nyamaropa communal land, Nyanga District, Zimbabwe. *Agriculture and Human Values*, 14, 353-372.

Migot-Adholla, S.E. (1991). Indigenous land rights systems in sub-Saharan Africa: A constraint on productivity? *World Bank Economic Review*, 5, 155-175.

Mollo, C.N. (2003). *Forest regulations and policy objectives in Cameroon: An economic analysis*. M.Sc. thesis. Department of Rural Economy, University of Alberta, Edmonton.

Moyo, S., Robinson, P., Katerere, Y., Stevenson, S., & Gumbo, D. (1991) *Zimbabwe's environmental dilemma: Balancing resource inequities*. ZERO, Harare, Zimbabwe.

North, D. (1990). *Institutions, institutional change and economic performance*. Cambridge: University Press.

North, D.C. (1993). The new institutional economics and development. *Economics Working Paper Archive* at WUSTL, Economic History, #9309002, Washington University. http://econwpa.wustl.edu.

Oates, W.E., & Portney, P.R. (2001). The political economy of environmental policy. *Resources for the Future Discussion Paper*, 01-55, Washington, D.C.

Pannell, D.J., & Nordblom, T.L. (1998). Impact of risk aversion on wholefarm management in Syria. *Australian Journal of Agricultural and Resource Economics*, 42(3), 227-247.

Pender, J. (1996). Discount rates and credit markets: Theory and evidence from rural India. *Journal of Development Economics*, 50, 257-296.

Pigou, A.C. (1940). *The economics of welfare*. London: MacMillan.

Place, F. (1995). *The role of land and tree tenure on the adoption of agroforestry technologies in Zambia, Burundi, Uganda, and Malawi: A summary and synthesis*. Land Tenure Center. University of Wisconsin-Madison.

Place, F., & Hazell, P. (1993). Productivity effects of indigenous land tenure systems in Sub-Saharan Africa. *American Journal of Agricultural Economics*, 75, 10-19.

Pope, R.D., & Just, R.E. (1991). On testing the structure of risk preferences in agricultural supply response. *American Journal of Agricultural Economics*, 73, 743-748.

Randall, A. (1987). *Resource economics: An economic approach to natural resource and environmental policy* (2nd ed). Toronto: John Wiley and Son.

Sen, A. (1995). Rationality and social choice. *American Economic Review*, 85(1), 1-24.

Sethi, R., & Somanathan, E. (1996). The evolution of social norms in common property resource use. *The American Economic Review*, 86, 766-788.

Sjaastad, E., & Bromley, D. (2000). The prejudices of property rights: On individualism, specificity and security in property regimes. *Development Policy Review*, 18, 365-389.

Sobel, J. (2002). Can we trust social capital? *Journal of Economic Literature*, XL, 139-154.

Stavins, R.N. (2004). Introduction to the political economy of environmental regulations. *Resources for the Future Discussion Paper* 04-12, Washington, D.C.

Stigler, G.J. & Becker, G.S. (1977). De Gustibus non est Disputandem. *American Economic Review*, 67(2), 76-90.

Street, D.R. (1990). Time rate of discount and decisions of Haitian tree planters. *Haiti Agroforestry Project*, South East Consortium for International Development (SEDIC) and Auburn University, SEDIC/ Auburn Agroforestry Report No. 25.

Sultan, F., & Winer, R.S. (1993). Time preferences for products and attributes and the adoption of technology-driven consumer durable innovations. *Journal of Economic Psychology*, 14, 587-613.

Zhang, D., & Pearse, P.H. (1996). Differences in silvicultural investment under various types of forest tenure in British Columbia. *Forest Science*, 42, (4), 442-49.

Zhang, D., & Pearse, P.H. (1997). The influence of the form of tenure on reforestation in British Columbia. *Forest Ecology and Management*, 98, 239-50.

Zhang. D. (1996). Forest tenures and land value in British Columbia. *Journal of Forest Economics*, 2(1), 7-30.

Zuhari, S.M.M. (1986). *Harvesting behavior and perennial cash crops: A decision theoretic study.* Unpublished Ph.D. thesis, Virginia Polytechnic University, Blacksburg, VA.

# CHAPTER 3

# MODERN ECONOMIC THEORY AND THE CHALLENGE OF EMBEDDED TENURE INSTITUTIONS: AFRICAN ATTEMPTS TO REFORM LOCAL FOREST POLICIES

MARITEUW CHIMÈRE DIAW

*Centre for International Forestry Research*
*Cameroon Regional Office, BP 2008, Yaoundé, Cameroon*
*Email: c.diaw@cgiar.org*

**Abstract:** Customary tenure, mingled with state law and occasional private titling, continue predominantly to govern African rural and forest lands. This is in spite of evolutionist theories that predicted its demise and colonial and post-colonial policies that tried actively to accelerate it. The chapter develops an anthropological conceptualization of the institutions of embedded tenure. With examples from Africa and various parts of the world, we highlight the factors that account for the flexibility, adaptability and resilience of this type of system. Embedded tenure has been able to cope with economic stress and hostile policies because of the unique way in which it nests private entitlements into the commons, and both of them into collective property and long-lasting social institutions. Two philosophical principles giving rise to three constitutional rights and four appropriations regimes make up its structure, while dynamic access and transformation rules govern the interlocking and transmutations of appropriation regimes across space and time. This opens three specific paths of agricultural change as well profuse right delegation and land transaction procedures. These have helped the system adapt to changing economic, demographic and social conditions since at least the 19th century. Reductionist economic analyses of non-market systems, including property rights, kinship, common property and non-wage systems, contributed to relegating the most innovative aspects of these systems to the limbo of imperfect markets. More interested in 'crafted' institutions, the CPR literature also failed to see or emphasize the theoretical and policy implications of the nesting of appropriation regimes in embedded tenure. Other analyses have tended to overemphasize the controversial role of traditional authority at the expense of a deeper institutional analysis of the embedded system. The chapter highlights the numerous policy mistakes that can derive from these forms of reductionism. We conclude that, to disentangle African forest policies from the social costs and inefficiencies of the past, it is necessary to integrate the complexity and validity of embedded tenure institutions and their demonstrated ability to adapt to legal pluralism and commodity markets.

*Kant and Berry (Eds.), Institutions, Sustainability, and Natural Resources: Institutions for Sustainable Forest Management, 43-81.*
© 2005 *Springer. Printed in Netherlands.*

## 1. INTRODUCTION

There is massive evidence across Africa that rural lands, including forests, continue to be predominantly governed by indigenous tenure principles, mingled with state law and occasional private titling (Bruce, 1998). Since at least the 1960s, numerous researchers have reviewed the systems of rights that govern African land and forest tenure to find that, far from disappearing, these systems, already complex in precolonial times, had further evolved into multidimensional constructs of econiches and overlapping rights[1]. This is consistent with observations made in other parts of the world (Brown, 1998; Doolittle, 2001; Fisher, 1989; Kant, 2000; Michon, de Foresta, & Levang, 1995; Sirait, 1997, Vandergeest, 1996); it also comforts Shashi Kant's finding[2] that property rights in tropical forest systems have evolved toward Pareto efficient pluralism rather than the singular private property optimum that economic theory had predicted.

This economic and legal pluralism[3] and the underlying resilience of indigenous tenure institutions (Diaw, 1997) are a formidable challenge to the theory of non-Western economic institutions. It has been argued that the evolutionist theory of land rights was not realized in the African case because of bad policy implementation and ambivalent, inconsistent and bureaucratic African land laws (Platteau, 1992b). We show in this chapter that African post-independence land and forest policies have actually been consistent with the evolutionist-modernist paradigm they inherited from the colonial era. Rather than looking at the superficial dimensions of those policies, we suggest instead investigating their ontological dimension. African land and forest policies have been rooted in epistemologies of modern transformations that considered indigenous tenures and other forms of non-Western economic 'otherness' as doomed to be replaced by higher forms of modernity. That this did not happen as predicted should be a powerful incentive for revisiting the theoretical parameters under which the debate on Western and non-Western forms of economic organization was originally framed.

This chapter begins with a review of evolutionist and New Institutional Economics (NIE) theories of property rights and their theoretical ramifications into the analysis of non-wage systems, common property and kinship in non-Western societies. This is followed by a historical sketch of the attempts made by colonial and post-colonial states in Africa to contain, dislodge, or replace customary tenure. The parallel between conventional economic views and African tenure policies highlights the syndrome of extraordinary treatment to which indigenous tenure systems have been subjected. We see in both cases an initial phase of scientific negation and policy exclusion, followed, after decades of failure, by a phase of limited recognition or rehabilitation. We argue that this way of 'walking backward' the path of understanding the social system generates epistemological bias and reductionism as well as a high social cost of policy change.

The two core sections of the chapter develop an anthropological conceptualization of the institutions of embedded tenure. Making ours Popper's falsificationism and the principle that theories should compete on the basis of 'always higher empirical content', we base this on an empirically-grounded comparative analysis of indigenous forest tenure; our examples and references are drawn from various parts of Africa and the world, including Cameroon where we collected solid empirical material over nine years of field work. Our analysis shows how interlocked private, common, and collective rights are embedded in long-lasting social institutions and allocation principles that explain the considerable resilience of indigenous tenure into modern times. This analysis also dispels naïve criticisms of "traditional" tenure, which overemphasize political discontinuities and the controversial role of authority in hierarchical models of 'traditional governance'. This has been generally done at the expense of a deeper institutional analysis of the system of rules –including 'rules of transformation'- that governs embedded tenure and its interaction with statutory law, markets and other externally driven forces of change.

The last, section revisits the policy implications of this analysis of embedded tenure. We start with a brief review of Karl Polanyi's theses and the fierce debates they triggered in the 1950s and 1960s among economic anthropologists and economic historians before reaching, much later, the mainstream of economic thought (North, 1977). We then proceed to contrast the peculiarity of tenures in feudal Europe – where this concept was invented – with embedded tenure forms that survived the development of capitalism. We provide examples from the recent history of forests, which show how the misunderstanding of the rules of interaction and transformation that guide embedded tenure can lead to policy mistakes. We also caution against new attempts to codify local property rights or to privatize the global commons without taking full stock of past mistakes and of what we should now know of the complex, fast changing and fine inner workings of embedded property rights systems.

We conclude by recognizing the social, political and economic pluralism of local forests as recent social and economic analyses (Kant, 2003) demonstrate. Within that broader framework, the attempts to disentangle African forest policies from the massive social cost and inefficiencies of the past should fully integrate the complexity and validity of embedded tenure institutions and their demonstrated ability to adapt to legal pluralism and commodity markets. This ability 'to change while remaining the same' should be a defining factor of theory and policy rather than a last resort or an afterthought.

## 2. THE 'EXTRAORDINARY TREATMENT' OF INDIGENOUS TENURE

The evolutionist theory of property rights and its revision by the New Institutional Economics are key markers of the debate on non-Western forms of tenure. Their theoretical extension into the analyses of non-wage systems, common property and kinship are also central, given their importance in embedded tenure dynamics. These theoretical analyses have strong parallels in policy. We review the broad lines of these policies since colonial times in order to see how they failed to take root in the African case.

## 2.1 The Evolutionist Theory of Property Rights

According to the evolutionist theory of property rights, agricultural systems are submitted to a general process of transition from communal forms of tenure to private land ownership. Under population pressure and market penetration, various changes take place in the relative prices of factors. At a certain point, land becomes alienable by private individuals. It thus acquires a collateral value and becomes an asset, which increases the supply of credit and allows the accumulation of capital in agriculture. According to Demsetz (1967), any externality comes from a potential gain of the sale of one set of property rights against another. If the exchange takes place, the externality is "internalized"; if not, there is market failure. The prohibition of exchange (the case of collective property) or the existence of prohibitively high negotiation costs (the case of 'common property') thus creates externalities prejudicial to investment and efficiency. This view has served to justify colonial and post-colonial policies that wanted to eliminate customary tenures as well as land reforms that sought to accelerate the pace of privatization in systems that do not fit these evolutionist conditions (Goodland, 1991, Harrison, 1987, World Bank, 1974). We will see later that these policies were in many ways unsuccessful.

## 2.2 Contractual Relations, Kinship, and the New Institutional Economics

In the 1960s, the frustration expressed in the post-war economic literature about "market failures" began to crystallize into an internal movement of reform of conventional microeconomics. With the publication of *The Problem of Social Cost* by Ronald Coase (1960), the hope of reducing the "black box" of these market failures by taking transaction costs and property rights into account began to take on an operational form within academic circles. This was when the issue of common property resources (Hardin, 1968) began to take shape, as did the emerging theory of collective action (Olson, 1965), both of which deal with the *prisoner's dilemma* and the problem of the *free rider* (see also Axelrod, 1983). The theorization of property rights (Demsetz, 1967), together with the neo-Marshallian (Cheung, 1968, 1969) and neo-institutional (Datta & Nugent, 1989; North 1990; Platteau, 1989, 1992a) theses on property rights and kinship formed part of this growing movement.

The New Institutional Economics (NIE); emerged and developed as an ambitious research program in the lakatosian sense (Alchian & Demsetz, 1972; Coase, 1984; Davis & North, 1971; North & Thomas, 1973; Williamson, 1975). As much by *percolation and translation* as by *coagulation*, this movement to restructure microeconomics annexed or rallied to its flag all the main themes of our time, including those of economic anthropology, dealt with in the form of a response to *"the challenge of Karl Polaniyi"* (North, 1977). The approach of *transaction costs*, linked to the central concepts of *bounded rationality*[4] and *opportunism*[5] (Williamson, 1975, 1985), provided the unifying framework needed for its analytical coherence. This framework was then massively applied to as diverse centers of interest as the mining and manufacturing industries, insurance, agricultural tenure, livestock, fisheries, the structure of firms and of the extractive industry, electoral coalitions, rent-seeking strategies, and the emergence and decline of civilizations.

Logically, the NIE revisited the initial analysis of property rights, criticizing it for its "mechanistic and technocratic bias" (Platteau, 1989) and for its 'naive approach' of institutional evolution: "it is absurd to argue that processes of institutional change 'optimize'" (Nelson, 1995, cited by Angelsen, 1997). According to Douglas North (1990), two contradictory forces define the path to institutional change and the different performances of societies: "increasing returns", responsible for the "exceptional success story of the Western World" in economic history (North & Thomas, 1973), and "imperfect markets characterized by significant transaction costs". The resilience of non-market forms of economic organization thus becomes interpretable in terms of "path dependency" generated by the inertia of mental constructs and institutional frameworks. Despite a greater sophistication, the policy implications of the analysis remain basically the same as with the initial theory. Private property rights alone can neutralize transaction costs instead of generating them (Baak, 1982); they must be instituted for economic development to occur. This gravely underestimates the role of power and politics in world history as well as the complexity of relational and innovative dynamics in non-market social systems.

The NIE's view of kinship and social networks in non-Western societies is consistent with North's thesis of path dependency. Basically, these pillars of the social system are construed as "*risk pooling* mechanisms" substituting for the inexistence of insurance markets, and as "*implicit insurance contracts*" in "primitive societies" facing the continuous threat of violence. This includes a diverse set of complex institutions such as reciprocal credit, village associations, gifts exchange, family enterprises, children adoption, and broader kinship, extra-domestic and intergenerational relations (Datta & Nugent 1989; North, 1990; Platteau, 1989). In the context of the evolutionist theory of property rights, Platteau (1992a) attributes this "*risk pooling*" function to the "extended African family". Presented as a "*collateral substitute*" for market imperfections, it supposedly slows down innovation and technology adoption "*because of the conservative mind of the elders*".

This metamorphosis of the African lineage into a "collateral substitute" concludes the process of integrating social otherness into the rationality of "insurances". This process is based on several 'absences', ranging from the absence of private property rights to the absence of writing and archival traditions or stable climatic conditions; it culminates in the "imperfection", "deficiencies", and "underdevelopment" of the market in its various states. This last 'absence' is, in a way, the basis and 'ontological truth' of all the others. It is because of market imperfections that the whole fabric of "traditional" societies becomes interpretable in terms of insurance networks; as if the logic of insurance was an intrinsic and innate fact of the social system. There is complete inversion of the relational history of the social system and the market (see sections 2-3).

*2.3 Common Property and Non-wage Systems*

This misunderstanding of non-Western systems has noteworthy ramifications in areas other than tenure and kinship. In a previous contribution (Diaw, 2002), we reviewed the history of the debates on "Common property" and non-wage systems such as the share system in fisheries and sharecropping in agriculture. From the concept of "rent dissipation" in common property fisheries (Gordon, 1953) to 'neo-marshallian' analyses of sharecropping (Bardhan & Srinivasan, 1971; Bell, 1977), we see a pattern where the existence or validity of these systems was first rejected before being laboriously reinserted into preset hypothetical-deductive models that could not express their practical rationality. From Hardin's (1968) tragedy of the commons to the realization that common property is not open access or "everybody's property" (Ciriacy-Wantrup and Bishop, 1975), the evolution of the common property debate is well known. The "productive inefficiency" of sharecropping was a direct result of the theoretical postulates of marginalism. Assuming mutual equality and equal shares, neither owner nor tenant would invest its resources beyond the point where the marginal product equals half (and not all) of the product. This was a theoretical impossibility. But this result could not stand by itself in light of the need to explain the continued existence of this system. It was revisited by Cheung (1968), who introduce new elements into the model and triggered a debate that raged for years between traditionalists (e.g., Bardhan and Srinivasan, 1971, Bell, 1977) and modernists (e.g., Newberry, 1974; Stiglitz, 1974). The share system was first discussed in the mid-50s at the round table of the International Economic Association (Turvey and Wiseman, 1956). Despite interesting survey data produced by Zoetweweij (1956), the discussion soon sidetracked on issues of *raison d'être*. This led to a recess of 20 years[6] that, ironically, saved it from the negationist stage that had marked the debate on sharecropping.

For both systems, however, the introduction of new elements (uncertain and imperfect markets, transaction costs, negotiation and contractual outcomes, minimization of market and opportunistic risks, etc.) into the marginalist matrix basically sought to retain the core of its maximizing assumptions. As a result, the share system could be no more than an "illusion" to be modeled as a long run proxy for wage labor (Anderson, 1982; Flaaten, 1981, Nugent & Platteau, 1989). In the case of sharecropping, Srivastava, (1989) showed that the new elements introduced to explain its existence imply some other source of "imperfection" and inevitably *"push the models into the realm of 'second' best"*[7].

The comparative anthropology of sharecropping (Robertson, 1980, 1987) and of the share system (Diaw, 1989, 1994), showed the complex, flexible nature of these systems, their specific mathematical coherence (in the second case), and their capacity to redistribute wealth and to develop as post-capitalist innovations. On the whole, however, and with the exception of the common property discussion, contributions from non-economic social sciences had little influence in shaping these issues. Morisset and Reveret's (1985) economic analysis of the Individual Transferable Quota, ITQ, is somewhat an exception; it uses anthropological contributions to highlight the historical incapacity of the vertically integrated firm to compete with coastal fisheries on a simple basis of market and production costs. Beyond the rent 'dissipation-restoration' narrative that justified state intervention in the fisheries, it had to be seen that these two forms of production have neither the same concepts of costs nor the same management principles. Whilst the capitalist enterprise has to pay salaries and remunerate its capital, coastal fisheries use family and/or share-rewarded labor. With the introduction of the ITQ, coastal fisheries lost adaptability and became more vulnerable to market forces.

### 2.4 The Colonial Roots of African Land Tenure Nationalism

Historically, in Africa, land tenure policies have taken place in contexts of bitter struggles between nation-States and rural communities and have generally worked against local controls over the environment. This global phenomenon left very few nations untouched. The succession of *"national domain"* laws, which marked the emergence of *land tenure nationalism* in Africa, in the 1960s and 1970s, is symptomatic of the phenomenon. These laws sought to reduce "traditional resistance" to the development and modernization of societies in accordance with the European model. They were intended to "break" the communal basis of land tenure systems - to *"detribalize"* them (Melone, 1972) - in order to establish the territorial basis considered essential to the "rational development" of the new nation-States. The State thus became the exclusive "manager", "guardian", "administrator" (Senegal, Côte d'Ivoire, Mali, former Upper Volta, Madagascar, Cameroon, Ghana, etc.), or the "owner" (Guinea, Mauritania, former Zaire, etc.) of the national estate (Diaw & Njomkap, 1998). A few countries, such as Kenya and, to a lesser extent, Uganda, developed strong privatization programs, while others, such as Ethiopia and Tanzania, attempted to replace customary tenure with sweeping villagization or land-to-the-tiller reforms (Bruce *et al.,* 1998a). Across the continent, post-independence tenure nationalisms thus appear as a systematic effort to dislodge or displace indigenous tenure in order to replace it with registered or state-administered land, as a transition toward state leasehold and freehold tenure. In almost all the countries concerned, these policies came up against strong grass-roots resistance and were affected by sporadic conflicts (Coquery-Vidrovitch, 1982; Fisiy, 1990; Melone, 1972; Tjouen, 1982). This brings to mind the problems encountered in the dissociation of the forest from agrarian systems in Europe, where these policies have their historical roots[8]. Their results, in Africa, differed greatly, however, depending on the country and the environment[9].

A major paradox in African land tenure nationalism is its origin in colonial tenure policies. It was Faidherbe, in 1865 in Senegal, who began the French policy of promoting private ownership through land registration techniques. This basically denied pre-existing communal land rights. Alexandre Tjouen (1982), who shares the non-tribal and developmentalist ideals of the 1974 land ordinances in Cameroon, traces their origin in the Imperial ordinance of June 1896, under the German colonial regime. Reinterpreting the treaty signed twelve years earlier with King Akwa of Dwala, this latter incorporated so-called vacant (*herrenlos*) lands into the lands of the German crown (*kronland*). This move opened the way to the distribution of millions of hectares of traditionally owned forests to German agricultural and forestry companies (Egbe, 1997; Tjouen, 1982).

The French and British legislations, which succeeded the German occupation in 1916-1919, made several modifications to technical aspects of the legal framework without changing, however, the basis of the new relationship established between the State, the land and local communities. On the whole, "freehold lands" were kept outside local control, despite the recognition of limited areas within which traditional rights remained valid. Under the French colonial system, this limited recognition was done through a recording procedure, *le régime de la constatation*, which later introduced an administrative certificate and then a land record book. In the British system, customary rights to *"native lands"* were recognized by the Forestry Ordinance of 1916 and by the Land and Native Rights Ordinance of 1927. This did not include the former *herrenlos* lands. All lands were also placed under the ultimate authority of the Governor of Nigeria who had "all-embracing powers of regulation and disposition" in the British territory (Anyangwe, 1984; Coquery-Vidrovitch, 1982; Egbe, 1997; Ngwasiri, 1984)[10].

Through this whole process, the pre-eminence of reasons of State and of land titling remained the basis of the arguments opposed to communities in matters of land tenure. This 'ontological divide' did not change with post-independence legislation.  It even became more radical, as the regime of « *constatation* » disappeared from Cameroonian law in 1966 to be replaced by the principle of « *mise en valeur* » (*making valuable use*), more in tune with the ideology of planned development and the normative target of individualizing tenure rights[11]. The case of Cameroon is but an example of a general process that virtually touched all African societies. African forestry legislation tightly linked to these land tenure laws and to land use planning concepts were even more state-centric. They separated the *forest-as-trees* from other agrarian uses and put legal limitations to traditional authority over trees and forest expanses. They also gave the state considerable powers to grant state concessions and felling rights to private entrepreneurs, and to establish various forms of protected areas in its eminent domain. It is only in the 1990s that these policies started to undergo new cycles of decentralization reform under the two-fork pressure of international institutions and traditional "encroachments" and "resistance to change". The new framework, however, maintained global legal coherence with the philosophy of State edification and the pre-eminence of private interests over forest exploitation.

*2.5 The Failure to Achieve the General Dissolution of Indigenous Tenure*

As a whole, colonial and post-colonial policies in Africa failed to achieve the widespread dissolution of indigenous tenure anticipated as a condition for 'civilization' and 'modernization'. Effective tenure changes can be observed in urban, peri-urban, and special-development areas. However, the general situation in rural and forest areas is one of an uneasy compromise between externally-imposed statutory law and indigenous tenure. The 1996 country profile of land tenure in Africa published by the Land Tenure Center (LTC) (Bruce, 1998), found that customary or community-based tenure was the *"de facto dominant tenure type"* in almost all sub-Saharan countries. The only exceptions were Cape Verde, South Africa and Namibia. In Kenya, customary tenure was found to be co-dominant with private ownership despite one of the most aggressive, long standing programs of privatization on the continent. The situation is basically the same in Senegal, which privatization scheme, the first in francophone West Africa, started in the 1830s. In Cameroon, this coexistence takes the form of a 'legal compromise', established in the 1930s, under the French regime (Diaw and Njomkap, 1998). Under this compromise, local tenure matters are managed under 'informally recognized' indigenous law at the communal and administrative levels of local jurisdiction. It is only when such questions reach the judicial system (after three stages of local arbitration) that they are taken over by the statutory system of law.

In their review of the tenure profiles of 22 West African countries in 1996, Elbow *et al.* (1998) identify four classes of policies toward indigenous tenure: (i) 'non recognition or abolition' (50 %), (ii) 'neutral recognition' (14 %), (iii) 'recognition aimed at replacement' (14 %), and (iv) 'zoning recognition' (23 %). This last category, which lumps together Guinea Bissau, a former Portuguese colony, and all former British colonies but Nigeria, concerns mainly the continuation of the colonial legacy of tribal authority lands and does not preclude strong replacement policies in some of these countries. In addition, the study found that little difference could be discerned in practice among these four categories of policies, despite differences in the degree of formal recognition of customary tenure. None of them offered real, proactive protection to indigenous tenure and all of them reproduced the dualism of past colonial policies.

From the viewpoint of theory, the "last best-last resort" status of indigenous tenure systems is primarily the result of theoretical hypotheses that have not been empirically proven. To fill this lack of empirical and statistical data, the LTC and the World Bank carried out two studies on the issue of "land tenure security" (Bruce & Migot-Adholla, 1994). Covering seven African countries, the studies tried to verify if the individualization of land rights indeed leads to increased investment and agricultural productivity. This field orientation led to important *"counter-intuitive"* findings; on the whole, no significant relationship could be found between private land ownership and the use of agricultural credit, land investment, land improvement, and agricultural yields (Bruce, Migot-Adholla & Atherton, 1994). It was even discovered, as in Kenya and Senegal, that national legislation and land registration were a cause of uncertainty and land insecurity rather than the opposite. Simukonda (1992) raises related issues in a separate study on Malawi.

This failure of the 'land tenure security' paradigm to successfully pass the test of empirical validation is vindicated by studies on the effect of land and tree rights on the adoption of alley farming–hedgerow intercropping–in West Africa, notably, in Nigeria, Togo and Cameroon (Adesina, Nkamleu, & Mbila, 1998; Lawry, Stienbarger, & Jabbar, 1995). The former study, based on Logit investment models using aggregated and disaggregated property rights variables and data from three provinces in Cameroon, rebuts views of 'land security' as precondition for adopting that technology. Rather than land rights –whether 'complete', 'preferential' or 'limited'-, it is tree rights and other institutional and agroecological variables, which influenced alley-farming adoption. This confirms that 'tenure security' cannot be simply equated with 'complete bundles' of marketable rights. Further, and with regard to the issue of technological innovations, 'security' does not logically refer to the right to sell the land (the support base of innovation). It is the right to access the land and the returns on investments made in it that is the real security stake. We will see in section 2 that this is guaranteed by the notion of "productive rights" in embedded tenure systems. We look at the profusion of transaction and investment forms that it enables in section 3.

The resilience of indigenous tenures at the heart of legal constructions that sought explicitly or implicitly to displace it, or to uproot and break it, needs to be explained beyond superficial policy differences across time, countries and political regimes. We demonstrate in the following section that tenure embeddedness in long-lasting and dynamic social institutions is the key to this resilience.

## 3. THE EMBEDDED TENURE

By 'embedded tenure', we refer to an appropriation regime where private, shared, and collective rights to natural resources are nested into each other and into larger social institutions based on kinship and descent. This section focuses on the underlying system of rules that makes up this system and gives it basic structure and coherence. In doing so, we recognize the rule-base (Nabli and Nugent, 1989; North, 1990) as well as constructed nature (Giddens, 1987; Long, 1992; McCay, 2002) of social institutions. It is not possible to understand how actors and structure interplay to produce institutional change and institutional resilience without decoding the historical and structural conditions that shape this interaction and give it meaning. Most recent works on land tenure and local institutions have emphasized with reason the negotiated, fluid and messy character of local tenure arrangements in contexts of uncertainty and legal plurality (Lavigne Delville, 1998; Leroy, Karsenty, & Bertrand, 1996; Mehta *et al.*, 1999). The point is well taken but should not lead to ignoring the redundancies and coded patterns, which inform on the structure and emerging properties of these systems. There is always some level of indeterminacy in human understanding of social systems. This is due to indeterminacy in the system itself – in relation mainly to the uncertain and open nature of interactive outcomes – as much as to our own cognitive and ideological limitations. In the case of non-Western systems, ideology and ethnocentrism have played a major role in preventing a full understanding of those systems. Leaving their ground rules and coding patterns floating in indetermination would thus only add to rather than resolve the analytical confusion inherited from decades of reductionism. In the following sub-sections, we unpack the key constitutive elements of embedded tenure, notably, its 'constitutional roots' in history and social reproduction and the philosophical principles that organize its conversions and shifts between nested property and access regimes. We will look at issues of systemic change, agency and politics in the next section.

### 3.1 Blood Rights and Civil Rights

In embedded tenure systems, the right to access, withdraw, hold, or possess natural resources generally comes from membership in a natural or putative group of descent. 'Blood rights', that is, rights that are acquired and exclusively transmitted through a bloodline are the first defining characteristic of embedded tenure. We know since Morgan (1877) that the opposition between 'gentile society', where government is exercised through a descent group, and 'civil society', where government is based on political citizenship, is key to understanding government systems in history. These two competing principles still coexist within modern nation-states where the right to acquire citizenship, to hold certain public offices, or to vote is subject to conditions of descent and/or residence varying greatly from one country to the next. In tenure issues, the question is not political access but resource access. The divide between disembedded (e.g., private property) and embedded tenure in modern nations has also shown to be much sharper and long lasting than the opposition between these different organizing principles of political order.

## 3.2 Migration and First Occupancy: The Making of Territorial Rights

Tenure institutions based on descent and blood rights are still significant or dominant forms of land use and natural resource allocation in tropical agrarian and forested landscapes. This includes virtually all sub-Saharan African countries, as well as numerous countries and regions of Asia and South America. Typically, the relationship between blood rights and land rights in embedded tenure was established through a historical process of first occupancy. In *Property and persuasion*, Carol Rose (1994) describes the transformative and jurisprudential processes through which first occupancy and first possession were socially contested and debated in the framing of property law in the United States. In the Congo Basin, historical rights of first possession were established through the large enclosure movement that marked Bantu expansion in the forest during the second millennium. This movement accelerated in the 17th to 19th centuries, inducing major re-compositions of ethno-cultural landscapes (Diaw, 1997; Vansina, 1990)[12]. As elsewhere around the world, original rights of first occupancy could be displaced by force through war or with the arrival of a stronger or larger group. This is the case, for instance, of the Lokasani of southern Equateur, in D.R. Congo, who were displaced from their first-occupant territories by the massive arrival of other Mongo groups around the 16th century. In some cases, these groups–Kundo, Ekonda and Ntomba, mainly–were themselves pushed eastward and southward under the pressure of Ngombe and other non-Mongo groups (field data, Feb, 2004). Such situations were widespread in the region. They were a major feature of the historical colonization of the Cameroonian rainforest by myriad Bantu groups in the 16th-19th centuries (Diaw, 1997). We consider these two types of situations –first occupancy and 'final occupancy'- as basically similar, insofar as they stage the historical establishment of permanent territorial rights, whether through force or peaceful means. Neighboring groups then went through the stage of recognizing each other's exclusive right of use, possession or alienation over the territory each of them controlled. Such territorial rights thus became permanent and secure 'constitutional rights' across a region or a series of connected regions. Externality and universality of social recognition are, indeed, attributes of constitutional rights. Secure and permanent group ownership is, in that sense, the first constitutional principle of embedded tenure.

### 3.3 Genealogical Land Rights: The Encoding of Tenure in Social Reproduction

It is predominantly through clearing and symbolic marking of a territory that people established ownership of forestland. As the first act in forest agriculture, the *droit de hache*, or 'axe right', is a constitutional right in African land history; it expresses the land right gained through clearing of the forest and marks the taking over of a new territory by a founder of lineage. To be constitutionally recognized, the designation of a territory demarcated by natural boundaries such as rivers and hills was supplemented by the physical marking of that territory through symbols of human occupancy and productive use. In the *lamanat* of the Senegambia, the *droit de feu* or 'right of fire', which led to massive deforestation in the Middle Ages, was the equivalent of the axe right in the Congo basin rainforest. It designated the land rights acquired through burning of forest patches. In Cameroon, the exercise of the axe right was reflected in the occurrence of plants such as the bush mango, *Irvingia gabonensis*, which is one of the few fallow species to thrive in regrowth forest. Its presence in primary forest is thus a sign of ancient human use and land appropriation. Other plant species such as "the red flower", *Dracaena spp.*[13], are also used to mark out hunting and gathering territories.

The axe right and other rights of first occupancy had fundamental genealogical and collective implications in African tenure history. In the Congo Basin, it was a full component of the dominantly segmentary and patrilineal social organization. In patrilineal descent, jural relations in kinship are confined and transmitted through male lines going back to an original ancestor. The segmentary institution is based on increasing genealogical ramification, whereby an original clan or lineage is segmented into lineages of progressive inclusiveness tied to the founding lineage. According to Vansina (1990), the patrilineal and segmentary system may have been adopted as early as the 13th century to become the most general blueprint for social coordination among forest peoples. During the Bantu historical migrations, segmentation was one of the main mechanisms whereby minor burgeoning « houses » split from original lineages to colonize new territories (Diaw, 1997). A site would be identified, marked by the axe or its substitutes, and all offspring of the founders would become entitled to the resources on that land on the basis of the access principles that we describe below.

A number of other social institutions regulating marriages and alliances, residence, inheritance, and social reproduction in general contribute to the embedding of tenure in society. Among the Bantu and, in a more mitigated way, among Bagyeli, Baka and Batwa pygmies, clan exogamy and virilocality (marriage out of the clan and in the husband's residence) are direct complement of the patrilineal system. Both principles are essential to the gendering of tenure and inheritance. It is because a clan is of the « same blood » or the « same womb » that it is necessary to marry outside the clan to avoid incest. Under the virilocal rule of residence, women are the ones who have to leave to be married 'elsewhere'. They cannot, therefore act as guarantor of the land patrimony and inherit immovable property in their lineage of birth. In her lineage of adoption, a widow has full inheritance right but can transmit her holdings only to her sons and daughters-in-law, not to her daughters. The result is a precarious status of female inheritance rights within both lineages of birth and adoption. However, this marginalization of women applies only to the inheritance system as a woman enjoys full 'productive rights' in her place of residence. This is why the matrimonial status of women is the institutional key to their tenure status in patrilineal-exogamic systems, not the tenure system in itself. This is very important for gendered policy options in this type of embedded tenure system.

We must clarify that the patrilineal/segmentary mode of organization is not universal and coexists with other forms of tenure embeddedness into kinship and descent. Among the Fanti of Ghana, the Kongo of Congo and Angola, or the Minangkabau in distant Sumatra, Indonesia, matrilineal descent is the fundamental conduit of social organization, including tenure. Similarly, among the Chewa, Yao and Ma'nanja of Central and South-central Malawi, it is matrilineal and uxorilocal (residence of the husband in his wife's home) rules, which influence the embedded tenure system (Bruce et. al., 1998b). The result is an extremely complex map of governance rules and reproduction principles across societies.

## 3.4 The Philosophical Principles of Embedded Tenure

Embedded tenure is based on two simple philosophical principles, which, as far as we know, seem to hold everywhere. The first is the embodiment of private rights into permanent collective rights; the second is the inalienable right of individuals to live off their own labor through natural resource entitlements. Three series of constitutional rights[14] derive from these two principles:

(i) *Genealogical rights.* They are responsible for the collective nature of land ownership and the striking resilience of the embedded tenure system in history. *Land, and the resources it supports, belongs collectively and organically to the dead, the living and the unborn* (Agbosu, 2000; Diaw, 1997). Use and access are grounded in blood rights based on kinship and descent. Land is a permanent patrimony, which cannot be alienated, except in stressful circumstances. This principle of 'exo-inalienability' of land (Verdier, 1971) is a direct product of history, as we have seen. Collective property is its foundation; that is, property in the strongest sense, as it applies to real, immovable assets on which the owners, collectively, have complete, exclusive, and permanent rights.

(ii) *Productive rights* constitute the second series of rights in the system. They are the level at which the collective foundation of the system is reconciled with private rights and the value attributed to human labor. Land use has a base in usufruct and all members of the community have a fundamental right to live off and benefit from the product of their own labor. Individual rights to land express this: the right to work, 'to create', to use or open the forest, to clear a long-abandoned fallow field or to plant trees in a young fallow. This series of rights apply to all members, male and female of the founding lineages in a community. They can be delegated to strangers or to other community members under conditions that we describe in section 4.

*The investment of labor into the resource is the fundamental appropriation principle in this series of rights.* The security and duration of tenure depends on the enduring physical evidence of the work embodied in the resource. This principle was responsible, in great part, for the major drive to establish cocoa plantations in Cameroon in the 1930s, with the original view of extending the duration of individual tenure on collective land rather than realizing financial gains (Leplaideur, 1985; Weber, 1977). Productive rights are thus, typically, *subsistence and development rights* realizable through individual *axe, use or planting rights* that result in the permanent or temporary conversion of a collective asset into an individually controlled resource.

(iii) *Succession and inheritance rights.* The whole system is based on lineal descent, marriage and residence. The place of first and second sons, of other children and of widows and divorcees, varies considerably across societies and traditions[15]. In exogamous virilocal systems, women have to leave the clan to marry in their husbands' residence. The collective nature of ownership thus means that women cannot inherit from their original lineage and can only enjoy productive rights in their place of residence. In South Cameroon, widows do inherit from their husband, however, and direct inheritance by unwed daughters is a growing phenomenon. In matrilineal and uxorilocal systems, women are the natural guardians of the collective heritage and male members of the lineage inherit through their mother's line. Both patrilineal and matrilineal rules of inheritance are intricately tied to the collective origin of ownership and productive rights; this is also true of the variable place of different family members in the system. These three principles –collective ownership, private appropriation and lineal inheritance- are the bases upon which a diversity of appropriation regimes are articulated.

*3.5 The Interweaving of Property Regimes*

In the process of transforming their collective endowments into private entitlements, community actors apply the principles of rights outlined above in different ways across the landscape. Different property regimes, which are both contiguous and nested into each other, thus emerge to form different configurations of resources and land uses. In the most sophisticated systems, these property regimes are intertwined in constant cycles of environmental conversion and social mutation in both time and space. Four property and access regimes can be distinguished in this system:

(i) *Collective property.* All classes of land use, except for certain open access domains of the primary forest and the aquatic space, ultimately belong to this property regime. The basic form of collective property is lineage ownership. However, cases where this type of ownership is subsumed within larger political or social structures of allocation (e.g. a kingdom) are frequent. In the basic lineage system, access is limited to members of the lineages owning the portion of territory concerned. Several lineages in the same community may have exclusive rights on their separate portions of land, while other areas may be shared among them. In that case, collective lineal ownership is nested into the wider communal ownership of the village territory. Strangers can be granted conditional access rights through the individual lineages or the community leadership structure. The permanence of the property status of the resource base is the key factor distinguishing this property regime from other appropriation regimes in the system. Land and other natural resources are submitted to active processes of conversion and transmutation involving different, subsidiary forms of tenure. These other appropriation regimes do not alter the outer boundaries and ultimate status of the collective land endowment but reflect the concrete actualization of private and social entitlements to the resources and the land. This means that, while collective property provides the structural framework for the embedded tenure system, these other appropriation regimes are the key to its structural dynamic.

(ii) *Open access*. Areas of open access, where no property rights have been established are relatively rare in agrarian and forest environments. In Southern Cameroon, such areas are mainly arid and exhausted land, roads and trails, and major rivers and water bodies. Non-appropriated forests are rare, even when the communities around them do not actively use them. In that sense, open access must be distinguished from the regime of 'open communal access' that we describe below. Some tree products may have exceptional open access status. Such is the case of the *esok (Garcinia lucida)* bark, an antidote to a whole range of poisons. It is usually mixed with palm wine in social gatherings and considered a free social good, which should be accessible to anyone (Diaw, 1997). However, with the increasing pressure on this resource and the development of market opportunities, there is an increasing tendency to limit free access and enclose it more firmly within the bounds of communal ownership. These communal bounds actually favor the actualization of private benefits on this type of resource.

(iii) *Private holdings*. This appropriation regime covers the establishment of private use rights resulting in durable domestic entitlements. It is the closest to private property. These private possessions remain the ultimate property of the group of owners and should not be alienated to outsiders. This is the typical domain of agriculture and other land or resource conversion activities. The fundamental transformation condition in this private regime, therefore, is 'enclosure' (Figure 3.1). The physical proof and enduring quality of personal investment in a portion of a common pool resource signifies that this portion has been 'privatized' within the collective system. The 'productive axe right', a subsidiary to the genealogical right we described earlier, is the main form taken by private enclosures in embedded tenure. In South Cameroon, women exert a *'fis right'* on fishing sites *(fis)*, analogous to the axe right. This *fis* right grants exclusive and permanent use right on the stretch of river she demarcated to a woman's family-in-law. The *fis* inheritance right goes to her daughter-in-law.

Tree planting gives *'planting rights'* on both the trees and the surrounding land patch and is increasingly used to establish private land rights in the system. In South Cameroon, several forest species, such as the wild mango, the moabi, the *njansang* (*Ricinodendron heudelotii*) or the *mvut* (*Trichoscypha acuminata*), can be taken out of the common pool, either through fallow planting or through enclosure in a new forest field. More rarely, these private rights are established on the tree or its product without affecting in the same way the land area surrounding it. We also found such case in Sumatra's *damar* agroforests, where common pool rights can be maintained around someone's personal tree (field data, 1997). Elbow et al. (1998) report that in Sierra Leone, migrant farmers are exceptionally allowed to plant trees, with the understanding that this is not a permanent entitlement and that they would be compensated when they leave. We give other examples of this in section 3. Exclusive rights to a 'tree product' such as honey in someone else's farm are also found in certain arrangements between Bagyeli pygmies and Bantu farmers in South Cameroon.

(iv) *Common pool access.* This definition is more exact than that of 'common property', since it refers to an access rather than a property regime. *This access regime is the pivotal element of the whole system in land-surplus forest areas.* All privately held resources are subjected to it at some stage in their lifetime. They had to be subtracted from a common pool before becoming private holdings. Given certain transformation conditions, they can evolve back into commons accessible to the whole group of owners. It is this regime that makes dynamic conversions of resource tenure possible in embedded tenure. It is characterized by two mutually constitutive dimensions: (i) laws of transformation that govern takings from a shared pool of biological resources; (ii) an access regime that is open to a category of persons and closed to others.

(a) The common pool. There are only two fundamental ways of transforming the initial common pool status of a resource into an individually owned product: extraction and enclosure (Figure 3.1). Extraction is the general case, as enclosure is but a form–indeed critical–of extraction. One unit of resource enclosed or subtracted by one individual from the *common pool* is not available to the next individual; the pool is common but appropriation is individual. When private taking from the pool is purely extractive, it does not affect the common status of the land base or the biological stock. A fish or a fruit has been taken out but the pool, the fishery or the forest patch remains accessible to others. When extraction is done through enclosure, that part of the pool is cut off and becomes inaccessible to other members. Agricultural fields, tree plantations or fishponds are typically subtracted through enclosure, whilst fish, wildlife and wild forest products are generally taken through simple extraction. Mobile wildlife resources, such as fish and game, have special bioecological characteristics, which further condition their tenure status. They cannot become a 'property', whether individual or collective, unless they are captured (the so-called 'rule of capture'). This was, before the ITQ, the main obstacle to private property rights establishment in marine fisheries (Diaw, 1989). The common pool can also be modified by addition -through fallow and secondary forest regeneration, for instance. Common pools in embedded tenure are collective natural reserves, which future thus depends on the balance of extractive and transformative uses. The form and pace of their transformation are conditioned by the biophysical nature of the resource and by the process of valuation through which social actors attribute different social values and economic incentives to alternative uses.

(b) Restricted pool access. This concept emphasizes two characteristics of the common pool status of resources: (1) The level of inclusiveness or exclusiveness of access in reference to membership; (2) The condition of appropriation or transfer from the pool to an individual or a group of individuals. In our Cameroon cases, access to the forests (*nfos afan*) of a corporate lineage is generally open to all members of that lineage but not to other people in the community. On the other hand, the 'high forest' (*fut afan*) is, in most areas, governed by a communal access regime open to all members of the territorial community (village or group of villages) but only to them. The levels of permission and restriction of access follows the boundaries of the group of owners. A case that clearly shows the different levels of restriction to access is that of non-timber forest products. As a general rule, their gathering[16] is free in the high forest (open communal access), restricted to lineage members in secondary forests (lineage property), and limited to landholders (households and individuals) in fallows and cocoa plantations.

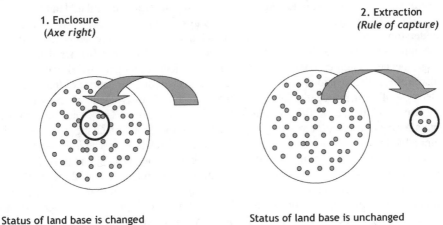

**1. Enclosure**
*(Axe right)*

**2. Extraction**
*(Rule of capture)*

Status of land base is changed                Status of land base is unchanged

*Figure 3.1. Appropriation Principles*

These restrictions in access membership are supplemented by restrictions related to transformation conditions. Regardless of the status of the land base, hunting is an activity open to all community members. Operational restrictions apply to trapping but not to projectile-based hunting in or around other people's fields. Fishing is generally open to all residents in the community, except on women's fishing grounds where these latter must authorize third party access. In agriculture, axe rights and planting rights are the conditions of appropriation. In aquaculture, the axe right applies only indirectly through the development of an agricultural field adjacent to the swamp area.

## 4. CHANGE PATHWAYS

There is a deep-rooted suspicion of 'traditional', 'customary' and 'community' concepts in contemporary discussions of African social systems and values. Institutions are not frozen in time, untouched and unchanged, and communities are not homogeneous entities devoid of conflicts, self-maximizing schemes and power politics. The very idea that customary tenure institutions could be valid vehicles of environmental governance at the age of virtual markets and globalization is profoundly counter-intuitive to many. In addition, the classic frontiers of social theory have expanded in last decades through the influence of deconstructivist, participatory, and actor-oriented analyses, which emphasize the constructed, fluid and "messy" nature of social processes and institutions (Giddens, 1987; Long, 1992; Long & van der Ploeg, 1994; Mehta et al., 1999; Sunderlin, 2003). The "myth of community" has been exposed in countless settings to offer differentiated, gendered perspectives on local governance (e.g., Biesbrouck, 2002; Guijt & Meera, 1999). Pluralist mediation approaches have pointed at the adaptive and negotiated nature of overlapping formal and informal legal orders (Anderson, 1998, Borrini-Feyerabend, Farvar, Nguinguiri, & Ndangang, 2000). Such an ebullient context calls for examination of how resilience, agency, and change interplay in embedded tenure.

We believe that social practice is always 'situated practice'; it cannot be understood out of the cultural, institutional and contextual elements that give it meaning and purpose. Institutions are 'guiding norms' of social order that enable coalescence and interpenetration of social practices and interests. However, they also reflect the tensions and paradoxes of socioeconomic and political orders. Because of their inherent dialectics, human interfaces always generate pressure and incremental change at the margins of social institutions. This leads to change, whether evolutionary or revolutionary. This section looks at change processes in embedded tenure. We start by clarifying the basic relation between the embedded system and the diverse polities that organize it in the political realm. We then proceed to describe the system's internal change rules across three paths of agricultural transformation. We end by looking at how external factors–non-kin immigration, money, land pressure and land transactions–play out through various modes of rights devolution and delegation; we also consider what this means in terms of change and innovation.

## 4.1 The Political Governance of Embedded Tenure

Political systems of customary land administration have been a visible feature of indigenous tenure system. There is, actually, a widespread tendency in the discourse on rural Africa to confuse the tenure system with the "traditional authority" of chiefs and rulers. From this confusion emerges two opposite but complementary themes; the first sees "traditional institutions" as inherently despotic, the second as inherently weak. The first takes its clues from strongly hierarchical systems such as the Swazi monarchy in Swaziland or the *lamidat* in North Cameroon; the second takes them from so-called "acephalous societies", which dominate in rainforest environments. In both cases, the political lenses through which customary tenure is analyzed prevent a deeper comparative understanding of its internal rules of behavior. Throughout the colonial and post-colonial periods, these two themes have fed policies that tried either to weaken or replace traditional authority or, at the opposite, to reinforce local hierarchies and autocratic elements in order to establish social control. Mamdani's *Citizen and Subject* (1996) is a landmark in the study of this fascinating question of power and authority in Africa. How political authority links up to the systemic foundations of embedded tenure is therefore an important point of research and theory.

There are basically two models[17] of customary land administration in the African political realm. The first is based on a political superstructure of "chiefly jurisdictions" going upward to the level of kings and paramount chiefs, and downward to homestead, compounds and family heads. The second is the family or lineage system, where "*land rights are normally enjoyed through voluntary occupation and exploitation of land without the necessity of a formal grant or allocation based on hierarchy or chiefly authority*." (Agbosu, 2000). Basically, the African pre-colonial State, from the huge West African empires of the 10th–16th centuries to the more scattered kingdoms of the 17th–19th centuries, superimposed itself on the lineage-based social organization. It linked up to them mainly through various tribute-paying and allocation mechanisms (Diaw, 1985) without really affecting the social make up of tenure. Rwanda, where the relation between the *ubukonde* lineage system and the *isambu-ikingi* system of political tenure was a conflictive opposition going back to the 16th century (Andre & Lavigne Delville, 1998) is somewhat an exception in that regard.

In a rich comparative analysis of the *Contradiction between Anglo-American and Customary Tenure Conceptions and Practices* in Ghana, Agbosu (2000) gives a summary of the underlying unity of vertical and horizontal systems of land control. Relying on an array of anthropological and legal works[18] spanning the 1950s-1970s, he comes to conclusions, which are in essence identical to those presented in this chapter:

> "it is possible to identify (…) a universal norm that underlies and determines the property rights, the beneficial enjoyment of rights in land, and the nature of such rights... Such a universal norm (…) is exemplified by the right of the individual member of a social group, such as the polity, the clan, the tribe, or the family, to beneficially enjoy property as a member of the group... This is (…) whether or not the land administration system and land rights are based on hierarchical structures or simply based on group membership."

In the political ideology of even the most hierarchical polities, the head of the system, whether it is the Swazi monarch, the *fon* in North-West Cameroon, or the Stool holder in Ghana, is never the 'owner' of the territorial estate. He holds it in trust for the entire group or nation, on behalf of the living as well as the ancestors and the future generations. As much as the clan or the lineage in horizontal systems, he represents a corporate juristic entity[19] that must guarantee the inherent right of each individual to access common assets and resources and preserve them as a collective patrimony.

## 4.2 Paths of Agricultural Transformation

The first change pathways in embedded tenure are determined by its generic make up and its internal appropriation and transformation rules. The four access and property regimes that we described in section 2 are substantially different from those of the CPR literature. In CPR analyses, open access, private, public and common property regimes form separate categories submitted to different institutional and legal orders. In embedded tenure, the property regimes are submitted to a unique set of appropriation and transformation rules. They can interlink to form a prism of nested rights applying to nested eco-niches across the landscape and over time. Several layers of rights are thus intertwined into different segments of the landscape.

Agriculture is the main driver of regime change in embedded tenure. It was historically built on long-term productive cycles implying systemic mutations and transmutations of the natural and social statuses of land over time (Figure 3.2). This cycling, characteristic of shifting cultivation systems, is neither linear nor unique, and leads to numerous variants. From our observations, we distinguish three fundamental paths of agricultural transformation within this dynamic system. These are illustrated by the three branches of the fork that appears after the fifth transformation ($t_5$) in Figure 3.2.

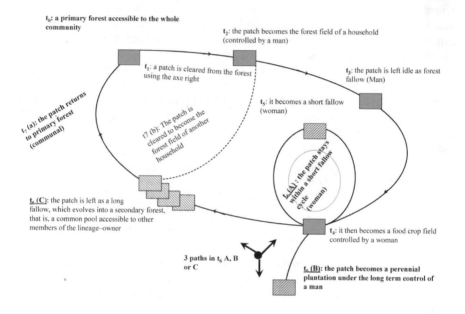

**Figure3.2.** *Cycles of Land Conversion and Mutation of Ownership*

The first path ($t_6$-A) is a path of intensification that can express land, population or productivity pressure as well as deliberate social strategies. Fallow cycles are shorter and private tenure of the land is stable and tends to become permanent or quasi-permanent. This path is widespread around the world and has led many to interpret it as the internal displacement of customary tenure by private property. This is not the case, even though this change reflects increased privatization of use. In the Gerze community of Gbaya in Southern Guinea, five exogamous patriclans coexist within an increasingly exiguous territory. Common pool areas shrunk decades ago into a single forest patch that was exceptionally preserved from human colonization by sacred and mystical attributes that scared people away. In the 1950s, the village took the collective decision to enter *en masse* into it, thus eliminating the last piece of commons in the territory. Fifty years later, collective property and exoinalienability of land is still the defining feature of tenure in the area. Though lineal authority and inherited lineage lands are increasingly concentrated in the hands of first sons, these latter do not have the power to alienate the land and continue to consider themselves as mere custodians of the inherited patrimony, the *nuna buluhënga*, '*the things of behind the father*'. The exceptional possibility to sell a piece of land exists but only as an extraordinary decision authorized by the village council of elders. Such examples of permanent and quasi-permanent land entitlements within the larger system of collective ownership are abundant. We observed it in as distant and disparate places as the community of Hobeni on the Wild Coast of South Africa and Karen and Hmong villages of the Doi Inthanon area in Thailand (Rasmussen et al., 2000). We saw it at play in Casamance (Senegal), Zimbabwe, D.R. Congo, and Sumatra,. Though they do not go into the detailed working of indigenous tenures, the 1996 African profiles provide other examples across the continent (Bruce, 1998).

($t_6$-B) is a second path of private rights consolidation through creation of perennial plantations. Again, this path is widespread around the world. The *damar* agroforests of the Pesisir (krui) area in Sumatra are a unique illustration of this. They feature the evolution of the private sphere in embedded tenure as well as its capacity to adapt to economic growth and sustainability challenges. The *damar* agroforest system evolved decisively around 1927. At that time, wild species of this dipterocarp family were being severely depleted in response to the increased European demand for industrial painting and varnish material between 1850 and 1920; this was in addition to the traditional Asian demand for the resin. Abusive tapping and increased encroachments and theft from external collectors posed an imminent threat to the wild *damar* and to the communities (Michon, de Foresta , & Levang, 1995). The communities of the Pesisir responded by creating the *damar* gardens. Within a span of 50 years, they reconstituted a complete forest bio-ecosystem within the traditional *landang* system of shifting cultivation (the complete $t_0$-$t_7$ cycle), effectively taking over 80 % of the national production in Indonesia. About 40 common species of trees and tens of additional species of large trees, treelets, shrubs, liana, herbs and epiphytes are now associated with the *damar* in this domesticated system. This has lead to the emergence of a range of new marketable commodities, including timber, rattan, medicinal and insecticide plants (Fay et al., n.d.; Michon *et al.,* 1995). To realize this, the Pesisir peasantry used its ethno-ecological knowledge and inventiveness to nest different species into complementary econiches in time and space. They also adapted the tenure system to the radical change introduced by the *damar* gardens.

In the wild, the first tapping constituted first possession of the tree but it did not affect the common pool status of the forest. Within the larger *ladang* system, private land rights were primarily acquired through forest clearing for rice cultivation. This is consistent with the basic tenure principles that we already outlined. The large-scale creation of damar plantations within the system was likely to induce major shifts in the balance between the commons and private holdings. According to Michon *et al.*, (1995), the *damar* gardens were created only after the customary system of law modified the appropriation rules and formally authorized the practice. Accordingly, the change did not lead to individual appropriation of the entire agroforest domain. Only the *damar* tree and resin really became private holdings. The other agroforest resources – fruits, fuel wood, sugar palm sap, or bamboo- remained available to the community, along various appropriation rules. In the *damar* garden that we visited in 1997, owners of *durian* trees were using nets to keep the fruits, highly valued in local markets, from falling on the ground. This was because any member of the group could, by right, collect a fruit from the ground. Despite the factors of growth that it liberated and the rearrangements that the system had to make within itself, the *damar* garden thus remains "inalienable lineage patrimony" in the Pesisir (Fay et al, n.d.).

($t_6$-C) is the traditional path of extensive shifting cultivation. It characterizes sustainable cultivation systems highly consumptive of space, which still dominate agricultural land uses in the Congo Basin and other land-rich tropical forest environments. It is along this path that the nesting and transmutations of land uses and appropriation regimes are the fullest. While the intensification paths show the system's important –though neglected- potential for economic growth and development, the extensive path allows the full realization of the equity principles of embedded tenure. In modern welfare economies, unequal property rights and unequal access to resources are attenuated through taxes and revenue redistribution schemes. In embedded tenure, equity is pursued through equal access rights to the resource and the possibility to redistribute this access according to the needs and capacities of social units (Diaw & Oyono, 1998). Even though, as shown below, donations and produce exchange as well as labor and land transactions are important, the system of wealth redistribution relies, primarily and wherever possible, on social cycles of allocation and reallocation of access to the primary source of wealth, the natural resource.

*4.3 How do 'Strangers' fit? African Land Transactions under Devolved and Delegated Rights*

There are wide membership differences among communities operating embedded tenure systems. Some are highly homogenous, while others host groups of different wealth and statuses, often including temporary and permanent migrants clustered in separate wards or quarters[20]. How strangers fit and how the status of land adjusts to social differences and to influxes of non-kin people into the system are important questions to address. They linked up to the wider question of land transactions in embedded tenure.

"Derived", "delegated" (Leroy, 1998, Lavigne-Delvigne and others, 2001), "attributed" (Oyono, 1998) or devolved rights are indirect use rights; the user is not the direct right holder; his/her right to use the resource transits through the established genealogical or productive rights of a member or a category of members of the tenure system (farm 'owner', chief, lineage, village council). Conversely, his/her use of the resource does not automatically lead to enclosure and personal appropriation, as is normally the case. In our definition, devolved rights include the rights temporarily or permanently transferred to non-native residents of a community as well as such right transfers among community members. Land sales themselves are often ambiguous forms of rights devolution, leading to succession disputes at the death of one of the contractors. The unifying characteristics of this type of use rights and land transactions are that (i) they are granted by a right holder to a non-right holder; (ii) the rights of both owners and user are secured by some form of provision or compensation, whether symbolic, in-kind or in cash; (iii) they take place under explicit or implicit conditions that serve to protect the system; (iv) they constitute a pressure on the system at the same time that they help it adjust to internal or external stress and play a role in maintaining it overtime.

There is a wide range of transactions that fulfill the above conditions. In a study of nine cases covering six West African countries (Lavigne Delville et al., 2001), the researchers found that Western transaction categories could not capture this diversity; recourse to original African languages was necessary to express the operational efficiency of the three dozen types of transaction they had found. What is the net effect of these transactions on the embedded tenure system? Most studies are not specifically designed to answer that question[21]. Yet, they bring important insights and generally conclude on the flexibility and adaptability of the indigenous tenure system. We explore the issue around three themes: (i) the conditions under which strangers have been able to acquire full tenure rights in the embedded tenure systems; (ii) the rationale of temporary or semi-permanent delegated rights and their effect on the system; (iii) the meanings and dynamics of land sales and other forms of land alienation.

### 4.4 Settlers' Incorporation in the Embedded Tenure System

There are numerous historic cases where settlers were able to gain full membership in another group's embedded system. This includes individuals who were donated or lent land with implicit or explicit inheritance rights to their descendants. In Burkina Faso, Matlon (1994) found that the likelihood of the user family becoming the 'owner' of the land increased with the duration of the loan across generations. In Cameroon, we found a few cases where the descendants did not even know of their ancestors' host status until they became entangled in land tenure conflicts demanding historical reconstitution. Full incorporation of larger family and settler groups are also historical facts. In our earlier Gbaya example in Guinea, four of the five owner-clans started as hosts of the founding Hunumu clan; during our visit, this was a theme of subdued complaints among Hunumu youth blaming the current land shortage on the 'excessive generosity' of their founding fathers. In the 16th century Senegambian lamanat, the incorporation of settler groups was a building block of the overall tenure system. The *lamaan,* holders of eminent right of fire, granted transmissible axe rights to incoming land clearing families. Until at least the 1950s, settler groups were still paying a symbolic fee to the *lamaan* when organizing the succession of a lineage head. This ritual bonding served to restate the origin of the system of rights while securing the land settlers' genealogical rights under the terms of the ancestral pact (Guigou, Lericolais, & Pontié, 1998). Lavigne Delville et al. (2001) distinguish these establishment processes from delegation procedures, but there is definite analytical contiguity between the two. Their examples from Burkina Faso of the *sigily* (full axe right to the migrants), *sisa sigily* (productive right on the tutor's land with interdiction to make physical investments) and *singely* (full but temporary use rights) indicate a clear semiotic and systemic continuity between establishment and delegation procedures.

### 4.5 The Social Rationality of Temporary Landholding Arrangements

Most modern delegation processes do not lead to full incorporation. From the literature, we generally see a kaleidoscope of arrangements where one party seeks access to productive land while the other uses its land entitlement as asset or collateral to access labor, cash, service, or productive investment. This is direct refutation of the ideological belief in private titles as only possible collaterals for credit, and productive investment. For the rural landholder, the transfer of rights is most often seen or presented as transient, rarely, even in cases of donations, quasi-donations and sales, as irrevocable alienation of family land. This is why many of these arrangements are contested at the death of the original right holder. For the land seeker, acquiring permanent rights over the land may be an overt or covert motive, but this is not always the case. Simple land loans, fixed-term leases, seasonal tenancy and share contracts are simple cases where the temporary nature of the delegation of rights is unambiguous.

Land loans are a bit trickier as it is sometimes difficult to distinguish them from outright gifts. They can also be short or long term, seasonal or open-ended; this has been a source of land grabbing problems in the context of some national legislations. In Senegal, for instance, land automatically goes by law to the user after a couple of years of continuous cultivation; this was a source of drastic drop of land loans in the mid 1970s. In South Benin, land loans are also declining because of the competition of arrangements more profitable to the 'landowner' (Lavigne Delville et al., 2001). In other land surplus areas, loans to strangers and kinfolks are still frequent[22] and secure. Arrangements are often made in front of witnesses or in public, with token payments by the borrower as recognition of the lender's ongoing landholding rights.

Share contracts are widely used in African agricultural systems to compensate for unequal endowments in land, labor, productive capital, and cash between social groups and geographic areas. Share tenancy forms vary greatly. They range from produce-sharing contracts on crop fields or perennial plantations to land–for-labor or land-for-service swaps (e.g. maintenance and guarding). *Share planting tenancies* express the inventiveness of these arrangements and their rationality within the embedded economy. We call "share planting" the tenancy arrangements under which the tenant establishes the plantation in return for a share of the plantation, that is, the productive capital itself (see Lavigne Delville et al., 2001). This definition excludes other tenancies on perennials where the tenant's labor and/or cash investment in the plantation is remunerated by a share of the product. Such cases correspond to the classic forms or sharecropping in African systems. They also share a similar historicity[23].

Share planting tenancies are different because of the sensitive role of tree planting in embedded tenure appropriation processes. This is why tree planting is often excluded from land delegation procedures. Some systems, however, found solutions that could accommodate tree planting by adjusting the tenancy arrangements and dissociating the ownership of the trees from the ownership of the land (Bruce, 1988). The *dibi-ma-dibi* ("I eat so that you can eat") in Togo, the *trukatlan,* in Côte d'Ivoire and the *abusa tenant in Ghana* (Lavigne Delville et al., 2001) are examples of such arrangements. In the first two, the share planter establishes the plantation and is entitled to its output during the development phase; Once it becomes fully operational, the share planter gets half or 2/3 (*abusa tenant*) of the plantation. His exploitation rights are guaranteed for the lifetime of the trees, that is, up to ¾ of a century and more for certain species (e.g. cocoa). There is no prescriptive acquisition, that is, no period of time after which the tenant's holding can automatically change into ownership.

Land pledges and land sales also express the flexibility and adaptability of the tenure systems. Like sharecropping, both have precolonial origins (Coquery Vidrovitch, 1982; Migot-Adhola, Bennet, Place, & Atsu, 1994). Possessory mortgage (Bruce, 1988) is the general case of African pledges in land. In the market form of mortgage, the land acts as security for the lender's money but the borrower remains in control. The exchange value of the land, not its use value, is the stake in the contract. In possessory mortgage, the lender takes hold of the land and uses it until the debt is repaid. In some variants (e.g. *awoba* in south Benin), the use of the land is a form of interest on the loan; in others, the lender's usufruct reimburses both principal and interest. The two forms can coexist in the same locality, sometimes, with the same name; this is the case of *"garantie"* contracts in center-west Côte d'Ivoire (Lavigne Delville et al., 2001).

*4.6 Land, Money, and Markets: Unexpected Meanings and Change Processes*

Observers of African land tenure systems have a hard time anticipating the long-term meaning of these land transactions in terms of systemic change. Most authors insist on individualization trends and the "novelty" of contractual relations while highlighting in the same breadth the system's flexibility and the "imperfect" commoditization of land[24]. The pluralism of property rights is an all embracing theme as well as the criticism of current tenure policies, which do not to integrate the dynamism of local tenure systems and land delegation procedures. But is land becoming a commodity or are the "new" forms of transactions self-contained adjustments of the embedded system? This question is often left in the limbo of "imperfect markets". This is because the historicity of these practices in no way fits the evolutionist expectations placed on African tenure systems. In addition, the position of private rights and the significance of individualization paths ($t_6$-A and $t_6$-B) within the embedded system have been thoroughly overlooked. Most land transactions and rights delegation are extensions of these individualization paths, which are integral parts of the system. Most also have precolonial roots and did not evolve in the linear fashion that was anticipated by evolutionist theories.

Late 19th century land sales in Ghana, for instance, did not come about because land had become scarce and was gaining market value; land started to sold because it was abundant and because the socio-political system was accommodating enough to absorb the interregional migrations triggered by a growing cocoa economy in the 1870s: "*Because most of the areas settled by the early migrant farmers were not inhabited by local people, it was not difficult to arrange for land purchases or other types of traditional leases*" (Migot-Adhola et al., 1994). Both patrilineal and matrilineal groups of migrants were involved in these purchases[25]. Land purchases were thus part of a process of political incorporation, broadly comparable to donations and axe rights, which included marriages, long-term lease and sharecropping arrangements.

Amanor and Diderutuah (in Lavigne Delville et al., 2001) note further that, following the initial 19th century royal land sales and land-based arrangements with migrant groups, *abusa* contracts evolved into an *abusa tenant* system. In this latter, local chiefs gave temporary clearing, planting and use rights to the migrants in exchange for the possibility to recuperate a fully-grown plantation afterward. This enabled them to take up lands that, otherwise, could have been allocated to other kinfolks. It was therefore an adaptive move away from land sales and a precursor of the delegation procedures that we discussed earlier. Starting in the 1980s, the migrant *abusa* workforce was progressively replaced by land-starved indigenous youth outside the land holding family. More recently, with increased scarcity of land and conflicts over internal modes of land allocation, this inward move reached the landholding group itself. Within the span of a century, the system has evolved from sales on land-rich communal areas to tightly kin-centered dynamics on increasingly scarce communal lands. This evolutionary process shows exactly the reverse of the change forecasted by evolutionist theories.

According to Bruce (1988), this is a general process still in progress in most tropical Africa. As crowding increases with population growth, fallow periods shorten, leading to a rush for the allocation of the remaining land. Once cultivation stabilizes, allocation decreases in importance and the rules of inheritance come into their own. This tends to shift the focus of social control of land toward the extended family. The critical decisions become those concerning how land will move from generation to generation in a "family" only two or three generations in depth. Migot-Adhola et al., (1994, p.117) make a similar analysis on Ghana: "*As population (grows) and the land frontier closed. It became more important for lineages to satisfy the resource needs of its members. This leads to increased lineage control over transfers by its members and increased restriction to alienation to non-lineage members*". Analyzing the Rwandan situation, Andre and Lavigne Deleville, (1998) observe the same pattern; with the progressive disappearance of common pastures, purchases are becoming the principal mode of family land acquisition in certain areas. This increases the competition and inequalities between domestic units involved in various strategies to augment family landholdings. Land gifts are often disguised sales in this fashion. These authors make a point to note that households' property rights do not thwart the permanence of collective control on the transmission of the land patrimony; such control does not prohibit sales or intensification.

The instrumentation of land sales as a domestic strategy is but one of the many coping mechanisms unfolding within embedded tenure systems across historical contexts and social situations. Various and changing forms of tutelage are used all over to keep migrants under the long-term bounds of communal systems (see e.g., Karsenty, 1996 on Côte d'Ivoire). Organized migrations of younger generations regulate structural overcrowding situations (see e.g., Gbaya above and Guigou et al., 1998, for the tenure system of the sereer Siin in Senegal). Women in some patrilineal systems use various forms of monetary transactions to circumvent their exclusion from inheritance. This is the case in south Benin where they are significantly involved in land lease, possessory mortgage, sharecropping and oil palm contracts (Pescay, 1998). In south Cameroon, we observed a few cases of divorcees combining the symbolic weight of a "purchase" with strong personal ties to a kinfolk to buy they way back in their village of birth[26]. Money is thus used as a legitimizing tool for transmissible female possession of lineage land. In other settings, other strategies are deployed. In Kenya, it is the traditional figure of the "female-husband" that widows used at times to fight for their daughters' right to inheritance (Mackenzie, 1994). The picture is rich and varied but it has remained so far within the folds of the embedded tenure system. Money and other social mechanisms have been incorporated, wherever needed, to expand rather than break the physical and institutional boundaries of the system; people have drawn differentially from the plural –and often contradictory– institutional and economic repertoire available to them in ways that could constantly be reprocessed by the system. Even when land became alienated to strangers, it was kept from becoming "free of all social claims by others" by tutelage and other forms of social control (Bruce, 1988; Karsenty, 1996; Leonard & Longbottom, 2000; Lavigne Delville et al., 2001); it also retained a family-based use value along the general lines of the embedded system. It is this structural capacity to change and, at the same time to contain change within its bound that defines the embedded system's resilience and its adaptability to constantly renewed internal and external demands.

## 5. POLICY, THEORY AND THE EMBEDDED ECONOMY

Sixty years ago, Karl Polanyi (1944, 1957) formulated the thesis that the formal categories of conventional economics where not valid for capturing the integrative principles of societies governed by reciprocity and redistribution rather than price-fixing market mechanisms. In *Trade and Market in the Early Empires* (1957), a collective work, which included his seminal paper on "*The Economy as an instituted process*", Polanyi argued that the study of the "changing place of the economy" in society requires an examination of its "substantive meaning" and its historical, empirically observable characteristics. This is what makes possible the identification of the "forms of integration"—reciprocity, redistribution and exchange—through which economies are institutionalized and gain stability, interdependence and recurrence of their elements. This thesis, at a time when Polanyi was already 70, was the departure point of fierce controversies among economic anthropologists and economic historians[27]; it passed virtually unnoticed among economists except in some US universities and in the circle of the old institutional school of Veblen, Commons and Ayres. It is not until the late 1970s that it was taken on by mainstream economic thought (North, 1977)[28].

Polanyi's analysis of non-market economies actually stemmed from his lifelong interrogation about the ultimate causes of the collapse of the early 19th century liberal economy and the disasters that befell Europe through the Second World War (Polanyi Levitt, 2003). In the *Great transformation* (1944, 1945), his most important work, he argues that the secret of the 19th century liberal economy was the disembedding of the economy from society and the invention of specific market institutions around which society became organized. In contrast to pre-capitalist society, social relations became embedded into the market instead of the market being embedded in social relations. But for Polanyi, the idea of self-regulating markets (the 'invisible hand') was essentially an utopian project that inevitably lead to massive social dislocation and counter-clock movements to reign in the market (Polanyi Levitt, 2003). This is how he interpreted the two Great Depressions and the decades of crises, wars and reorganizations that marked capitalism between the 1870s and the 1950s. This tension remains a defining feature of 21st century globalization, with multiple clashes of national and regional interests, 'market fundamentalism' and 'alter-globalization' in global forums. This, and the social struggles that underlie local policy choices across rural tropical environments, gives fresh relevance to the debate initiated half of century ago by Polanyi on the occultation and 'naturalization' of these social choices by mainstream economic theory.

In his work on *Anglo-American and Customary Law* in Ghana, Agbosu (2000) highlights the origins of the 'twin artificial doctrines of estates of tenure' in feudal services. Tenures emerged in feudal Europe in exchange for different classes of services—military, agricultural (socage) or religious—rendered by a pyramid of land grantees flowing up from the actual tenants to tenants-in-chief and the overlord. When feudal services became commuted into monetary payments, conditions became favorable for the evolution of the freehold estate. Landed property became a marketable commodity, along with new skills and forms of property. These conditions are fundamentally different from those of the embedded tenure where the product of labor is strictly personal and where universal blood rights establish both the boundary and the underlying principles of the system. With the long view of history, the syndrome of "extraordinary treatment" to which indigenous tenure has been subjected in order to replicate the conditions of freehold emergence in Europe appears fundamentally misguided.

There is an obvious 'mutual construction' between modernization theories and tenure policies, each feeding into the other to blur the picture of local change dynamics. It is therefore legitimate to ponder over the continuing risk of reductionism associated with the 'rehabilitation' of local institutions in current forestry reforms. It must be reminded that the interest for indigenous forest systems stems in great part from the environmental crises of the 70's, which exposed the vital link between population and the environment. Despite genuine ethical concerns, the larger mobilization of the international community was mainly triggered by efficiency considerations related to global interests in conservation. The shifts in policy came mainly from realization that local people had 'access power' on their environment and could not be indefinitely excluded from decision-making without serious environmental backlash. The understanding of indigenous systems remained largely shallow and an afterthought of policies. As a result mistakes have been made and continue to be made in the attempts to develop new community-based approaches of forestry.

The history of community forestry is an interesting case in that regard. The first generation of social forestry projects (India, Kenya, Yemen, Malawi, Pakistan, Haiti, Zimbabwe, etc.) were based on the idealistic assumption that reforestation and 'basic needs' objectives would be better achieved by massive planting of fuel wood by 'communities' on communal lands. This option resulted in a series of setbacks, which were only made good by a return (Haiti, India, Tanzania) to smaller social units, mainly family farms, more appropriate for this type of activity (Cernea, 1991). As a logical backlash from these experiences, it was concluded that community action was ineffective and that individuals and households were more relevant units for achieving community forestry goals (Arnold, 1991).

But this conclusion overlooked the fact that the first generations of projects were elaborated in a context of arid, semi-arid and deforested areas, while the current focus of community forestry has largely moved to maintaining existing tropical forest ecosystems. From the standpoint of customary tenure systems, as we saw in this chapter, *planting* belongs, *as farming,* to the realm of *productive and development rights*, that is, to the private sphere in embedded tenure. Tree tenure being tied to land tenure in almost all such systems (see Elbow et al., 1998, for West Africa), planting automatically generates private appropriation of the surrounding land[29]. This is not the case of forests, which belong predominantly to common pool access regimes. In the same way that, under customary principles, large communities were not the best-fit entities for tree farming objectives, the management of common pool forests cannot be vested in individual farm households, that is, in a segment of potential forest claimants.

In Cameroon, a good deal of the problems faced by the community forest reform of 1994, and of fiscal decentralization in the years 2000 can be traced to the limited understanding of these governance rules. The legal associations favored by the legal process were crafted for economic projects and collective action but the social mandate to address local property rights issues were vested in the lineages. As a result, there was lots of free riding and conflicts. On the other hand, the resource-based redistributive system that we outlined in section 2 was not designed for processing large tax revenues at the community level. This naturally led to various forms of mismanagement and to backlash re-centralization (Bigombe Logo, 2002; Diaw, 2002; Diaw & Oyono, 1998,).

It must be noted that the CPR theoretical tradition has mainly focused on 'crafted' institutions (Ostrom, 1990, 1992), that is, organizations and rules purposively designed to address collective action dilemmas in resource management. Embedded tenure institutions, at the opposite, were not "created" to achieve particular management performance. As we saw, they evolved out of a process structured by natural human demands (e.g., produce and reproduce) and long term social goals (e.g., organize a system of descent prohibiting incest, and a system of resource use based on universal access across households and generations). "Crafted" and embedded tenure institutions do not have the same cultural history or the same scope, capabilities and functions. The CPR literature did not emphasize this distinction that proved crucial in cases such as the ones mentioned in Cameroon.

In their review of tenure regimes – Les maîtrises foncières – Leroy, Karsenty, and Bertrand (1996) give the example of the Salomon islands, where a reform that had made possible the development of community logging had devastating effects on communal cohesion, because of the same type of misguided assumptions. These examples, and many others, must be meditated in light of the on-going efforts to develop rural land plans (Benin, Côte d'Ivoire, Burkina Faso, Guinea) and new forestry codes, as in the Democratic Republic of Congo. In the first case, there is a risk that the registration procedures could introduce more rigidity in the dynamic systems of land rights that we saw. In the second, the forest zoning implies the demarcation of separate logging, conservation, and community areas. Experiences in the region indicate that, in practice, such supposedly discreet units are inevitably confronted with overlapping claims and conflicts, unless properly negotiated with local actors.

There is a considerable policy risk entailed by scientific and bureaucratic assumptions about local tenure institutions. The idea that the world is facing a global tragedy of the commons that will be fixed by privatizing "the global commons" introduces some perspective of equity between poor and rich countries through various entitlement transfers (Chilchinsky, 2004). We saw, however, that the local dimension of environmental entitlements is not reducible to state-based policies and that collective, common, open and private regimes of appropriation interact differently in local forest settings than is generally assumed. The virtualization of property rights and their delocalization toward global allocative markets and institutions will also come with their own measure of reductionism and structural problems. The question is complex and will certainly need further exploration, but there are strong reasons for caution. The reinsertion of local populations in new 'participatory' or global schemes of resource management being clearly a *move by default*, its success ultimately depends on how well policy makers and their technical advisors can avoid the epistemological trap of reductionism. Such a task requires an epistemological break and a serious attempt to understand the design principles of resilient forest institutions. The exposure of those principles, *through field-grounded research and appropriate social methodologies*, is a prerequisite because of the financial, environmental and social costs of potential failure.

## 6. CONCLUSION: FOR NEGOTIATED AND INVENTIVE SOLUTIONS

We tried in this chapter to outline the working principles of embedded tenure and the fine processes that explain its considerable flexibility and resilience over time. We showed that embedded tenure has important universal features despite the significant political differences that may distinguish lineage-based systems from more hierarchical polities. We pointed at evidence that these systems can sustain economic growth and social change without having to dislocate or be replaced by private property. We also highlighted the mutual construction of conventional privatization theories and policies that sought, without decisive success, to accelerate the predicted demise of these systems. Such processes are not limited to forests, as we see them in larger agrarian settings as well as in state regulation of fisheries.

Our main argument is that the historical and epistemological distortions, which have facilitated the marginalization of embedded tenures and other non-market systems from world accumulation processes need to be transcended. The policy and theoretical focus should aim at understanding the dynamic rationality of these systems and at working with local actors in ways that reinforce their institutions rather than postulate their demise as a precondition of development and modernity.

In the case of on-going forestry reforms, a clear consensus already exists on the need for stakeholders' negotiation. It must only be stressed, again, that this involves a *negotiation of meanings*, wherein indigenous conceptions and structuring of the natural and social world would be fully understood *for what they are* and integrated as legitimate components of the negotiating framework itself. This importance of *'social fitness'* for induced social change (Cernea, 1991) highlights the responsibility of social sciences with regard to the fitness of their own theories and to the task of transforming social knowledge into tools for action.

## NOTES

[1] See e.g., Agondjo-Okawe, 1970, Adeyoju, 1976, Fortman and Bruce, 1988, Bruce and Nhira, 1993, Mackenzie, 1994, Rocheleau and Edmunds, 1997, Diaw, 1997, Diaw and Oyono, 1998, Agbosu, 2000, Lavigne Delville et al., 2001, Robiglio, Mala and Diaw, 2003, and other references throughout this chapter.

[2] Kant, S., 2003 – Economic theory of emerging forest property rights. Presentation at the World Forestry Congress, September. See also Kant and Berry, 2001.

[3] See e.g., Fitzpatrick, 1983, Santos, 1987, Merry, 1988, and Fisiy, 1990, on legal pluralism

[4] According to this idea, developed from the works of Simon (1962, 1969), the *intent of rationality* of economic agents is bounded by limitations in their information and cognitive competency.

[5] Opportunism, as Williamson (1985) puts it, is strategic self-interest seeking -"with guile". It weakens contractual arrangements by generating externalities, in the form of both ex-ante and ex-post (enforcement) costs, among which, "adverse selection" (bad risks mistaken for a good ones) and "moral hazard" (cheating) problems, initially identified in the study of insurance markets (Holmstrom, 1979).

[6] Jon Sutinen developed the first comprehensive analysis of the share system in 1979.

[7] In fact, and as Albert Berry (pers. com., 2004) correctly points out, the mainstream view of "second best" as something inferior and problematic is questionable: "real world economists do recognize that we are always in the second (or lower) best, i.e. that the first best of the neoclassical theory is unattainable. The literature on uncertainty, transaction costs and so on does not lead one to the conclusion that these obstacles to the theoretical first best can be eradicated by policy... As soon as that qualitative point is accepted, all theoretical second-bests are necessarily candidates for the best actually attainable".

[8] In France, this was only achieved in the 19th century, at the outset of protracted conflicts between the forestry administration and rural people, and through the transformations of the industrial revolution (Karsenty, 1995). See also, Buttoud, 1997.

[9] In general, the effect of those policies is weaker in the rainforest, where low population pressure and strong clan structures reinforce indigenous institutions. Tjouen (1982) notes that Cameroon's 1963 Decree-Law, which allows strangers to acquire ownership titles on customary lands has an effect only in urban centers. In the countryside, the *"unshakable position of customary chiefs"*, immigrants' consciousness of the legitimacy of indigenous rights and their fear of *"the reaction of the dead, which translates into a succession of deaths"*, are such that potential beneficiaries do not dare claim their new 'rights'.

[10] This separation of "freehold lands" from native reserves had serious consequences on the viability of social organization in Bakweri areas, as reported by a 1925 colonial report (Tjouen, 1982). It triggered a strong movement of protest for the restitution of customary lands. In the aftermath of WWII, this lead to the creation of the Cameroon Development Corporation (CDC), a government agency, which role is to manage huge plantations in the name of the "natives". This prefigured the "nationalist" form of agro-industrial transformation of customary forests.

[11] Land registration became the unique mode of recognizing ownership rights and required, in the case of a collective request, to list all community members and to prove the *"mise en valeur"*.

[12] People, then, moved in various ways and for a multitude of reasons, through oil stain' expansion as well as 'flyover' jumps across space. In those times, a Bantu chief could lead up to 3 or 4 migrations during his lifetime; migrations could take the form of short distance shifting cultivation rotations, or result from the displacement of large clusters of villages over distances as great as 200 kilometers (Alexandre, 1965; Vansina, 1982; Leplaideur, 1985; Santoir, 1995; Bahuchet, 1996).

[13] According to N. Tchamou (pers. com., March 1997), there is no natural difference in the spatial distribution of the various species of *dracaena* in South Cameroon. This confirms that the use of the red *dracaena* as a marker of territory, or to demarcate cocoa plantations, is not an innocuous side-benefit of natural vegetation succession but the result of a conscious strategy of land rights establishment.

[14] We do not include here 'derived' and 'delegated' rights (next section), which are not constitutional, but are devolved to third parties by constitutional right holders.

[15] Among the Gerze of south Guinea, for instance, it is the first son who inherits the collective assets of the family, while the Bamileke of West Cameroon give this position to the second son. Among the Beti, Fon and other groups of Southern Cameroon, assets are more horizontally distributed among all male children.

[16] There is no single concept in Bulu for the gathering (*akole*) of forest product. The verbs used are tied to the particular act undertaken in the process: *atoe esok*, to take off (the bark of) the *esok; asang ndo'o*, to crack the *ndo'o*. This is also true of fishing where the term *nop,* sometimes used as a generic, means in reality line fishing.

[17] In the first category, we can cite the hierarchical polities of North, West and North-West Cameroon, the Swazi kingdom, the Ashanti and other Akan groups in Ghana, the Lebu of Senegal, the Lozi of Zambia or the Yoruba in Nigeria. In the second group, we find the Western Bantu of Congo, DRC, Gabon, Cameroon and Rwanda, the Joola of southern Senegal, the Ibo in Nigeria or the Northern Ewe and other communities of Upper and Northern Ghana (see, e.g., Agondjo-Okawe, 1970, Vansina, 1990, Diaw, 1997, Subramanian, 1998, Andre and Delville, 1998, Agbosu, 2000, Mope Simo, 2000).

[18] These works (see Agbosu, 2000: 9-16) cover *the political systems and laws of property of the Ashanti, Ewe, Yoruba, Benin, Ibo*, Lozi, Tonga, Sotho, and other ethnic nations of West, East and Southern Africa.

[19] Agbosu, 2000; see also Diaw, 1997, for the concept of "corporate lineage".

[20] In some areas such as the ones around Mbandaka and Wendji Secli in DR Congo (field data, 2004) these migrant groups can become the dominant population while remaining under the tutelage of the host community. Across the rainforest, however, clan- or lineage-based homogeneity is predominant as revealed by a 1996 focus survey of 471 villages of Center and South Cameroon (Diaw, 1997).

[21] In particular, key social characterization data on the status, relationship and origin of the land taker are

not always collected, as well data pertinent to the operation mode and long-term future of the new social unit within or outside the communal system. These follow-up questions would have helped address issues related to (i) the level of higher-order communal control exerted on land permanently acquired by non-kin and (ii) the long-term relation between the exchange and use values of land acquired in this way.

[22] Leonard and Longbottom (2000) cite studies from Burkina Faso showing that refusing land to a person in need was socially unacceptable, except in limited cases. In that context, fully endowed kin members could borrow land in order to adjust to uncertain fertility cycles and avoid overusing their family fields.

[23] The seasonal sharecropping *nawetaan* in the Senegambia, for instance, was a by-product of the colonial expansion of the groundnut economy and has recently declined along with that economy. In similar ways, the *abusa,* in 19th-20th century Ghana, and the *busan*, its 1960s-70s variant in Center Côte d'Ivoire, stimulated the expansion of the cocoa and coffee economies in those regions. These latter are mainly plantation maintenance contracts in which the tenant gets half to 1/3 of the cocoa produced. In Benin, *lema* share tenancies are contracted on oil palm plantations as well as tomato fields (respectively, 1/3 and 2/3 of the shares to the sharecropper).

[24] Bruce (1988:43) warns for instance that to think of these land transactions in market terms, "we must learn to handle more informal, less impersonal land markets than those of the Western economists". In south Benin, Pescay (1998) notes trends toward individualization of land rights and cases of "Western-type private appropriation" but is "surprised" that the model of private property organized actually remains exceptional in rural areas. "Andre and Lavigne Deleville, (1998) observe a similar pattern in Rwanda: "even at a population density of 300-500 inhabitants/km², this individualization remains partial".

[25] In many cases, the migrants bought far more land than they needed, which suggests the desire to the establish land reserves for their heirs, including sons rather than nephews in some matrilineal groups.

[26] Money combines here with legitimate blood ties and favorable circumstances (e.g. a close uncle with no male heir) to overcome the triple taboo of land sales, female inheritance, and female return from a severed exogamic marriage.

[27] Among the former, the opposition was mainly between the 'substantivists' and the 'young formalists' (as opposed to Malinowski and Firth, the founders of this sub-disciplinary field) who believed that the principles of 'maximizing behavior' applied to all human activities. Among economic historians, Polanyi's theses stimulated new research, mostly to prove him wrong, which ended up vindicating his central argument about "the misuse of modern concepts" and the "uncritical assumptions about the existence of market conditions" in ancient economies (Finley, 1955, and Oppenheimer, 1964, cited by Humphreys, 1969).

[28] While Polanyi challenged the validity of market categories for societies governed by reciprocity and redistribution, the NIE submits those societies to the transaction costs framework. Authors such as North and Myers (1982), actually see conceptual borrowings from other social science as useless. They consider that economic analysis is self-sufficient to explain the existence of 'other allocation systems', insofar as those are considered, in the context of ill-defined market, as "rational responses to certain types of transaction costs in the exchange of resources" (Myers,1982:274).

[29] There are rare exceptions to this principle, but these are deliberate innovations as we saw (Sierra Leone case, share planting tenancies, *damar* gardens, etc.).

## REFERENCES

Adesina, A.A., Nkamleu, G., & Mbila, D. (1998). Land and tree property rights and farmers' investment in alley farming in Cameroon. For review, *Land Economics*.

Adeyoju, S.K. (1976). Land tenure and tropical forestry development. *FAO committee on forest development in the tropics,* 4th session. Rome 15-20 Nov, 36 pp.

Agbosu, L.K. (2000) Land law in Ghana: Contradiction between Anglo-American and customary tenure conceptions and practices. *Working Paper 33*, Land Tenure Center, University of Wisconsin-Madison.

Agondjo-Okawe, P.L. (1970*).* Les droits fonciers coutumiers au Gabon (Société Nkomi, Groupe Myene). *Revue Juridique et Politique, Indépendance et Coopération*, XXIV, 4, 1135-1152.

Alchian, A.A., & H. Demsetz (1972). Production, information costs and economic organization. *American Economic Review,* 62, 777-795.

Alexandre, P. (1965). Protohistoire du groupe Béti-Fang. Essai de systématisation. *Cahiers d'Études Africaines*, 5, 503-560.

Anderson, L.G. (1982). The share system in open-access and optimally regulated fisheries. *Land Economics*, 58(4), 435-449.

Anderson, T.L., & HILL, P.J. (1984) Privatizing the commons: An improvement? *Journal of Southern Economic Review*, 51, 438-450.

André, C., & Delville, P.L (1998) *Changements fonciers et dynamiques agraires*. Le Rwanda, 1900-1990. In P. L. Delville (Ed.), *Quelles politiques foncières pour l'Afrique rurale. Réconcilier pratiques, légitimité et légalité* (pp 157-182). Paris : Karthala – Coopération française,.

Angelsen, A. (1997). *The evolution of private property rights in traditional agriculture: Theories and a study from Indonesia*. WP1997:6, Chr. Michelsen Institute, Bergen.

Anyangwe, C. (1984). Land tenure and interests in land in Cameroonian indigenous law. *Cameroon Law Review*, 27, 29-41.

Arnold, J.E.M. (1991). *Foresterie communautaire. Un examen de dix ans d'activité*. Rome: FAO.

Axelrod, R. (1983). *The evolution of cooperation*. New York: Basic Books.

Baak, B. (1982) Testing the impact of exclusive property rights: The case of enclosing common fields. In R.L. Ransom, R. Sutch, G.M. Walton (Eds.), *Exploration in the new economic history* (pp. 257-272). New York: Academic Press.

Bahuchet, S. (1996). La mer et la forêt : Ethnoécologie des populations forestières et des pêcheurs du sud-Cameroun. In A. Froment, I. De Garine, C. Binam Bikoi, J.F. Loung (Eds), *Anthropologie alimentaire et développement en Afrique intertropicale : Du biologique au social* (pp. 145-154). Actes du colloque tenu à Yaoundé (1993). Paris: ORSTOM.

Bardhan, P.K., & Srinivasan, T.N. (1971). Crop sharing tenancy in agriculture. A theoretical and empirical analysis. *American Economic Review*, 61, 1.

Bell, C., (1977). Alternative theories of sharecropping.  Some tests using evidence from Northeast India. *Journal of Development. Studies*, 13(4), 317-346.

Biesbrouck, K. (2002). New perspectives on forest dynamics and the myth of 'communities': Reconsidering co-management in tropical rainforests in Cameroon. *IDS bulletin*, 33(1), 55-64.

Bigombe Logo, P. (2002). Économie politique de la performance de la fiscalité forestière décentralisée au Cameroun : Logique d'État et gestion locale en question. Washington : WRI, CIFOR.

Borrini-Feyerabend, G., Farvar, M.T., Nguinguiri, J.C., & Ndangang, V.A. (2000). *Co-management of natural resources: Organising, negotiating and learning-by-doing*. Kasparek Verlag, Heidelberg (Germany): GTZ and IUCN.

Brown, T. (1998). *Cadastre, corruption, and the commons: The privatization and persistence of the commons in North-East Nepal*. Paper presented at the International Association of Common Property Conference, Vancouver.

Bruce J.W. (Ed.) (1988). A perspective on indigenous land tenure systems and land concentration. In R.E. Downs, S.P. Reyna (Eds), *Land and Society in Contemporary Africa* (pp. 23-52). Hanover, London: University Press of New England.

Bruce J.W. (Ed.) (1998). Country profiles of land tenure: Africa, 1996. *LTC Research Paper 130*, Land Tenure Center, University of Wisconsin-Madison.

Bruce, J.L.F, & Nhira, C. (1993). Tenures in transition, tenures in conflict: Examples from the Zimbabwe social forest. *Rural Sociology*, 58(4), 626-642.

Bruce, J.W., & Migot-Adhola, S.E. (Eds.) (1994) *Searching for land tenure security in Africa*. Dubuque: Kendall/Hunt Publishing Company.

Bruce, J.W., Migot-Adhola, S.E. &  Atherton J. (1994). The findings and their policy implications: Institutional adaptation or replacement. In J.W. Bruce, & S.E. Migot-Adhola (Eds.), *Searching for land tenure security in Africa* (pp.251-265). Dubuque: Kendall/Hunt Publishing Company.

Bruce, J.W., Subramanian, J., Knox, A., Bohrer, K., & Leisz, S. (1998a). Synthesis of trends and issues raised by land tenure country profiles of greater Horn of Africa countries. In J.W. Bruce (Ed.), *Country profiles of land tenure: Africa, 1996* (pp. 137-200). LTC Research Paper 130, Land Tenure Center, University of Wisconsin-Madison.

Bruce, E.J., Cloeck-Jenson, S., Knox, A., Subramanian, J., & Williams, M. (1998b) Land tenure country profiles, southern Africa, 1996. In J.W. Bruce (Ed.), *Country profiles of land tenure: Africa, 1996* (pp. 201-282). LTC Research Paper 130, Land Tenure Center, University of Wisconsin-Madison.

Buttoud, G. (1997). The influence of history in African forest policies: A comparison between Anglophone and Francophone countries. *Commonwealth Forestry Review*, 76(1), 43-46.

Cernea, M. (Ed.) (1991). *Putting people first. Sociological variables in rural development* (2nd Edition). The World Bank, Oxford University Press.

Cheung, S. (1968). Private property rights and sharecropping. *Journal of Political Economy,* 76, 1107-1122.

Cheung, S. (1969) *The theory of share tenancy.* Chicago: University of Chicago Press,.

Chilchinsky, G. (2004) *Property rights and the efficiency of markets for environmental services.* Chapter 7, in this volume.

Ciriacy-Wantrup, S.V., & Bishop, R.C. (1975). 'Common property' as a concept in natural resource policy. *Natural Resources Journal,* 15, 713-27.

Coase, R.H. (1960). The problem of social cost. *Journal of Law and Economics,* 3, 1-44.

Coase, R.H. (1984). The new institutional economics. *Journal of Institutional and Theoretical Economics,* 140, 229-231.

Coquery-Vidrovitch, C. (1982). Le régime foncier rural en Afrique Noire. In E. Lebris, E. Le Roy, F. Leimdorfer (Eds.), *Enjeux Fonciers en Afrique Noire* (pp. 65-84). Paris : Karthala.

Datta, S.K., & Nugent, J.B. (1989). Transaction cost economics and contractual choice : Theory and evidence. In J.B. Nugent, M.K. Nabli (Eds), *The new institutional economics and development. Theory and applications to Tunisia* (pp. 34-79). Amsterdam, New York, Oxford, Tokyo: North-Holland.

Davis, L.E., & North, D.C (1971). *Institutional change and American economic growth.* New York: Cambridge University Press.

Demsetz, H. (1967). Toward a theory of property rights. *American Economic Review,* 57, 347-359.

Diaw, M.C., & Oyono, R.P. (1998). Dynamique et representation des espaces forestiers au sud-Cameroun : Pour une relecture sociale des paysages. *Arbres, Forêts et Communautés Rurales,* 15-16, 36-43.

Diaw, M.C., & Njomkap, (1998). *La Terre et le Droit : Une anthropologie institutionnelle de la tenure foncière et de la jurisprudence chez les peuples bantu et pygmées du Cameroun méridional forestier.* Yaounde: Inades-Formation.

Diaw, M.C., 1985 - La pêche piroguière dans l'économie politique de l'Afrique de l'Ouest. Les formations sociales et les systèmes de production dans l'histoire. 45e congrès des américanistes, Bogota, 1-7 juillet, 38 pages.

Diaw, M.C. (1989). Partage et appropriation : le système de part et la gestion des unités de pêche. Cahiers Sciences Humaines, 25(1-2), 67-87.

Diaw, M.C. (1994). La portée du partage. *Les implications théoriques et épistémologiques du système de parts pour l'étude de l'altérité en économie.* Unpublished Ph.D Dissertation, Laval University, Québec.

Diaw, M.C. (1997). Si, Nda bot and Ayong: Shifting cultivation, land use and property rights in southern Cameroon. *Rural Development Forestry Network Paper* 21e.

Diaw, M.C. (2002). L'altérité des tenures forestières : les théories scientifiques et la gestion des biens communs. *Informations et Commentaires,* 121, 9-20.

Doolittle, A.A. (2001). From village land to "Native Reserve": Changes in property rights in Sabah, Malaysia, 1950-1996. *Human Ecology,* 29(1), 69-97.

Egbe, S. (1997). *Forest tenure and access to forest resources in Cameroon. An overview.* Forest Participation Series 6, International Institute for Environment and Development, IIED, London.

Elbow, K., Bohrer, K., Furth, R., Hobbs, M., Knox, A., Leisz, S., & Williams, M. (1998). Land tenure country profiles, West Africa 1996. In J.W. Bruce (Ed.), *Country profiles of land tenure: Africa, 1996* (pp. 1-136). LTC Research Paper 130, Land Tenure Center, University of Wisconsin-Madison.

Fay, C.C., de Foresta, H., Sirait, M.T., & Tomich, T.P. n.d. A Policy breakthrough for Indonesian farmers in the Krui damar agroforests. *Agroforestry today,* in press (2000).

Fisher, R.J. (1989). Indigenous systems of common property forest management in Nepal. Working Paper No. 18, East-West Center, Honolulu, Hawai.

Fisiy, C. (1990). Peasant resistance to land law reform. *XIV Congress of the European Society for Rural Sociology,* July, Giessen.

Fitzpatrick, P. (1983). Law, plurality and underdevelopment. In D. Sugarman (Ed.), *Legality, ideology and the state* (pp. 159-181). London, New York: Academic Press.

Flaaten, O. (1981). Resource allocation and share-systems in fish harvesting firms. *Resources paper 72, PRNE Resources Discussion Series.* The University of British Columbia, Vancouver.

Fortman, L., & Bruce, J. (1988). *Whose tress? Proprietary dimensions of forestry*. Boulder: Westwiew Press.

Giddens, A. (1987). *Social theory and modern society*. Oxford: Blackwell.

Goodland, R. (1991). Tropical deforestation. Solutions, ethics and religion. *Environment Working Paper 43*, The World Bank.

Gordon, H.S. (1953). An economic approach to the optimum utilization of fisheries resources. *Journal of the Fisheries Research Board of Canada*, 10, 442-447.

Guigou, B., Lericolais, A., & Pontié, G. (1998). La gestion foncière en pays sereer siin (Sénégal). In L. Delville (Ed.), *Quelles politiques foncières pour l'Afrique rurale? Réconcilier pratiques, légitimité et légalité* (pp. 183-196). Paris : Karthala – Coopération française.

Guijt, I., & Meera, K.S. (Eds.) (1999). *The myth of community. Gender issues in participatory development*. London: Intermediate Technologies Publications.

Hardin, G. (1968). The tragedy of the commons. *Science*, 162, 1243-1248.

Harrison, P. (1987). The greening of Africa. London: Paladin Grafton Books.

Holmstrom, B. (1979). Moral hasard and observability. *Bell Journal of Economics*, 10, 74-91.

Humphreys, S.C. (1969). History, economics, and anthropology: The work of Karl Polanyi. *History and Theory*, 8(2), 165-212.

Kant, S. (2000). A dynamic approach to forest regimes in developing economies. *Ecological Economics*, 32, 287-300.

Kant, S. (2003). *Economic theory of emerging forest property rights*. Conference Proceedings (the XII World Forest Congress, Quebec City, Canada, September 21-28, 2003): C-People and Forests in Harmony, pp.207.

Kant, S., & Berry, R.A. (2001). A theoretical model of optimal forest resource regimes in developing economies. *Journal of Institutional and Theoretical Economics*, 157(2), 331-355.

Karsenty, A. (1996). Marchandisation imparfaite de la terre en Côte d'Ivoire. In E. Le Roy, A. Karsenty, A. Bertrand (Eds.), *La sécurisation foncière en Afrique (p. 22). Paris : Karthala*.

Lakatos, I. (1978). *The methodology of scientific research programs*. Cambridge: Cambridge University Press.

Lavigne Delville, P. (Ed.) (1998). *Quelles politiques foncières pour l'Afrique rurale? Réconcilier pratiques, légitimité et légalité*. Paris: Karthala – Coopération française.

Lavigne Delville, P., Toulmin, C., Colin, J.P., Chauveau, J.P. (2001). L'accès à la terre par les procédures de lélégation foncière (Afrique de l'Ouest rurale) : Modalités, dynamiques et enjeux. Paris : IIED, GRET, IRD RÉFO.

Lawry, S., Stienbarger, D., & Jabbar, D. (1995). Land tenure and the adoption of alley farming in West Africa. pp. 464-471 In B.T. Kang, A.O. Osiname, A. Larbi (Eds.), *Alley farming research and development: Proceedings of the international conference on alley farming*, 14-18 September 1992, Ibadan, Nigeria.

Leonard, R., & Longbottom, J. (2000). *Land tenure lexicon: A glossary of terms from English and French speaking West Africa*. London : IIED.

Leplaideur, A. (1985). *Les systèmes agricoles en zone forestière: les paysans du Centre et du Sud Cameroun*. Paris: CIRAD-IRAT.

Leroy, E., Karsenty, A., & Bertrand, A. (1996). *La sécurisation foncière en Afrique : Pour une gestion viable des ressources renouvelables*. Paris: Karthala.

Long, N., & van der Ploeg, J.D. (1994). Heterogeneity, actor and structure: Towards a reconstitution of the concept of structure. In D. Booth (Ed.), *Rethinking social development. Theory, research and practice* (pp. 62-89). Harlow: Addison Wesley Longman Ltd.

Long, N. (1992). From paradigm lost to paradigm regained? The case for an actor-oriented sociology of development. In N. Long, A. Long, (Eds.), Battlefields of knowledge. The interlocking of theory and practice in social research and development (pp. 17-43). London, New York: Routledge.

Mackenzie, F. (1994). Un-customary laws: Issues for research into land rights, Kenya. *Paper prepared for the Conference on Gender and the Politics of Environmental Sustainability*, Centre for Africa Studies, university of Edinburgh, 25-26 May.

Mamdani, M. (1996). *Citizen and subject. Contemporary Africa and the legacy of late colonialism*. Princeton: Princeton University Press.

Matlon, P. (1994). Indigenous land use systems and investments in soil fertility in Burkina Faso. In J.W. Bruce, S.E. Migot-Adhola (Eds.), *Searching for land tenure security in Africa* (pp. 41-69). Dubuque: Kendall/Hunt Publishing Company.

McCay, B. J. (2002). Emergence of institutions for the commons: Contexts, situations, and events. In E. Ostrom, T. Dietz, N. Dolsak, P.C. Stern, S. Stonich, and E.U. Weber (Eds.), *The Drama of the commons* (pp.361-402). Wahsington, DC: National Academy Press.

Mehta, L., Leach, M., Newell, P., Soones, I., Sivaramakrishnan, K. & Way, S.A. (1999). *Exploring understandings of institutions and uncertainty: New Directions in natural resource management*. IDS Discussion Paper 37, Brighton.

Melone, S. (1972). *La parenté et la terre dans la stratégie du développement*. Yaounde/Paris: Klinksienck.

Merry, S.E. (1988). Legal pluralism. *Law and Society Review*, 22, 869-896.

Michon, G., de Foresta, H., & Levang, P. (1995). Stratégies agroforestières paysannes et développement durable : Les agroforêts à damar de Sumatra. *Nature, Science et Société*, 3(3), 207-221.

Migot-Adhola, S.E., Bennet, G., Place, F., & Atsu, S. (1994). Land, security of tenure, and productivity in Ghana. In J.W. Bruce, S.E. Migot-Adhola (Eds.), *Searching for land tenure security in Africa* (pp. 97-118). Dubuque: Kendall/Hunt Publishing Company.

Mope Simo, J.A. (1998). Systèmes fonciers coutumiers et politiques foncières au Nord-Ouest Cameroun. Pp.81-102 In P. Lavigne, P. Delville, C. Toulmin, S. Traoré (Eds.), *Gérer le foncier rural en Afrique de l'Ouest. Dynamiques foncières et interventions publiques*. Paris/St. Louis (Sénégal): Karthala – URED,.

Morgan, L.H. (1971) (1877). *La société archaïque*. Paris: Anthropos.

Morisset, M., & Reveret, J.P. (1985). Les quota individuels dans l'agriculture et la pêche : une analyse critique. In *agriculture et politiques agricoles : transformations économiques et sociales au Québec et en France*. Co-edition L'Harmattan/Boréal Express.

Nabli, M.K., & Nugent J. B. (eds), (1989). The new institutional economics and economic development : An introduction. In J.B. Nugent & M.K. Nabli (Eds.), *The new institutional economics and development. Theory and applications to Tunisia* (pp 3-33). North-Holland: Elsevier Science Publishers.

Newberry, D.M.G. (1974). Cropsharing tenancy in agriculture: Comment. *American Economic Review*, 64(6), 1060-1066.

Ngwasiri, N.F. (1984). The impact of the present land tenure reforms in Cameroon on the former West Cameroon. *Cameroon Law Review*, 27, 73-85.

North, D.C., & Thomas, R. (1973). *The rise of the western world. A new economic history*. Cambridge: Cambridge University Press.

North, D.C. (1977). Non market forms of economic organization. The challenge of Karl Polanyi. *Journal of European Economic History* (Fall).

North, D.C. (1990). *Institutions, institutional change and economic performance*. Cambridge: Cambridge University Press.

Nugent, J., & Platteau, J.P. (1989). *Contractual relationships and their rationale in marine fishing*. Miméo, University of Southern California/Faculté Notre-Dame-de-la-Paix, Namur.

Olson, M. (1965). *The logic of collective action*. Cambridge: Harvard University Press.

Ostrom, E. (1990). *Governing the commons*. New York: Cambridge University Press.

Ostrom, E. (1992). *Crafting institutions for self-governing irrigation systems*. San Francisco: Institute for Contemporary Studies Press.

Oyono, P.R. (1998). *Création et promotion des forêts communautaires au Cameroun. Aperçu général, contraintes socio-anthropologiques et alternatives théoriques*. Yaoundé: WWF-CPO.

Pescay, M. (1998). Transformation des systèmes fonciers et « transition foncière » au Sud-Bénin. In Lavigne Delvigne, P. (Ed), *Quelles politiques foncières pour l'Afrique rurale ? Réconcilier pratiques, légitimité et légalité* (pp 131-156). Paris : Karthala - Coopération française.

Place, F., Roth, M., & Hazel, P. (1994). Land tenure security and agricultural performance in Africa: Overview of research methodology. In J.W. Bruce, S.E. Migot-Adholla, *Searching for land tenure security in Africa* (pp. 15-39). Dubuque: Kendall/Hunt Publishing Company.

Platteau, J.P. (1989). La contribution de la nouvelle économie institutionnelle pour l'analyse des relations contractuelles et des formes organisationnelles dans le secteur de la pêche maritime. In «La Recherche face à la pêche artisanale», Contributions Provisoires, Livre 2 : 749-764. Montpellier : ORSTOM-IFREMER,.

Platteau, J.P. (1992a). Small-scale fisheries and the evolutionist theory of institutional development. In I. Tvedten, B. Hersoug, (Eds.), *Fishing for development. Small-scale fisheries in Africa* (pp. 91-114). Uppsala: The Scandinavian Institute of African Studies.

Platteau, J.P. (1992b). - Formalization and privatization of land rights in Sub-Saharan Africa: A critique of current orthodoxies and structural adjustment programmes. *Research Seminars in Rural Development Studies Institute of Social Studies*, 50 pp.

Polaniyi, K. (1944). *The origins of our time: The great transformation*. New York: Rinehart.

Polaniyi, K. (1957). The Economy as an instituted process. In K. Polaniyi, C.M. Arensberg, H.W. Pearson (Eds.), *Trade and markets in the early empires*. Glencoe: The Free Press.

Polanyi Levitt, .K. (2003). The English experience in the life and work of Karl Polanyi. *Paper for Conference Proceedings, Polanyian Perspectives on Instituted Economic Processes, Development and Transformation*. ESRC, Center for Research on Innovation and Competition, University of Manchester, October 23-25.

Rasmussen, J.N, Kaosa-Ard, A., Boone, T.E., Diaw, M.C., Edwards, K., Kadyschuk, S., Kaosa-Ard, M., Lang, T., Preechapanya, P., Rerkasem, K., & Rune, F. (2000). *For whom and for what? Principles, criteria and indicators for sustainable forest management in Thailand*. Study Report, The Danish Forest and Landscape Research Institute/Chiang Mai Univ.

Robertson, A.F. (1980). On sharecropping. *Man*, 15(3), 411-29.

Robertson, A.F. (1987). *The dynamics of productive relationships. African share contracts in historical perspective*. Cambridge, London: Cambridge University Press.

Robiglio, V., Mala, W.A., & Diaw, M.C. (2003). Mapping landscapes: Integrating GIS and social science methods to model human-nature relationships in Southern Cameroon. *Small-scale Forest Economics, Management and Policy*, 2(2), 171-184.

Rocheleau, D., & Edmunds, D. (1997). Women, men and trees: Gender, power and property in forest and agrarian landscapes. *World Development*, 25(8), 1351-1371.

Rose, C.M. (1994). *Property and persuasion: Essays on the history, theory, and rhetoric of ownership*. Boulder, San Francisco, Oxford: Westview Press.

Santoir, C. (1995). Les groupes socio-culturels. In C. Santoir, A. Bopda (Eds.). *Atlas regional du sud Camreoun. MINREST/Institut national de cartographie* (pp. 15-18). Paris: ORSTOM Edition.

Santos, B.S. (1987). Law: A map of misreading. Toward a postmodern conception of law. *Journal of Law and Society*, 14(3), 279-302.

Simon, H.A. (1962). The architecture of complexity. *Proceedings of the American Philosophical Society*, 106, 467-482.

Simon, H.A. (1969). *The sciences of the artificial*. Cambridge: MIT Press.

Simukonda, H.P. (1992). Land tenure change on customary land as a strategy for agricultural and rural development: A critical analysis of its relevance in a Malawian case. *Journal of Rural Development Hyderabad*, 11(4), 377-391.

Sirait, M.T. (1997). *Simplifying natural resources: A descriptive study of village land use planning Initiatives*. In W. Kalimantan, M.A. Thesis, Ateneio de Manila University, Philippines.

Srivastava, R. (1989). Tenancy contracts during transition : A study based on fieldwork in Uttar Pradesh (India). *Journal. of Peasant Studies*, 16(3), 339-95.

Stiglitz, J.E. (1974). Incentives and risk sharing in sharecropping. *Review of Economic Studies*, 41, 126.

Subramanian, J. (1998). Swaziland country profile. In J.W. Bruce (Ed.), *Country profiles of land tenure: Africa, 1996* (pp. 161-165). LTC Research Paper 130, Land Tenure Center, University of Wisconsin-Madison.

Sunderlin, W. (2003). *Ideology, social theory and the environment*. Lanham, Boulder, New York, Oxford: Rowman and Littlefield Publishers.

Sutinen, J. (1979). Fishermen remuneration systems and implications for fisheries development. *Scottish Journal of political economy*, 26(2), 147-162.

Tjouen, A.D. (1982). *Droits domaniaux et techniques foncières en droit Camerounais (Étude d'une réforme législative)*. Paris: Economica.

Turvey, R. & Wiseman, J. (Eds.) (1956). *Proceeding of a round table organized by the International Economic Association*. FAO, Rome.

Vandergeest, P. (1996). Mapping nature: Territorialization of forest rights in Thailand. *Society and Natural Resources*, 9, 159-175.

Vansina, J. (1982). The peoples of the forest. In D. Birmingham, P.M. Martin (Eds.), *History of Central Africa* (Vol. 1, pp. 75-117). Longman.

Vansina, J. (1990). *Paths in the rainforests. Toward a history of political tradition in Equatorial Africa*. Madison: The University of Wisconcin Press.

Verdier, R. (1971). Evolution et réformes foncières de l'Afrique noire francophone. *Journal of African Law*, 15(1), 85-101.

Weber, J., (1977). Structures agraires et évolution des milieux ruraux : Le cas de la région cacaoyère du Centre-sud Cameroon, Paris: ORSTOM,.

Williamson, O.E. (1975). *Markets and hierarchies: Analysis and antitrust implications*. New York: Free Press.

Williamson, O.E. (1985). *The economic institutions of capitalism*. New York: Free Press.

World Bank, (1974). Land Reform. World Bank Development Series. Washington, DC: The World Bank.

Zoeteweij, H. (1956). Fishermen's remuneration. In R Turvey, J. Wiseman (Eds.), *The economics of fisheries* (pp. 18-41). Proceeding of a Round Table organized by the International Economic Association, FAO, Rome, September.

# CHAPTER 4

# ORGANIZATIONS, INSTITUTIONS, EXTERNAL SETTING AND INSTITUTIONAL DYNAMICS

SHASHI KANT

*Faculty of Forestry, University of Toronto*
*33 Willcocks Street, Toronto, Ontario, Canada M5S 3B3*
*Email: shashi.kant@utoronto.ca*


R. ALBERT BERRY

*Department of Economics, University of Toronto*
*150 St. George Street, Toronto, Ontario, Canada M5S 3G7*
*Email: berry2@chass.utoronto.ca*

**Abstract:** To study the dynamics of forest regimes, an institutional analysis framework which takes account both of factors internal to the institutions and organizations as well as of the external setting - the social, environmental, economic (including markets) and international factors – is developed. Adaptive efficiency, an efficiency measure different from allocative efficiency, is suggested for institutional changes that are path-dependent rather than just price or market-dependent. The framework is used to analyze the dynamics of Indian forest regimes. The main feature of those dynamics has been incremental path-dependent change, the exception being the sudden shift from the dominance of community regimes in the pre-British period to that of state regimes in the British period. The dominant factors in this pattern of incremental change have varied markedly over time. In pre-colonial India the inertia of the informal institutions played a major role. At the outset of the colonial period, "organisational energy" was directed at the dismantling of the existing institutions. But, later many self-reinforcing mechanisms contributed to path-dependent changes. In post-colonial India, self-reinforcing mechanisms at the level of the Legislative Wing (LW) and "organisational inertia" of the Executive Wing (EW) dominated the process of institutional change for a time. But, later the "organisational energy" of the LW, the external setting, and "organisational surges" of the EW allowed more rapid change. The adaptive efficiency varied – higher in decentralized regimes of pre-British India and recent regimes and lower in the centralized regimes of British India and the first four decades of independent India. Organisational inertia has been one of the main factors impeding institutional changes towards adaptive efficiency. Hence, policy and management prescriptions for sustainable forest management, in these countries, should address institutional and organisational aspects in an integrative manner.

*Kant and Berry (Eds.), Institutions, Sustainability, and Natural Resources: Institutions for Sustainable Forest Management, 83-113.*
© *2004 Springer. Printed in Netherlands.*

## 1. INTRODUCTION

Neo-classical economic theory, when applied to natural resources, is generally focused on technological developments but ignored the institutional structure that shapes the interactions between policy makers, resource managers and resource users. The set of institutions that serve to order the actions of those involved with forest resources is commonly termed the forest (resource) regime (Young, 1982, p.15). In terms of forest regimes, discussion in the neo-classical framework has been limited to private forest regimes versus state controlled forest regimes. However, over the last decade or so there has been an increasing recognition of the importance of institutions as a determinant of economic performance, and need to extend the institutional discussion beyond private and state regimes. One of the important elements of that discussion is the process of institutional changes termed as "institutional dynamics". Institutional economists, such as Coase (1960), Commons (1961), Ayres (1962), Veblen (1975), Schotter (1981), Bromley (1989), North (1990), and Setterfield (1993) have discussed the issues related to institutional dynamics. However, these economists have mainly emphasized the role of existing institutions, their inertia, and market forces as critical factors in institutional dynamics, and the role of organizations, their inertia, and external factors other than market forces have not been adequately incorporated into the thinking of institutional economists[1].

In the context of new paradigm of sustainable forest management, an understanding of the dynamics of forest regimes has become critical, and it has generated a huge literature on dynamics of forest regimes. However, most of these discussions are in the framework of public policy analysis, where mainly the role of government has been analyzed. Some examples of these discussions are McCarthy (2000) and Cashore (2001). Kissling-Naf and Bisang (2001) used property right approach and public policy analysis, and Kant (2000) included socio-economic factors in their analyses of forest regimes. In the case of India, since the early eighties, many scholars[2] have discussed the dynamics of forest regimes. Some have focused on forest regimes specifically; others have treated those regimes as a part of a broader environmental analysis. Guha, Gadgil, and Shiva - probably the most prolific writers on the topic - are highly critical of the "technocratic" state which disregards indigenous knowledge systems and cultural practices. They call for the replacement of state management of forests (and other natural resources) by community-level management. To analyze institutional changes from 1976 to 1994, Vira (1995) uses the concept of *relative autonomy*, in which the state is an arena of social conflict among social groups which are not political or economic equals. He explains changes in forest regimes in terms of the shifting configuration among ten forest-dependent groups (including state agencies). Rangan (1997) does not see the state as the powerful and predatory monolith visualized by Guha, Gadgil, and Shiva, operating independently of markets and civil society; in his view natural resource management policies are affected by a wide range of groups and processes.

Most discussions of the dynamics of forest regimes, however, do place the state at the centre of the economic and political processes leading to institutional change. These discussions completely disregard the role of institutional factors in the dynamics of institutions, discussed by institutional economists. In addition, state power is not a simple concept; it is exercised through a variety of institutions with their own organizational structures (Pathak, 1994). The interests of various social groups and the demands stemming from economic, social, and environmental forces are translated into new institutions and actions through the operations of existing institutions and state organizations. But, the role of (state) organizations has not attracted the desired attention of either institutional economists or political scientists in the discussions of dynamics of institutions, specifically dynamics of forest regimes[3].

To understand the process and path of institutional change, and to draw lessons on how to channel such change towards efficient outcomes, requires consideration both of factors internal to the institutions and organizations as well as of the external setting- the social, environmental, economic, and international factors which form the context for change. In this chapter, we develop such a framework and use it to explain the dynamics of forest regimes in India. The framework is discussed in the context of the dynamics of forest regimes, but the main features of the framework – organizations, institutions, external setting, and their interactions – will remain the same irrespective of the context.

Our framework for the analysis of forest regime dynamics has its roots in institutional economics, as explained in Section 2, and is presented in Section 3. Section 4 deals with the main features of forest regimes in India during the pre-colonial, colonial, and post-colonial periods, the dynamics of forest regimes, and the nature of changes and factors contributing to those changes. Section 5 reviews the impact of recent changes in forest regimes towards community-based regimes. Finally, some policy implications are drawn.

## 2. AN OVERVIEW OF INSTITUTIONAL ECONOMICS

Among economic analyses giving serious attention to institutions a distinction has emerged between the Old Institutional Economics (OIE) and the New Institutional Economics (NIE). The former, associated with authors such as Commons (1961), Ayres (1962), and Veblen (1975), is characterized by an holistic approach stressing the idea that individual behavior and phenomena cannot be explained without taking due account of the context. This perspective gives considerable emphasis to institutions relative to the activities and choices of individuals in the determination of economic outcomes (Setterfield, 1993). The NIE, associated with authors such as Coase (1960), Schotter (1981) and Williamson (1985), emphasises the importance of the self-interested behavior of individuals and posits that, during the evolution towards a market economy, institutions arise because they are valued by rational economic agents. Bromley (1989) argues that such positive valuation of institutions may be related not only to their contribution to allocative efficiency or to a desired redistribution of income but also to profit-seeking unproductive activities.

Both variants have been criticized for being unidirectional - the OIE for overlooking the impact of individual behavior on institutions, and the NIE for overlooking the impact of institutions on individuals' behavior. To overcome these shortcomings, Setterfield (1993) has suggested a model of institutional hysteresis characterized by the short-term exogeneity and long-term endogeneity of institutions. In the short-term, due to a degree of, it is the institutional setting which mainly guides economic activities. In the longer-run, however, institutional changes come about through pressures from the current patterns of economic activity-- pressures that are also usually counterbalanced to some extent by the forces of institutional inertia. In other words, long-run institutional changes are evolving, not-necessarily-optimal, path-dependent phenomena[4], unlike the standard equilibrium metaphor of mainstream economic theory.

Another new variant of the theory of institutional change has recently been posited by North (1990), with foundations in the theory of technological change proposed earlier by David (1985) and Arthur (1988). David described a form of path-dependent technological change beginning with a set of accidental events, and identified strong technical inter-relatedness, scale economies, and irreversibilities due to learning and habituation as the main factors contributing to path dependency. Arthur (1988) linked path dependency to the increasing returns economy, which was seen as characterized also by multiple equilibria, and the related possibility of inefficiency and lock-in. He identified four generic sources of self-reinforcing mechanisms: large set-up or fixed costs; learning effects; coordination effects; and adaptive expectations. These concepts have been used to explain the choice of AC electricity (David and Bunn 1987), the selection of light-water nuclear reactors, and the gasoline engine (Arthur 1989), as well as the FORTRAN computer language and VHS videotape formats (Arthur 1991). This literature on technological change draws a number of parallels to the broader process of change, including, mostly implicitly, institutional change. However, North (1990, pp. 92-104) incorporated explicitly in his explanation of institutional change a group of concepts associated with increasing returns and imperfect markets - path dependence, lock-in, and existence of inefficiencies. North argues that increasing returns are an essential ingredient to technological as well as institutional change, and all four of Arthur's self-reinforcing mechanisms apply, although with somewhat different characteristics. North also observes that the perceptions of actors play a more central role in institutional than in technological change. North (1990, pp. 97-98) used the example of the Northwest Ordinance to illustrate a path-dependent pattern of institutional evolution.

In summary, both Setterfield's model of institutional hysteresis and North's theory based substantially on increasing returns point to the path-dependent nature of institutional change. The partial counter-balancing of the external forces for institutional change, which may be continuous in nature, by internal factors (such as institutional inertia and self-reinforcing mechanisms), results in path-dependent incremental institutional change. Many economists have used the concept of path-dependent evolution of institutions in a variety of different fields and drawn related policy implications; examples include electric power in the city of Chicago (Throgmorton and Fisher 1993), investments in fossil fuel conservation (England 1994), environmental decline (Goodstein 1995), and urban sprawl (Atkinson and Oleson 1996). However, in this framework, it is assumed that there is no effect or role of organizations in shaping institutional changes, and there are no interactions between institutions, organizations and external setting. Next, we propose a model of the institutional (forest regime) dynamics in which all three components and their interactions are included explicitly.

## 3. A FRAMEWORK FOR ANALYSING THE DYNAMICS OF INSTITUTIONS (FOREST REGIMES)

Institutions refer to the rules, norms, codes etc., whether formal or informal, which define the rights, privileges and obligations of various groups under a regime. Organizations are physical manifestations of institutions, designed by their creators to achieve certain objectives. An organization is a collection of functions carried out by people who are influenced by organizational culture, norms, and practices and who in turn influence the implementation of institutions and the pattern of institutional change (Sastry, 1997). Organizations do not operate in a vacuum, but are continuously subjected to external forces, commonly referred to as the external setting. These three essential elements -institutions, organizations, and external setting - and the interactions among them, determine the dynamics of institutions - forest regimes. We now discuss each in turn.

### 3.1 Institutions

Forest regimes, like other institutional structures, include both formal and informal elements. Formal institutions involve formal rules that operate at a minimum of two levels - rules for making the rules and operational rules. For India the constitution constitutes the first or upper level. Various levels of operational rules can be distinguished, with overall forest policy at the top and legislation such as the Indian Forest Act, government orders, and guidelines[5] by the central government to translate broad policy decisions into actions forming a second level. The follow up state government acts, orders, and directions constitute a third level, and corresponding/resulting orders and directions by the head of the state forest department a fourth. Other levels may be present depending on the complexity of the hierarchy. Our focus here is on the first level of operational rules - the changes in forest policy, and within that the particular issue of inclusion/exclusion of local people in forest management, and the process of change between regimes where they are included and regimes where they are excluded.

A necessary condition for the effectiveness of formal institutions is reasonable compatibility with informal institutions (North, 1990; Kant & Cooke, 1998). The informal institutions of a local user group with respect to forest resource use/management are embedded in broader informal institutions, such as those relating to the management of other natural resources (water and pastureland), religious places, and schools; these in turn are part of the group's culture (Kant, Singh, & Singh, 1991). Since cultural change tends to be very gradual, this embeddedness means that changes to the informal institutions associated with forest regimes are also likely to be incremental and that formal forest resource institutions will be subject to the inertia of these related informal institutions. Formal forestry institutions are linked both vertically- with other institutions at different levels but involving the same resources or issues, and horizontally-with institutions involved in other relevant areas such as general administration, tax administration, etc. Institutional integration via these links may make the costs of/or impediments to change in any given area prohibitively high; even if one institution is in favor of change, others with which it interacts may not be, so that, as a whole, the group of relevant institutions may demonstrate strong inertia against change. These change-retarding forces may involve "frequency dependency effects", whereby the strength of a particular set of institutions depends upon the frequency with which they have held sway in the past. Complexity of institutions can also generate inertia against change. The forces against institutional change generated by such features as integration, complexity, repetition etc. are termed "institutional inertia" and may arise from informal institutions - "informal institutional inertia" - or from formal institutions - "formal institutional inertia".

## 3.2 Organizations

Organizations are created to pursue certain objectives identified by their creators. In the case of business organizations, owners (shareholders) are the creators. In the case of government organizations in a Parliamentary Democracy, the Legislative Wing (LW) of the state is the creator, which also defines the broad objectives, while the Executive Wing (EW), composed of organizations such as the forest department, is responsible for carrying out the designated functions; the interaction of the formal institutions of forest management with the informal institutions of local user groups occurs largely through the EW (forest department).

In principle, an organization should work efficiently to achieve the objectives of the creators. But, its members may develop their own goals in addition to or even in conflict with those of the creators; the resulting conflicts are well documented in the literature on the "principal-agent problem" (Jensen & Meckling, 1976). Not infrequently the creators want to reform the organization, but meet resistance in the form of the attitude of managers, self-reinforcing mechanisms related to the informal institutions (or culture) of the organization, or organizational structure. Organizational resistance based on forest managers who are unwilling to subjugate their own interests to those of the owners is referred to as "attitudinal inertia". Normally, the creator (the LW) develops prescriptive rules (codes of conduct) for interactions among the members of an organization, but over time the members develop informal institutions governing their day-to-day behavior, attitude, and interactions with each other--an "organizational culture". The self-interest of the forest managers and the associated organizational culture give rise to many self-reinforcing mechanisms which impede institutional change, like the irreversibilities due to learning and habituation referred to by David (1985). Resistance created by these mechanisms is termed "cultural inertia". Sometimes it is not so much the attitudes or the culture of the organization but its structure which impedes institutional change; we refer to it as "structural inertia". Organizational resistance to institutional change, fed by "attitudinal inertia", "cultural inertia", and "structural inertia", is termed "organizational inertia (OI)".

The degree of organizational inertia varies from case to case. Sometimes, even where the gap between the organizational culture and the intent of the creators has become wide, there are individuals not fully immersed in that organizational culture, whether because of their short period in the organization, their prior experience from other organizations, their social background, or their particular life goals. They may, for example, be concerned with the external image of the organization and will respond to external pressures such as the needs and demands of forest-dependent groups. They contribute to what we call "organizational energy (OE)" for institutional change. Normally such energy is low. But if some innovative but risky experiment beyond the boundaries of the existing formal institutions is undertaken by a few of its members and meets with initial success this induces other members to join them, and may generate enough organizational energy to change existing institutions. We refer to such events as "organizational surges (OS)". In other situations "organizational energy" may reflect a sort of takeover, as where organizations implanted by foreign rulers or local organizations acquired by multi-national companies have enough "organizational energy" to dismantle the existing institutions.

In the case of government organizations, the LW (the creator) also exhibits the characteristics of an organization. However, there are critical differences between the LW and the EW. First, in a parliamentary democracy, the LW is directly responsible to the people while the EW is not. Second, the tenure of the members (elected) of the LW is only a few (four or five) years, so that a newly elected LW may have a different ideology from that of its predecessor; tenure of the EW members is typically much longer, say 30 to 35 years where job security is high. Third, the role of leadership in the LW is more prominent than in the EW. Hence the degree of "organizational inertia" is likely to be less and the degree of "organizational energy" greater in the LW than in the EW, and individual leaders can be a source of much "organizational energy".

*3.3 External Setting*

The external setting of any forest regime is shaped by social, economic, political, environmental, and international factors as well as by various forest-dependent groups, and by the interactions between these factors and groups. Environmental groups will play a major role in bringing environmental issues to the forefront. Economic factors such as liberalization will change the nature of dependence of forest industries on forests, and social movements and social awareness may change the outlook of local user groups towards forest management. International factors such as the forest policies of the United Nations, the World Bank and foreign governments, along with changing governance systems (centralization versus decentralization) are part of the external setting.

*3.4 Interactions between External Setting, Organizations, and Institutions*

Elements of the external setting normally interact with the top formal (policy) level of forestry institutions through the LW; if the energy they create breaks the inertia there, it may lead to a re-examination of the objectives of forest management. Normally, such a re-examination will be done by the EW, and its outcome will reflect the balance between the organizational energy of the LW and the inertia of the EW. In some cases, "organizational surges" may reduce the OI of the EW and/or enhance the OE of LW, so that significant changes are accepted by the EW. In exceptional cases, such as the presence of very strong leadership, the LW may direct the EW to pursue new social objectives without any review by the latter; in the opposite case, if formal objectives are changed at all they may be defined ambiguously to accommodate the conflicting demands/interests from the changed external setting on the one hand and the managers of the organization on the other.

In any case, change of formal objectives is only a first step toward institutional change, which can be grouped in two categories – path-dependent incremental change and path-independent discontinuous change. Normally, an institutional structure is comprised of a variety of formal rules, enforcement procedures, and informal norms, and institutional change takes the form of marginal adjustment to this complex institutional structure. Consider, for example, a community forest regime which consists of decision-making rules, boundary rules, exclusion rules, harvesting-quantity rules, harvesting-period rules, penalty rules, and conflict-resolution rules. Suppose that, at some point, the harvesting period is extended from three to four months. Such change is incremental because only a marginal change has been made to the overall structure of the community forest regime, and it is path-dependent because it has been influenced by the history of the regime. A path-independent discontinuous change is a radical change in the existing regime structure, as where the stare terminates the community forest regime, and imposes a state forest regime. In such a case, the formal rules of the community regime are replaced by new formal rules, so change is discontinuous, and there is no role for history as a determinant of the character of the new regime, so it is path-independent. Wars, revolutions, conquest, and natural disasters are the main sources of discontinuous institutional change (North, 1990, p. 89), but such change has been observed in the absence of these factors also. Privatization of a majority of government forests during the early nineties in New Zealand is an example of path-independent discontinuous change.

Institutional change usually comes up against "institutional inertia", which plays a role of a constraint parallel to that of the "organizational inertia" of the EW. Normally, the combination of institutional and organizational inertia limits institutional change to the incremental and path-dependent variety. Only in exceptional cases, when the OE of the LW is extremely high, institutions change in a discontinuous and path-independent way. In certain special cases, such as entry of multinationals or imposition of institutions by foreign rulers, change, although temporally discontinuous and path-independent, will be continuous and path-dependent relative to the prior experience of those organizations newly present on the scene. Analysis of institutional changes which are path-dependent, rather than just price or market-dependent, calls for efficiency measure different from the neo-classical concept of allocative efficiency applied in that other case.

## 3.5 Adaptive Efficiency - An Efficiency Measure of Institutional Change

Given the complexity of institutional change, there is no guarantee, nor even a general presumption, that outcomes will systematically be desirable ones. One factor which may help to shape the path of institutional change in positive ways is learning over time by individuals, organizations, and societies and the diffusion of that learning. Societal advance thus depends on the capacity of its institutions to induce learning processes which lead to beneficial institutional change. Institutional arrangements that help a society to acquire knowledge and learning, to induce innovations (e.g. by encouraging such learning mechanisms as trials and experiments), to undertake risk and creative activities, and to resolve problems and bottlenecks contribute to such learning; North (1990, p.80) referred to this quality of institutions as "adaptive efficiency". North also clearly points out that we may not know all the aspects of adaptive efficiency but the institutional structure that allows the trials, experiments, and innovations will be adaptively efficient compared to those structure which does not allow these elements. Similarly, institutional structures that have incentive mechanisms for learning by doing that will lead individual agents to evolve systems gradually different from the existing ones will be adaptively efficient. A similar idea is imbedded in Hayek's (1960) argument that the society that permits the maximum generation of trials will be most likely able to solve its problems through time. Hence, adaptive efficiency encourages the development of decentralized decision making processes that allows societies to maximize the efforts required to explore alternate ways of solving problems (North, 1990). In this process, agents learn from failures and try to eliminate errors. However, these errors may not only be probabilistic, but also systematic, due to ideologies that may give people preferences for the kinds of solutions that are not oriented to adaptive efficiency (North 1990). On other hand, rigid institutional arrangements which leave no scope for these processes will be adaptively inefficient. On the similar lines, institutional changes that are based on the existing norms, behavioral patterns, moral codes will be adaptive efficient while the institutional changes that attempt to replace the existing norms etc. by a formal set of rules that are in-coherent to existing norms will be adaptive inefficient. In other words, complementarity of formal and informal institutions will lead to adaptive efficiency, while non-complementarity to adaptive inefficiency. At this stage it would be difficult or impossible to define a measure of adaptive efficiency in quantitative terms, and thus parallel to static allocative efficiency; we opt instead for a three-point qualitative scale (high, medium, low) in the discussion of Indian forest regimes.

## 4. THE DYNAMICS OF INDIAN FOREST REGIMES

The colonization of India by the British had a marked impact on forest regimes, such that a logical periodization for our analysis is pre-colonial, colonial, and post-colonial. Though details of the pre-colonial forest regimes are limited, its inclusion adds continuity and completeness. Both the colonial and the post-colonial periods have two distinct sub-periods, so our history of forest regimes involves five periods in all; their main features are presented in Table 4.1. The above discussion of the LW and the EW is applicable only to the post-colonial democratic period.

*Table 4.1.* The Main Features of the Dynamics of the Indian Forest Regimes

| Period | Pre-colonial | Colonial (up to 1864) | Colonial (1864-1947) | Post-colonial (1947-1980) | Post-colonial (1980-2004) |
|---|---|---|---|---|---|
| **Dominant Regime(s) and their nature** | Community (Informal) | State (Semi-formal) & Community (Informal) | State (Formal) | State (Formal) | State (Formal) & Community (Formal, semi-formal & informal) |
| **Nature of Change of Forest Regimes** | Temporal Path-dependent & Continuous | Temporal Path-dependent & Continuous | Spatial Path-dependent & Discontinuous | Temporal Path-dependent & Continuous | Temporal Path-dependent & Continuous |
| **Main Features** | 1.Absence of Formal Organization<br><br>2. Stable External Environment<br><br>3.Non-exclusion of Local People Decentralized Institutions | 1.Absence of Formal Organization<br><br>2.Only Semi-formal State Institutions<br><br>3.Non-exclusion of local people | 1.Creation of Formal Organization<br><br>2.Imposition of Formal and Centralized State Institutions<br><br>3.Exclusion of Local People | 1.Stable External Environment<br><br>2.Centralized State Institutions<br><br>3.Exclusion of Local People | 1.Social movements, emergence of non-government organizations, and local-level actions<br><br>2.Recognition of Decentralized Institutions<br><br>3.Inclusion of Local People |
| **Main Factors Contributing to the Regime Dynamics** | 1. Informal Institutional Inertia | 1.External Environment (colonization)<br><br>2. Informal Institutional Inertia | 1.Organizational Energy<br><br>2. Self-reinforcing Mechanisms (Set up costs, learning effects, & limited resistance) | 1.Self-reinforcing mechanisms at the LW level (adaptive expectation & set-up costs)<br><br>2. Self-reinforcing mechanisms at the EW level (continuation of Indian civil Service & organizational inertia of the forest department) | 1.External Environment<br><br>2. Organizational Energy of the Legislative Wing<br><br>3. Organizational Surges of the Executive Wing |
| **Adaptive Efficiency** | Medium | Marginal decrease from the previous period (Medium) | Reduced from the previous period (Low) | No change from the previous period (Low) | Start increasing |

*4.1 The pre-British Period - Temporal Path-dependence due to Informal Institutional Inertia*

In ancient India, learning and culture were mainly seen as a product of hermitage in the solitude of the forests (Mookerji, 1950). Indian epics such as Vedas, Puranas, Ramayana, and Mahabharat thus placed a very high importance on forests. According to Puranas, trees not only provide physical products such as timber and fruits, but also help ancestors to find a way to heaven (Dwivedi, 1980, p.7). The forest dependence of people was institutionalized through a variety of cultural and religious mechanisms such as sacred groves, temple gardens, and worship of some trees. For the people of these local communities, destruction of forests meant the end not only of material items but also of spiritual benefits necessary for eternal life.

The welfare of subjects was a prominent motto of the rulers of this period. As Chanakya (Kautilya's Arthasatra, translated by R. Shamasastry, 1929, p. 38), a revered teacher and the principal adviser to the king Chandra Gupta Maurya, opined: "In the happiness of his subject lies his (the king's) happiness; in their welfare his welfare; whatever pleases himself he shall not consider good, but whatever pleases his subjects he shall consider as good." Chanakya elaborately discussed and suggested (in Arthashastra, written during 325-273 BC) how forest management could contribute to the welfare of the subjects. He created three categories: (i) reserve forests, for the recreational use of the king and to meet the state's needs for construction timber and elephants for defense purposes; (ii) forests donated to eminent Brahmans for religious learning and for the performance of penance; and (iii) forests for the subsistence needs of the public (Dwivedi, 1980, p.9). Though the classification does reflect the existing social hierarchy, it clearly recognized the needs of the public as well as those of the rulers and the elite. The dictums of Chanakya were followed by the Mauryan Empire and continued in practice until at least the 8th century AD (Jha, 1994, p.21). Most of the forests, which were owned by the rulers, except those donated to Brahmans and those reserved for the exclusive use of the state, were under a community regime. Local decision-making decided which trees to use for firewood and other purposes, when to harvest, and how the forest products were distributed among households; decisions were taken in general meetings, rather than by local notables alone. At this point, the territory of post-Independence India consisted of hundreds of relatively small kingdoms and principalities, a fact which implied a smaller distance between ruler and ruled that emerged later.

In medieval India (800 AD to 1526 AD), and especially the subsequent Mughal period (1526-1756), the priority attached to the welfare of subjects declined (Upadhyaya, 1991). The former period saw a gradual trend towards centralization in the sense that the kingdoms became on average larger, through a process of conquest. It also saw an increase in the share of rulers who were non-Indian. Many Sultans, their courtiers, and senior subordinates enjoyed an increasingly luxurious life at the expense of their subjects (Jha, 1994, p.22) and, though the forest area remained adequate--forests were not commercialized and the public was not excluded in principle, direct involvement of the rulers on behalf of the public was reduced. Still, Sultans such as Ala-ud-Din Khalji (1296-1316) showed concern for public welfare by taking up social amenities programs such as roadside plantations. The Mughal period (1526-1756) saw re-unification and integration of states (Upadhyaya, 1991), and an increase in the importance of forest products due to urban development. There is no record of the Mughal rulers returning to classification-based forest management, but they did lay substantial emphasis on social amenities as well as on trade; large-scale roadside plantations and mulberry block plantations (for the silk trade) were developed during this period, mainly for the benefit of the public. On the whole, the Mughal rulers took a serious attitude to the forests, with the result that a number of forest products were available to fulfil both their needs and those of the public (Jha, 1994, p. 27). But the now-greater centralization of government in the region again tended to distance the rulers from the ruled.

In short, forests were an integral part of life in the pre-British period and forest regimes were governed mainly by conventions reflecting a vision of fair distribution of benefits among all sections of society.[6] Even though, forest land was owned by the rulers, community regimes in forest products, either explicit or implicit, were dominant, and the public's welfare had a considerable though declining weight. Even under the non-Indian rulers of the medieval and Mughal periods the forest regimes did not come into enough conflict with the social structure to force it to change. Their dominant feature remained the community-based informal institutions, and "informal institutional inertia" due to the embeddedness of forest regimes in other social institutions contributed to the path-dependence (here, continuity) of forest regimes during this period. There were no formal organizations dedicated exclusively to forest management, and hence the contribution of organizational factors to institutional dynamics was minimal. The forest regimes (informal institutions) were by nature decentralized; local people were the forest managers and decision makers, and they had the freedom to experiment, learn from failures, and make changes to the existing institutional arrangement. Hence, we judge the adaptive efficiency during this period to have been at least at the medium-level[7].

*4.2 The First Forest Policy Phase of the British Period (Up to 1864) - Path-Dependence due to Informal Institutional Inertia and Lack of Organizational Energy*

The British brought to India an attitude towards forests based on their own specific history of drawing down their own forest resources as well as those of Ireland, southern Africa, and the north-eastern United States to obtain timber for shipbuilding and iron smelting, and to get land for agriculture (Guha, 1996). Troops and settlers in seventeenth-century Ireland had cleared forests to deny cover to Irish rebels (Rangarajan, 1996, p.16). Throughout the seventeenth and eighteenth centuries, forest dwellers in England were locked in struggles with Crown officials and landlords over control of forestlands (Thomas, 1983, pp.194-195). The agenda of "agrarian progress" led to the breaking-up of the common tenurial system in Ireland and the Scottish highlands (Bayly, 1989, pp.123-4). Soon after their arrival in India, the British rulers extended land under cultivation as a way of consolidating their control, and sought military advantage against their foes by denuding the countryside (Rangarajan, 1996, p.17). Extension of agriculture and strategic denudation were of course not new to India (Pouchepadass, 1995); the British only increased the pace of these processes, and the objective of forest conversion to agricultural land became revenue-generation rather than the subsistence needs of the local people, as in ancient India. A very significant new pressure came from the contemporary strategic and commercial imperatives of the British empire (Rangarajan, 1996, p. 19). The shortage of timber in Britain, and the isolation of Britain from the Baltic supply lines during the Revolutionary and Napoleonic wars between 1793 and 1815 forced the empire to look to alternative sources of wood for shipbuilding. In the late eighteenth and early nineteenth centuries Indian forests were mainly used by the British to meet the requirements of the Royal Navy, on whom the safety of empire depended (Smythies, 1925 cited in Guha, 1983)). In the middle of the nineteenth century, after the Indian mutiny in 1857, a strategic priority of the empire became rapid troop movement within India. On the commercial front, expanding imperial trade was high on the agenda. To meet these strategic and commercial objectives the British began construction of a huge Indian Railways network, and the railway ties (sleepers) came from the Indian forests.

In summary, during this first phase of British rule forest regimes aimed to secure economic, political, and strategic advantages for the empire. Since forest resources were understood to be inexhaustible, local users were not in principle excluded from the resource. Some new semi-formal institutions were introduced by the British, while the informal institutions of local communities continued to exist except in a few cases where they gave way in the face of large scale harvesting by the rulers. The dynamics of forest regimes were mainly influenced by the external setting (colonization) and by the rulers' strategic and economic considerations. The lack of formal forest organizations, and therefore of forest-specific organizational energy, together with informal institutional inertia due to embeddedness in the social structure contributed to the path-dependant continuation of informal decentralized community regimes. There may have been some decrease in adaptive efficiency due to the introduction of semi-formal institutions and shift of control to the new rulers, but in the community regimes it continued to be the same as in the pre-British period.

*4.3 The Second Forest Policy Phase of the British Period (1864 – 1947): Spatial Path-Dependence (or Temporal Discontinuous Path) due to Organizational Energy and Self-reinforcing Mechanisms (Positive Feedback)*

Forest degradation, due to the reckless harvesting practices of the first phase of British rule (Pearson, 1969), eventually forced the colonial authorities to recognize that Indian forests were not inexhaustible. Scattered steps were taken early in the nineteenth century to ensure the timber supply for shipbuilding, but only in 1862 did the Governor General call for the establishment of a forest department to ensure the sustained satisfaction of the enormous demand for railway sleepers (Webber, 1902, cited in Guha, 1983). That department's creation in 1864 signaled a new phase, characterized by state control and the increasing exclusion of local people from forest use. The first Forest Act of 1865 empowered the state to declare any land covered with trees or brushwood as government forest and to set the rules for its management. At this point the government's right was still subject to the condition that it not abridges the existing rights of the local people. Mr. Dietrich Brandis, the first Inspector General of Forests, came from Germany, the leading European nation in forest management, and was sensitive to such existing reflections of indigenous Indian forestry as the sacred groves and the frequently competent management by the Indian rulers. He argued for a parallel system of communal forests for village use, separate from the state forests. But Mr. B.H.Baden-Powell (the British head of the revenue department) advocated total state control over all forest areas as the only check on individual self-interest and short-sightedness (Guha, 1996). Baden-Powell's view prevailed, leading to the 1878 Indian Forest Act, which put restrictions on the public's access to forestland and produce. Though the Act did countenance the provision of village forests, this option was exercised only in a few isolated cases (Guha, 1996). The first general statement of forest policy by the British Government (in 1894) further weakened local rights, as reflected in a shift of terminology away from traditional "rights" to "rights and privileges"[8]. The policy emphasised the need for state control and use of forests to augment government revenue. The reserve forest area was expanded at the expense of that allocated to villagers' use (Guha and Gadgil, 1989). Though the British defended their changes in terms of efficiency, arguing that well defined property rights would increase production, in practice these steps were less about clarification than about abolishing the rights of local people established through conventions developed over long periods of time. The new regime not only entailed the predictable welfare losses to the people, but also had negative efficiency implications due to the high transactions costs involved in excluding them.

The Indian Forest Act of 1927 incorporated the main features of the National Forest Policy of 1894, empowering the government to declare any piece of land to be state forest and recognizing only the rights and privileges of persons--not of communities. The shift from indigenous management systems to state control of forests incorporated in these two documents has been identified by some observers as the first step towards forest conservation or scientific management (Tucker, 1988). The preparation of management plans (referred to as 'Working Plans') was initiated on a major scale in 1884, and large forest areas in many provinces such as United Provinces, Central Provinces, Madras and Bombay Presidencies were brought under such plans (FRI, 1961b, pp. 91-97). The increased degree of scientific (silvicultural) professionalism of the forest service was also evident from problem-specific efforts such as the regeneration of Sal forests in early 1920s (Tucker, 1988). However, the basic principle of the working plans - sustained yield management - and other silvicultural principles were frequently superseded in the economic, political, and strategic interests of the empire, especially during the two World Wars. In World War I approximately 1.7 million cubic meters (mostly teak) were exported annually and the indigenous resin industry proved to be a great boon at a time when American and French supplies were unavailable (Guha, 1983). The Second World War saw an even more extreme "mining" of the Indian Forests. Timber management was placed on an emergency basis, with supplies and prices of timber strictly controlled by the Wartime Mobilization Board and the Forest Department (GOI, 1944). During this period, no management plans were followed, the only limit to harvesting was the supply of labor, and fellings were estimated to be six times annual yields (GOI, 1948). There is similar evidence of profligacy with respect to hunting practices. The new laws restricted small-scale hunting by tribal peoples but continued to facilitate large-scale hunting by whites[9].

To implement and manage its new forest regimes, the colonial government had created a large bureaucracy--the forest department. Mr. Brandis, the first Inspector General of Forests, was responsible for its establishment, including the forest service, forest training, and research. This German botanist recommended the selection of Imperial Forest Service officers from Europe and their training there. He was of the view that:

> "Attention should particularly be paid to scientific requirements, especially in natural sciences, and they should be competent to survey a forest and to plan and construct forest roads. Although climate and vegetation in India are different, yet the fundamental principles of forest management are the same everywhere and persons, whose practical experience is supplemented by scientific education will be able to apply these principles in the forests of another country" (FRI, 1961a, p. 105).

Although he did recognize some of the merits of community involvement in forest management, Brandis only went so far. While appreciating the social, cultural, and economic setting of local user groups, he failed to incorporate it in the formal forest regime (forest policy, forest law, structure of the forest department, training of forest officers, and forest research). While the officers were being trained in Europe, preparation of subordinate staff began in 1878 at the Central Forest School at Dehradun. In 1926 the training of forest officers was started at the Indian Forest College, Dehradun; this and later Indian schools were headed and managed by European specialists. The main objective of the training was to provide basic skills in engineering and natural sciences to fulfill the empire's demands from Indian forests; social science inputs were totally missing from the programs. This lacuna contributed to the isolation of forest officers from local communities, and also to their belief that local communities could not manage forest resources efficiently.

A Forest Research Institute was created at Dehradun in 1906. It was to become the colonial world's premier research station and the model for later centres in Britain's tropical colonies (Tucker, 1988). Its main objective was to provide support to the economic, political, and strategic interests of the empire. In the early twentieth century this meant the antiseptic treatment of inferior timber species for use as railway sleepers, which made the use of chir and blue pines possible on a commercial scale in 1912, and in the following year led to the reserving of extensive pine forests (Guha, 1983). Another research area was the industrial uses of Minor Forest Products (MFPs), for such items as resin and turpentine, kutha, myrobolans, and essential oils. India was the only source in the empire for these products, and their export contributed greatly to empire revenues (Guha, 1983). During the Second World War, forest research was dedicated to finding new substitutes for valuable species that were no longer available, thus promoting the harvesting of those species still left in the forests.

The colonial forest regime influenced the attitude of those princely state rulers who had maintained their own identity under British rule and who still administered about half of Indian territory. They observed the commercial benefits from forests enjoyed by the British, and started leasing forest tracts to the latter. Later some of them appointed their own forest officers, trained in colonial institutes together with the Imperial Forest Service officers, to manage their tracts. Revenue generation became the main objective of forest management in these states as well, and brought with it a similar exclusion of local people.

The progressive diminution of rights and the consequent loss of control over their forest resources evoked a sharp reaction from some forest communities against both the British government and the local rulers, as witness the Rampa rebellion of 1879-80 in Andhra Pradesh, the 1918 militant revolt by Santhals in Midnapur district of the Bengal Presidency, the 1916 and 1921 social movements in Uttar Pradesh Himalayas, and the 1940 revolt by the Gonds and Kolams tribes in the Adilabad district of Hyderabad (Guha and Gadgil, 1989). The milder rebellions were crushed by the British, but their response to some of the stronger, more protracted ones was to yield to local demands, as in the Uttar Pradesh hills where, in the 1920s, the concept of Van Panchayat—a community based forest management system, was accepted and forest areas were identified for management by local people. Similarly, Forest Co-operatives were established in Himachal Pradesh. In some cases, instead of handing over the forest to communities, arrangements were designed to provide land for agricultural crops by adopting agro-silvicultural systems such as Taungya.

In summary, in this second phase the British rulers sought to and largely succeeded in displacing all prevailing concepts of societal rights to forest-based benefits with the objective of maximizing the economic gains to the empire-- industrial development in Britain and expansion of the colonial foundation in India. In other words, this was the period of an extension of the concept of the "state" of the British empire to forest resources of India[10]. The establishment of a formal forestry organization, exclusion of people from forest use, and the superposition of the formal state forestry institutions over the informal community institutions were the main features. The prior British experience of forest exploitation for military and agricultural purposes, and of establishing private property rights, contributed to the "organizational energy" needed to dismantle the existing informal forestry institutions in India. The principle of private property rights - exclusion of all people except the owner - was applied in a modified form in which instead of individuals those rights resided in the state. In the beginning, therefore, it was the experience picked up elsewhere that encouraged the shift towards exclusionary forest regimes. What constituted a discontinuous change of regime in India was at the same time part of a pattern of path-dependence within British-ruled territories. However, once the British had introduced these regimes in India, the initial set-up costs, learning effects from the experience in India itself, limited resistance by Indian people in most of the areas, and increased demands that the forests meet the empire's strategic, political, and economic needs, contributed to positive feedbacks, and worked as self-reinforcing mechanisms. These feedbacks, further strengthened by the existing research and training activities, resulted in a step-by-step reinforcement of the exclusion of local communities in every subsequent forest act/policy statement between 1864 and 1947. In a few specific situations, such as Van Panchyats of UP and the Forest Cooperative of Himachal Pradesh, strong opposition from local elements forced the new formal institutions to incorporate features of the pre-existing informal ones.

Overall, however, in the process of centralization and formalization of forestry institutions, the elements of adaptive efficiency present in the forest regimes prior to the colonization were attenuated. Trial and error, experiments, and corrections were now limited to the departmental research laboratories alone, and were based only on departmental perspective. These experiments were also limited to the technological aspects of forestry, with the institutional aspects left out. There was no role for the communities, no experimentation at the local level, no lessons learned from communities, and success was not judged from a community perspective. Local communities were not free to modify the centralized institutional arrangements at their level to best meet local requirements. In conclusion, adaptive efficiency of forest regimes was lower than that during the pre-British period.

### 4.4 The First Forest Policy Phase of Independent India (1947-1980): Temporal Path-Dependence due to Organizational Inertia of the LW and the EW and Formal Institutional Inertia

The 1894 Colonial Forest Policy provided the basis for independent India's first (1952) forest policy; the fundamental idea of that earlier approach--that the state should administer the forests, was presumed to hold good. Formally the new policy recognized the protective, ecological, and societal values of forests as sources of social welfare. However, such recognition at the top institutional (forest policy) level did not translate into the changes needed at the lower level institutions (the Indian Forest Act and other down-the-line operating rules). De facto, the perceived national interest was given priority over local village interests and the former was interpreted in a very narrow sense which gave little weight to avoiding the destruction of forests. In extinguishing local rights, the new Indian government continued along the British path, bringing more and more forests under state control. After independence, the princely states were formally brought under Indian forest law. State control was also extended to include the power to regulate the collection of grass and other forest products in village forests, to prescribe their management practices, and to take up direct management of private forests (Alcorn and Molnar, 1996).

The major difference vis a vis the colonial period was that, whereas the British used the forests to promote the industrialization of Britain (a goal which, inter alia, called for the construction of the Indian rail network), the states of independent India engaged in a sort of competition to provide low-cost raw material to forest based industries such as pulp and paper in order to attract those industries to locate in their region. The idea of forest production as a commercial activity gained ground. Although commercialization and industrialization were only accorded secondary importance in the 1952 policy statement, in fact India's heavy emphasis on industrial development was a prime determinant of forest history over the next few decades. This contradiction between the policy, in which other goals were reasonably prominent, and operations at the forest management level, where they were not, reflects the fact that changes in the first level institutions (forest policy) did not trickle down to the lower-levels of operational rules, an example of "formal institutional inertia".

In 1970, the Government of India appointed a National Commission on Agriculture (NCA) to examine and make recommendations for improvement and modernization of that sector. Since forestry was still lodged administratively in the Ministry of Agriculture, it was included in the mandate of this commission. An accident of administrative structure thus implied that forestry policy would be set by way of a report on agriculture. The commission's terms of reference included a very general charge, together with a list of twenty-four specific items, of which only one dealt with forestry (GOI, 1976a). Though committees were constituted for each item and sub-item with experts in the respective fields, none of the NCA's Chairman, Secretary, or five full-time members was a forestry expert, and only one of ten part-time members was. Nonetheless, the NCA report resulted in a second round of major changes in the forest policy of independent India. It emphasized that production of industrial wood should be economically defensible in terms of cost and returns. This led to large-scale plantations of fast growing tree species, which replaced the existing slow-growing native (so-called inferior) species. Though the commission argued for economic efficiency (achieving good value of outputs relative to value of inputs), the resulting decisions focused only on technical efficiency (good ratio of physical outputs to physical inputs); price efficiency (making sure that output and input combinations duly reflected output and input prices) was never a criterion for these new plantation activities. In fact this state activity was plagued by numerous economic distortions, including supply of raw material to industries at subsidized prices and adoption of harvesting rotations based on purely silvicultural rather than economic criteria. There may have been some gains in the production of industrial wood, but the conversion of natural forests to industrial plantations has been highly criticized for its environmental costs. The NCA recommended a new organizational structure to manage these commercial activities on business principles and to attract institutional finance (GOI, 1976b). In response, Forest Development Corporations were set up in almost all states; they were, however, manned by forest officers rather than business managers. Although the NCA foresaw this problem and recommended the training of forest managers in business management skills, the establishment of an institute to impart these skills took almost two decades--a good example of organizational and formal institutional inertia, and when the first batch of graduates emerged there were no openings for them in the forest corporations, a reflection of "attitudinal inertia" in those organizations.

The NCA accepted the British presumption that free access by local populations would lead to the destruction of the forests, and recommended instead a National Social Forestry Program to meet the fuel, fodder, and small timber requirements of rural people through plantations on non-forest lands (GOI, 1976b). In the late seventies and eighties, social forestry programs were started in many states, Gujarat and Uttar Pradesh being the pioneers. Foresters and donor agencies promoted fast-growing species for large and fast biomass production irrespective of their fuel wood or fodder qualities. Ironically, when the product was found wanting by the villagers, the output of the social forestry program wound up contributing raw material to industry (Chowdhry, 1989). In many cases poor people were actually hurt by the conversion of common lands to these plantations. A program of free distribution of seedlings also benefited the big farmers and industrial units that took up plantation activities. Though the Social Forestry Program was supposed to benefit poor people, at best only marginal trickle-down effects reached them; distributional disparities probably increased rather than being reduced. The foresters lacked the initiative (through "attitudinal" and "cultural" inertia) needed to make the shift from a forest department program to a more people-oriented one.

During this period, the focus of training and research, like the overall policy orientation and the basic organizational structure of the forest department, remained essentially unchanged, though facilities were extended. The course curriculum, studied by the Indian Forest Service (and other) officers, was a carry-over from the British period. Though the nomenclature of positions was changed for Social Forestry projects, and some state level training institutions were set up to meet those projects' staffing needs, the curriculum and mode of training remained the same as in the other training institutes. On the research front, though many new centers were started by the central and state governments, including some attention to social forestry, the main focus remained silviculture and forest products.

In summary, in spite of the natural redefinition of forest policy objectives in terms of the national welfare, the first four decades of independent democratic India saw almost no significant institutional change either of a general character or with respect to the inclusion of communities. Forest-dependent groups were hoping and expecting that an independent Indian government would address their forest requirements more seriously. This optimism on their part bred a patience which contributed to stability in the external social setting for policy. The national focus on economic development through industrialization, which was supported by forest policies of international organizations such as FAO[11], facilitated continuity. As illustrated in the above discussion, "formal institutional inertia" and "organizational inertia" were the main elements underlying that continuity, with many self-reinforcing mechanisms contributing to the "organizational inertia".

In the case of the LW, the main mechanism was the adaptive expectations [12] of the members of the LW based on existing organizations and beliefs, and fuelled by the forest managers trained mainly in natural sciences such as forest botany, silviculture, and forest management.[13] Though the function of forestry organizations had supposedly been changed from "channeling goods and services to the empire" to "provision of goods and services to the population", adaptive expectations impeded recognition of the inappropriateness of the existing forestry organization to the new objective. Also impeding change at this level was the expectation of large imminent payoffs from the existing forestry (institutions and organization) system. During the last phase of British rule, while local populations were being excluded from the forests, large investments were made in the design and establishment of the new forestry institutions (state system) and in forestry organization, in terms of notification of reserve forests, settlement of the rights of local people, training of forest managers, and forest research. The dismantling of the previous forest regimes and organizations and their replacement had involved significant costs, from which the Indian government did not expect a commensurate payoff.

At the EW level, the first self-reinforcing mechanism was the continuation without significant alteration of the Indian Civil Service (now known as the Indian Administrative Service, IAS), a most powerful organization. It has wielded a strong influence over policy formulation, in many cases overriding the wishes of local people represented through their elected representatives. The "attitudinal inertia" of the members of this service, generated through their experience under British rule, impeded changes to the forest regimes designed under British rule. The forest managers, who had been trained to benefit the empire by excluding the local population, suffered a similar case of "attitudinal inertia". Meanwhile, the forest department suffered from "cultural inertia" also. Since the forest officers, trained under the British Empire, were responsible for training new recruits, their views tended to be imparted to the latter; this tendency was heightened by the typically military style of the training programs. Along with training in technical subjects came a heavy dose of organizational culture, within which it was inappropriate for a junior officer to disagree openly with the decision of a senior, even if convinced that it was completely incorrect. New officers inherited not only the colonial thinking about forest regimes, but also a perceived self-interest in managing the forests for the state, not for the local communities. Having developed the rulers' habit of treating the public as serfs, they found the role of servant to society in an independent country a long stretch. Finally, the "structural inertia" of the forest department may have been increased, at least marginally, by the addition of several new levels to its hierarchy. Hence, all three components - attitudinal, cultural, and structural, contributed to an "organizational inertia" in the forest department that strongly resisted any departures from the existing forest management practices.

As with the forest regimes prior to independence, there were no built-in mechanisms or provisions for institutional experiments, innovations, adaptations, or inclusion of communities in the learning process; learning was limited to the technological aspects of forestry, through departmental laboratories. Hence the adaptive efficiency of forest regimes remained at the same (low) as prior to independence.

*4.5 The Second Phase of Independent India (1980-2004): Dominance of the External Setting, Organizational Energy of the LW, and Organizational Surges in the EW*

The period 1980-2004 differs markedly from the previous one. Even though several important processes had started in seventies, their main effects were felt in the eighties. Swami Chidanandji, a spiritual leader who attended The United Nations Conference on the Human Environment held in Stockholm, 1972, launched a movement for community forest rights in the Himalayan region, where the Sarvodaya (Brotherhood) spirit created by Sarala Behn, a disciple of Mahatma

Gandhi, was already prevalent; it later came to be known as the Chipko (hugging the trees) movement (Bahuguna, 1987). In response to the long struggle of these hill villagers, Gandhian leaders like Jayaprakash Narayan and Kaka Kelekar, and ecologists like Salim Ali made an appeal to stop tree felling in the Himalayas (Bahuguna, 1987). All the major political parties included protection of the environment in their 1980 election manifestos. In March of that year the World Conservation Strategy, an international outcome of the Stockholm conference, was launched in New Delhi under the leadership of Prime Minister Indira Gandhi. In April Mrs Gandhi invited Chipko leaders for a discussion of their demands. As a follow-up to these discussions and to the launching of the World Conservation Strategy, the Government of India brought in the Forest Conservation Act of 1980, which put strict legal restrictions on the conversion of forestland to non-forestry purposes and placed responsibility for decisions on such transfers under the direct control of the central government. A Himalayan region ban on the felling of green trees for commercial purposes at altitudes above 1000 meters followed in April 1981. In 1982 the Centre for Science and Environment (CSE) published The State of India's Environment - 1982, which provided an environmental blueprint dealing, *inter alia*, with the forests. In 1982, the Society for Promotion of Wasteland Development (SPWD), a national non-governmental organization, was established. After replacing his mother on her death, Rajeev Gandhi in 1985 recognized the importance of forests by creating a separate Ministry of Environment and Forest, and acknowledged the significant role of non-government organizations in forestry by constituting the National Wasteland Development Board to bring wasteland under production through a peoples' program. The first chairperson of the latter board was Mrs. (Dr.) Kamala Choudhry, also the chairperson of the SPWD. The key role of NGOs was thereby formalized at the highest level, and they became part of the decision making process. In 1989, Mrs. Menaka Gandhi, an environmentalist, became the Minister of Forests and Environment and strengthened the role of NGOs in forest policy decision-making.

The impact of these evolving environmental, social, and political factors has been complemented by that of population pressure on the forests, which has risen dramatically since independence. Over the period 1901-1951, India's population increased from 238 million to 361 million (at an annual growth rate of 0.84%); after independence, it increased to 846 million in 1991, at a rate of 2.15% per year. The resulting pressure by local people on the forests accentuated the conflict between the formal institutions of forest management and the local, informal institutions of user groups. Local communities reacted to growing scarcities of forest products with forest protection activities; in the late 1970s and early 1980s, initiatives of this sort sprung up in thousands of villages all over the country (especially in Bihar, Orissa, Madhya Pradesh, and Gujarat). By the 1980s many local communities were challenging the authority of forest officials and their management systems. Environmental groups and non-governmental organizations emerged as new power centers with strong bargaining positions, and used the political system and public awareness to challenge the existing forest regimes. Politicians, including the Prime Minister and the Forest Minister, supported the inclusion of NGOs and other public groups in policy making. In isolated cases, bold and innovative forest officers also supported the cause of local people by involving them in forest management, against the traditional legal practices.

With these inputs and under the above pressures the government finally accepted (at least in principle) the failure of forest regimes which excluded local people. The second (1988) forest policy of independent India is a clear departure from the earlier (1952) one in terms of the rights accorded to local poor people, specifically tribals and scheduled castes, which take priority over all other rights to forests such as those of industries looking for raw material. It emphasizes that customary rights and concession holders should be motivated to identify themselves with the protection and development of the forests from which they derive benefits. Following the 1988 Forest Policy, the Government of India in June 1990 issued detailed directions to state governments for the design of forest management programs in collaboration with local people--now known as Joint Forest Management (JFM). For the first time in independent India, a new policy was followed by detailed guidelines to transform changes at the policy-level into the corresponding changes at the lower-level operational rules. By late 1998, 20 state governments had also issued enabling orders on JFM, the next lower level of operational rules, and around 21,000 Forest Protection Committees were managing about 2.5 million hectares of forests (Thomas, 1998). Depending on the origin of these forest protection committees, some forest areas are under community regimes (though the ownership of land remains vested in the state)[14] while others are under joint regimes. Forest communities and forest departments are now working in close collaboration to develop forest regimes based on the principle of partnership. Despite this return towards community-based forest regimes, state regimes remain important and still account for a majority of total forest area.

On the training front, Rajeev Gandhi intervened personally in 1985 to initiate one-week refresher courses for members of all India's Services, including the Forest Service. Courses for the latter group were aimed at providing exposure to aspects of forest management beyond the traditional technical inputs. Unfortunately, their impact was quite limited; they were provided only to officers of the Indian Forest Service and their brevity contributed to their frequently being viewed as short vacations at government expense. However, the new forest policy in 1988 together with the emergence of joint forest management did bring some remarkable changes in training and forest research. In the JFM states, training of forest officers and subordinate staff in participatory management tools has been started with the support of NGOs. Some social science and management institutions have also initiated training courses for forest officers in different aspects of social and management sciences. Many research centers other than the traditional forest research institutes have begun to study a variety of matters related to joint forest management, such as ecological, economic, institutional, and gender issues. During the evolution of JFM, some funding agencies (especially the Ford Foundation), have played the role of catalyst by organizing national workshops for sharing the ideas and supporting the research and training initiatives.

In summary, the external setting, the "organizational energy" of the LW coming from individual leaders, and the "organizational surges" of the EW are the main factors that contributed to the important forest regime changes during this period. Social movements like Chipko, emergence of non-government organizations, and local-level initiatives from user groups, together with the external factors like the UN conference were among the pressures for change. In addition, a new development paradigm, based on decentralization and people's participation, also gained popularity during this period, and found global acceptance, including institutions of the United Nations system and the World Bank (IADB, 1991; World Bank, 1991). Hence, global trends in decentralization fully supported the changes in forest regimes. The two Prime Ministers - Mrs. Indira Gandhi and Mr. Rajiv Gandhi - and the Forest Minister - Mrs. Menka Gandhi, contributed to the "organizational energy" of the LW. The local-level experiments by some forest officers contributed to the "organizational surges" of the EW. All these factors together were able to overcome to some extent the organizational inertia of the FD and the institutional inertia of the existing forest regimes, resulting in institutional changes that may appear sharp and discontinuous. Nevertheless several considerations incline us to describe the process as incremental. First, the idea of involving the communities in forest management is not new, community regimes having existed at least in isolated cases throughout India's forest history. The social forestry program was started in the seventies; there was a modest de-facto shift towards community regimes even before the 1988 policy was enunciated. Finally, the 1988 policy has of course not converted all state regimes into community regimes. Effects of the 1988 policy like the changing attitudes of some forest officers towards community involvement in forest management appear revolutionary, but have not been general and have not occurred overnight. They are the outcome of a continuous process of criticism of forest officers by the general public, an increasing role of non-government organizations in bringing the forest department's anti-people attitude to the forefront, and the experimentation and learning of some daring and innovative forest officials. There are thus more elements of continuous incremental change than of revolutionary transformation. In addition, at the operational-divisional level, the transformation of state regimes to community or joint regimes is still dependent on the perceptions and attitudes of the local forest officer, and here the pattern of path-dependent incremental change is particularly apparent.

One important development since the 1988 Forest Policy seems to be the emphasis on adaptive efficiency. The behavior of both the forest department and the local communities suggest that a process of positive adaptation is underway. It rests on the willingness of the department and the communities to acquire knowledge and learning, to introduce innovations, to undertake risks and creative activities of all sorts, and to resolve problems and bottlenecks as they emerge. The apparent origin of the new (joint) forest regime is the risky innovations adopted by some Divisional Forest Officers (DFOs) in the state of West Bengal--risky both in the sense that they were experimental --hence their effects could not be predicted, and in the sense that the DFOs could have been reprimanded or punished for this deviation from standard departmental practice. In Purulia district in 1972 a new DFO faced immense pressure from illegal harvesting of the forests by local communities. Initially he worked with the state police officers to raid villages and local fuel wood market centers and to arrest fuel woodcutters, but this created considerable unrest in the area and tension between the forest officials on the one side and the local people and politicians on the other. He responded by suggesting to the communities that they take on protection responsibilities in return for a share of the fuel wood and minor forest products. In another area of south-west Bengal, Arabari, the local DFO[15] offered a 25% share of the sal timber and rights to all non-timber forest products including leaves, medicinal plants, fiber and fodder grasses, mushrooms and fruits in return for forest protection by communities (Poffenberger, McGean, & Khare, 1996). These were daring moves by the DFOs since they lacked legal authority to venture into such partnership programs. When it later came to formalization of the new joint management regimes, the different states developed a variety of provisions, in consultation with non-government organizations, defining the categories of forests to be covered, the participants from the community, the management unit, the representation of different sectors in the executive committee, the power of the committee, the benefit-sharing arrangements, etc. It was felt that the reorientation and training of forest officers was perhaps the key to the attitudinal and institutional changes necessary to support JFM over the long-term. The forest departments have been very receptive to this idea. Social learning and local innovations by community members and forest officials are also contributing to the adaptive efficiency of the new regimes. In 1994, consultations between the DFO of Harda Forest Division, Madhya Pradesh, and the local communities of Malpone village resulted in an innovative management plan that focused on nested silvicultural prescriptions (i.e. prescriptions consistent with the local socio-economic and cultural milieu), dividing the forest and its management both by species and by canopy tiers (Campbell & Rathore, 1995). In such exercises, community members and forest officials together develop ways of combining traditional knowledge and modern scientific methods of forest management. Forest officers have learned a variety of "ethno-silvicultural" techniques such as seed sowing in Euphorbia bushes by drawing on the indigenous knowledge base (Campbell & Rathore, 1995). The local communities also search for answers to the various challenges confronting joint regimes, such as convincing other people to join and achieving an equitable distribution of benefits. When some herdsmen in Badagorada village of Orissa let their cattle graze in nearby forests, four school teachers and 150 students from the local school went to the village and lay prostrate in front of the herdsmen's houses. The latter apologized and promised not to repeat the offence (Kant et al., 1991). Such examples indicate that the adaptive efficiency of the forest regimes has been improving during this period, and should in the near future surpass the level of adaptive efficiency of the pre-colonial period.

## 5. OUTCOMES OF SHIFTS TOWARDS COMMUNITY-BASED FOREST REGIMES

Overall outcomes of community-based forest regimes, created from 1988 on, have been encouraging. Country level quantitative data on the impacts of these new forest regimes are not available, but an observed decrease in the rate of deforestation and an expansion of dense forest area (crown density equal or more than 40%) during the last decade point to a positive impact from this innovation. During the period from 1961 to 1987, forest area decreased from 783,962 sq km to 642,041 sq km while in the period 1987 to 1995, it declined only to 639,600 sq km (GOI, 1988 & 1996); the annual rate of deforestation was cut to 0.05% in the latter period from 0.78% in the former. In addition, dense forest area has been continuously increasing since 1987, from 361,412 sq km in 1987 to 385,756 in 1995 (GOI, 1996), an annual increase of 0.82%. Our belief that the shift to community-based regimes has been a major contributor to these outcomes is supported by numerous micro-level studies showing multi-dimensional positive impacts of community based regimes. For example, in the three districts - Bankura, Midnapore, and Purulia - of the Southwest Bengal, where large-scale community-based forest management systems were initiated in the early 1980s, forest cover increased from 14.94% in 1984 to 17.96% in 1988, and further to 19.22% in 1991 (Pattnaik & Dutta, 1997). Lal, Bahuguna, Uddin, & Hussain, (1995) observed substantial improvement in biodiversity and regeneration in forest areas under community -based forest management systems of Bankura district. Similarly, in the Jamboni range of Midnapore district, biodiversity increased by almost four times in forest areas under community-based regimes (Poffenberger & McGean, 1996). Within a period of 3 years of the initiation of community- based regimes, annual household returns from non-timber forest products increased by 200% in many villages of Midnapore district (Malhotra & Poffenberger, 1989). These returns from non-timber forest products have not only increased the total income of the households but have also decreased income disparities between rich and poor people (Kant, Nautiyal, & Berry, 1996). In Andhra Pradesh, degraded Teak forest areas been brought under joint forest management systems have responded favorably, and the concerned forest protection committees received an extra income of Rs30,000 ($750) to Rs75,000 ($1600) from the teak billets in Rajamundary Forest Circle (Mukerjee, 1997). In the Budhikhamari area of Orissa, production of Sal (Shorea robusta) seeds increased by five times in the five-year period of community based forest management regimes (Mishra, 1994).

## 6. POLICY IMPLICATIONS

The main feature of the evolution of forest regimes in India has been incremental path-dependent change, the main exception being the sudden shift from the dominance of community regimes in the pre-British period to that of state regimes in the British period. The dominant factors in this pattern of incremental change have varied markedly over time. In pre-colonial India the inertia of the informal institutions played a major role. In colonial India, the first major force for change was the "organizational energy" which dismantled the existing institutions. But, later many self-reinforcing mechanisms such as initial set-up costs, learning effects, and limited resistance contributed to path-dependent changes. In post-colonial India, first self-reinforcing mechanisms at the level of the LW such as adaptive expectations and large set-up costs, and "organizational inertia" of EW dominated the process of institutional change. But, later the organizational energy of the LW, the external setting, and organizational surges of the EW became dominant. The level of adaptive efficiency varied directly with the degree of decentralization of forest regimes – higher in the decentralized regimes of pre-British India and recent regimes and lower in the centralized regimes in British India and the first phase of post-colonial India.

Such features of forest regimes, and the outcomes of the post-1988 experience, suggest a number of implications for forest management in countries whose forest histories have considerable in common with India's. First, neither the traditional political-economy framework nor the new institutional economics provide an adequate base for the analysis of forest regimes; organizations, institutions, external setting and their interaction have to be included in a more explicit way. Second, the design of forest regimes should take account of path-dependency and associated factors, since this is so manifestly present in most actual experiences. Third, the concept of adaptive efficiency is central to the evaluation of the institutions and organizations constituting forest regimes. Fourth, the concept of property rights, as applied in neoclassical economics, is not sufficiently subtle to explain the success or failure of forest regimes. In India, the British began the process of defining property rights from 1865; by the first Indian Forest Act of 1927 those rights were clearly laid out. But, the results were unsatisfactory, mainly because of non-complementarities between formal and informal institutions which led to adaptive inefficiency. Well-defined property rights have been less important than adaptive efficiency of the forest regimes. Fifth, outcomes of forest regime changes depend on the process of implementation, which is highly sensitive to the perceptions of members of forestry organization and organizational culture. "Organizational inertia" has been one of the main factors impeding institutional changes towards adaptive efficiency in India. Institutional change alone, without complementary change in the attitude of members of forestry organizations and organizational culture, will not provide the desired results. As a result, policy and management prescriptions for sustainable forest management, in these countries, should address institutional and organizational aspects in an integrative manner.

## NOTES

[1] North (1990) is an exception. However, his discussion is in the context of an economic and not a state organization, and his focus is on the stock of knowledge and investment and not on organizational inertia discussed in this paper.

[2] These include Bahuguna 1982, 1987; Bhatt 1988; Corbridge & Jewitt, 1997; Gadgil & Guha, 1992, 1994, 1995; Guha, 1983, 1986, 1989; Guha & Gadgil, 1989; Jewitt, 1995; Nadkarni, Pasa, & Prabhakar, 1989; Pathak, 1994; Pouchepadass, 1995; Rangan, 1995, 1997; Shiva, 1989, 1991; Shiva & Bandyopadhyay, 1986; and Vira, 1995.

[3] Vira (1995) included the forest bureaucracy as one of the forest-dependent groups in his schema, but the role of a central organization, the forest department, in the dynamics of forest regimes has not been

addressed adequately.

[4] Please refer to Section 3.4 for a discussion of path-dependent incremental changes.

[5] In this hierarchical categorization, the legal distinction between acts, orders, and guidelines is not our concern. In the court of law, a government order that is in contravention of the existing forest act may not be sustained. But if the executive branch of government, which is presumably aware of the act, issues an order that contravenes it, this is likely to mean that it wants to change the existing law. Such anomalies, and the possible conflict to which they may give rise, can occur when amending an act is a lengthy process. Eventually such amendments bring the act into line with the new orders of the government.

[6] The concept of "fairness" involves, inevitably, a degree of subjectivity. In this case the distribution of benefits was fair to the less advantaged groups in the sense that it did make provision for everyone's receiving certain benefits (in particular firewood and some other forest products), and in the sense that it was generally perceived as fair by the affected groups.

[7] Scholars have substantially divergent views about certain aspects of forest regimes in the pre-British period. Gadgil, Guha, and Shiva focus on community-regimes and argue that communities were aware of environmental issues and of the scarcity of the resources, and accordingly managed forests in a sustainable manner. They visualize the pre-British situation as in many ways an ideal one. Nadkarni, Rangan, and Pouchepadass are in varying degrees critical of this view. Pouchepadass (1995) points out that, like the British, the previous rulers used the forests for strategic purposes, including maintenance of the army and extension of agriculture. But he agrees that colonization gave a new dimension to this phenomenon: "and, finally, to serve their own interests, they set up everywhere an increasingly efficient framework of government control, which gradually denied the local populations free access to their traditional natural resource bases, at a time when their numbers were beginning to increase. Although the ecological stresses and traumas resulting from European colonization were not by any means the first events of their kind in the tropics, the scenario for the first time were modern, representing the onslaught of commercial and industrial capital on the natural resources of the world at large." Nadkarni et al. (1989, p. 32) writes: "However, there was a distinction between exclusive private rights over forest land and community rights over its produce. There was no alienation of the locals from the forest in spite of the state ownership of forest land during the pre-capitalist stage." Ribbentrop (1900), a British Inspector General of Forests, has given a detailed account of forest regimes just prior to the arrival of British. "Where the population had settled in joint village communities, any forest or waste land that fell within their boundaries was, as a rule, considered as common property. The cultivators living in un-united villages never had any proprietary rights except in the areas actually under cultivation, though they had, in some instances, doubtless acquired prescriptive rights of user. Though it was then known that the state had inherited extensive proprietary rights in the forests of India from the rulers by whom the territories were ceded, the actual status of the property and its extent were uncertain. This condition of things was probably quite in accordance with the state of society previous to British occupation, when every one was accustomed, without let hindrance, to get what he wanted from the forest." What seems clear enough, from these accounts, is the frequency (dominance) of community regimes in the pre-British period, and their decline thereafter. Whether the communities evinced a high level of environmental awareness is harder to judge, and beyond the scope of this paper.

[8] Until this 1894 policy, there had been only one level of forest institutions in British India, represented by the forest acts. The 1894 forest policy statement added a new level, the top one in the terminology of this paper.

[9] One British planter killed four hundred elephants in the 1860s in the Nilgiris (Guha & Gadgil, 1989).

[10] In the medieval period (roughly AD 500 to 1500), government in Europe was highly decentralized and divided among persons, groups, and orders. Community was extremely localized, and the manor was the basic unit of society. The lord was the largely independent ruler of this domain. However, tension among the localized power centers eventually become too great, the medieval order started deteriorating in the twelfth century, and its collapse was complete by the seventeenth century, at least in western Europe, to be succeeded by the typical political organization of modern life, the "nation state" emerged. The 18th and 19th century witnessed the extension of nation states and bureaucratic management to other parts of the world (Hitchner & Harbold, 1992). Hence, new Indian forest regimes were just an extension of the concept of the "state" of the British empire to forest resources of India.

[11] A seminal paper "The Role of Forest Industries in the Attack on Industrial Development" by Jack Westoby (1962), an FAO forest economist, provided a basis for that institution's policies for at least a

110                                    SHASHI KANT & R. ALBERT BERRY

couple of decades.
[12] In economics, the phenomenon whereby an increased role of the market enhances belief in its continued prevalence is an example of "adaptive expectations". In the case of institutions, an increased prevalence of contracting based on a specific institution reduces uncertainties about its permanence (North, 1990, p.94).
[13] Similar expectations on the part of the government were also reflected in the continuation of such core components of the general administration as the Indian Civil Service and the Indian Police Service.
[14] For a detailed discussion of forest regimes in India see Kant, 1996, and Kant & Berry, 2001.
[15] Dr. A. Banerjee, who was DFO in this area, graduated from the University of Toronto in 1969. The more open, less bureaucratic, and more innovative environment of a western university may have contributed to Dr. Banerjee's inclination to try something new when back in India.

## REFERENCES

Alcorn, J.B., & Molnar, A. (1996). Deforestation and forest-human relationships: What can we learn from India. In L. Sponsel, R. Bailey, T. Headland (Eds), *Tropical deforestation: The human dimension* (pp. 99-121). Cambridge: Cambridge University Press.
Arthur, W.B. (1988). Self-reinforcing mechanisms in economics. In P.W. Anderson, K.J. Arrow, D. Pines (Eds.), *The Economy as an Evolving complex system* (pp.9-31). Reading, MA: Addison-Wesley.
Arthur, W. B. (1989). Competing technologies, increasing returns and lock-in by historical events. *The Economic Journal*, 99: 116-131.
Arthur, W. B. (1991). Positive feedback in the economy. *Scientific American*, February 1991, 92-99.
Atkinson, G., & Oleson, T. (1996). Urban sprawl as a path dependent process. *Journal of Economic Issues*, XXX(2): 609-615.
Ayres, C.E. (1962). *Theory of economic progress.* New York: Schocken Books.
Bahuguna, S. (1982). Let the Himalayan forests live. *Science Today*, March, 41-46.
Bahuguna, S. (1987). Chipko: The people's movement with hope for survival of humankind. In *The Chipko message*. Tehri-Garhwal, India: Chipko Information Centre.
Bayly, C.A. (1989). *Imperial meridian: The British Empire and the world.* London: Longman.
Bhatt, C.P. (1988). The Chipko movement: Strategies, achievements, and impacts. In M.K. Raha (Ed.), *The Himalayan heritage* (pp.249-265). New Delhi: Gian Publishing House.
Bromley, D. (1989). *Economic interests and institutions.* Oxford: Basil Blackwell.
Campbell, J.Y., & Rathore, B.M.S. (1995). *Evolving forest management systems to meet peoples needs innovating with planning and silviculture.* Ford Foundation, New Delhi, Internal Working Paper.
Cashore, B. ( 2001). *In search of sustainability: British Columbia forest policy in the 1990s.* UBC Press, Vancouver.
Chowdhry, K. (1989). Social forestry: Roots of failure. *Indian Journal of Public Administration*, 35, 437-443.
Coase, R.H. (1960). The problem of social cost. *The Journal of Law and Economics,* 3, 1-44.
Commons, J.R. (1961). *Institutional economics.* Madison: University of Wisconsin Press.
Corbridge, S., & Jewitt, S. (1997). From forest struggles to forest citizens? Joint forest management in the unquiet woods of India's Jharkhand. *Environment and Planning*, A., 29, 2145-2164.
David, P. (1985). Clio and the Economics of QWERTY. *The American Economic Review*, 75(2), 332-337.
David, P., & Bunn., J. (1987). T*he economics of Gateway technologies and network evolution: lessons from electricity supply history.* Stanford Centre for Economic Policy Research.
Dwivedi, A.P. (1980). *Forestry in India.* Dehradun: Jugal Kishore and Co.
England, R. W. (1994). Three reasons for investing now in fossil fuel conservation: technological lock-in, institutional inertia, and oil wars. *Journal of Economic Issues,* XXVIII (3):755-775.
Forest Research Institute (FRI). (1961a). *100 Years of Indian Forestry* (Vol. I). Dehradun: FRI.
Forest Research Institute (FRI). (1961b). *100 Years of Indian Forestry* (Vol. II). Dehradun: FRI.
Gadgil, M., & Guha, R. (1992). *This fissured land: An ecological history of India.* Berkeley, CA: University of California Press.
Gadgil, M., & Guha, R. (1994). Ecological conflicts and the environmental movement in India. *Development and Change*, 25(1), 101-136.

Gadgil, M., & Guha, R. (1995). *Ecology and equity: The use and abuse of nature in contemporary India.* London: Routledge.

Goodstein, E. (1995). The economic roots of environmental decline: property rights or path dependence. *Journal of Economic Issue,* XXIX (4):1029-1043.

Government of India (GOI) (1944). *Second report on reconstruction planning.* Delhi: Government Press.

Government of India (GOI) (1948). *India's forests and the war.* Delhi: Government Press.

Government of India (GOI) (1976a). *Report of the National Commission on Agriculture, Part 1.* New Delhi: Government Press.

Government of India (GOI) (1976b). *Report of the National Commission on Agriculture* (Part 1I, Vol. 9). New Delhi: Government Press.

Government of India (GOI) (1988). *The state of forest report 1988.* Dehradun: Forest Survey of India.

Government of India (GOI) (1996). *The state of forest report 1996.* Dehradun: Forest Survey of India.

Guha, R. (1996). Dietrich Brandis and Indian forestry: A vision revisited and reaffirmed. In M. Poffenberger, B. McGean (Eds), *Villages voices forest choices* (pp.86-100). Delhi: Oxford University Press.

Guha, R. (1983). Forestry in British and post-British India: A historical analysis. *Economic and Political Weekly,* October 29, 1983.

Guha, R. (1986).Commercial forestry and social conflict in the Indian Himalaya. *Forestry for Development Lecture Series,* Department of Forestry, University of California, Berkeley.

Guha, R. (1989). *The unquiet woods: Ecological change and peasant resistance in the Indian Himalaya.* Oxford: Oxford University Press.

Guha, R., & Gadgil, M. (1989). State forestry and social conflict in British India. *Past and Present,* 123, 141-177.

Hayek, F. A. (1960). *The Constitution of Liberty.* Chicago: University of Chicago Press.

Hitchner, D.G., & Harbold, W.H. (1992). Modern government. New York: Dood, Mead and Company.

Inter-American Development Bank (IADB). (1991). *Economics and social progress in Latin America.* Washington, D.C.: The IADB.

Jensen, M.C., & Meckling, W.H. (1976). Theory of the firm: Managerial behaviour, agency costs and ownership structure. *Journal of Financial Economics,* October, 305-360.

Jewitt, S. (1995). Europe's others? Forestry policies and practices in colonial and postcolonial India. *Environment and Planning D, Society and Space,* 13, 67-90.

Jha, L.K. (1994). *India's forest policies: Analysis and appraisal.* New Delhi: Ashish Publishing House.

Kant, S. (1996). *The economic welfare of local communities and optimal resource regimes for sustainable forest management.* Ph.D. Thesis. The University of Toronto, Toronto.

Kant, S. (2000). The dynamics of forest regimes in developing economies. *Ecological Economics,* 32(2000), 287-300.

Kant, S., & Cooke, R. (1998). Complementarity of institutions: A prerequisite for the success of joint forest management. Paper presented in the *International Workshop on Community-Based Natural Resource Management,* Washington D.C., May 10-14, 1998.

Kant, S., & Berry, R.A. (2001). A theoretical model of optimal forest resource regimes in developing economies. *Journal of Institutional and Theoretical Economics,* 157, 331-355.

Kant, S., Singh, N., & Singh, K. (1991). *Community based forest management systems; Case studies from Orissa.* New Delhi: ISO/Swedforest.

Kant, S., Nautiyal, J.C., & Berry, R.A. (1996). Forests and economic welfare. *Journal of Economic Studies,* 23(2), 31-43.

Kissling-Naf, I., & Bisang, K. (2001). Rethinking recent changes of forest regimes in Europe thorugh property-rights theory and policy analysis. *Forest Policy and Economics,* 3(3-4), 49-111.

Lal, J. B., Bahuguna, V.K., Uddin, H., & Hussain, S. (1995). Ecological studies of Sal forest in Bankura, North Division. In S.B. Roy (Ed.), *Experiences from participatory forest management* (pp.171-177). New Delhi: Inter-India Publications.

MacCarthy, J.F. (2000). The changing regime: Forest property and reforms in Indonesia. *Development and Change,* 31(1), 91-129.

Malhotra, K. C., & Poggenberger, M. (1989). Forest regeneration through community protection: The West Bengal experience. Proceedings of the *Working Group Meeting on Forest Protection Committees,* West Bengal Forest Department, June 21-22.

Mishra, G. (1994). The Budhikhamari model of participatory management of the forest ecosystem in Orissa. *The Indian Forester,* 120(7), 597-601.

112                                SHASHI KANT & R. ALBERT BERRY

Mookerji, R.K. (1950). *Hindu civilization*. Bombay: Bhartiya Vidya Bhawan.
Mukherjee, S.D. (1997). Is handing over forests to local communities a solution to deforestation? Experiences in Andhra Pradesh (India). *Indian Forester*, 123(6), 460-471.
Nadkarni, M.V., Pasa, S.A., & Prabhakar, L.S. (1989). *The political economy of forest use and management*. New Delhi: Sage Publications.
North, D.C. (1990). *Institutions, institutional change, and economic performance*. Cambridge: Cambridge University Press.
Pathak, A. (1994). *Contested domains: The state, peasants, and forests in contemporary India*. New Delhi: Sage Publications.
Pattnaik, B.K., & Dutta, S. (1997). JFM in South-west Bengal: A study in participatory development. *Economic and Political Weekly*, December 13, 1997, 3225-3232.
Pearson, G.I. (1869). Sub-Himalayan forests of Kumaon and Garhwal. In *Selections from the records of the Government of the North-Western Provinces* (Second Series, Vol. II). Allahabad: Government Press.
Poffenberger, M., McGean, B., & Khare, A. (1996). Communities sustaining India's forests in the twenty-first century. In M. Poffenberger, B. McGean (Eds.), *Village voices, forest choices: Joint Forest Management in India* (pp.17-55). Delhi: Oxford University Press.
Poffenberger, M., & McGean, B. (Eds.) (1996). *Village voices, forest choices: Joint Forest Management in India*. Delhi: Oxford University Press.
Pouchepadass, J. (1995). Colonialism and environment in India: Comparative perspective. *Economic and Political Weekly*, August 19, 2059-2067.
Rangan, H. (1997). Indian environmentalism and the question of the state problems and prospects for sustainable development. *Environment and Planning A*, 29, 2129-2143.
Rangan, H. (1995). Contested boundaries, state policies, forest classifications, and deforestation in the Garhwal Himalayas. *Antipode*, 27(4), 343-362.
Rangarajan, M. (1996). *Fencing the forest: Conservation and ecological change in India's central province 1860-1914*. Delhi: Oxford University Press.
Ribbentrop, B. (1900). *Forestry in British India*. Calcutta, India: Office of the Superintendent of Government Printing.
Sastry, M.A. (1997). Problems and paradoxes in a model of punctuated organizational change. *Administrative Science Quarterly*, 42(3), 237-275.
Schotter, A. (1981). *The economic theory of social institutions*. Cambridge: Cambridge University Press.
Setterfield, M. (1993). A model of institutional hysteresis. *Journal of Economic Issues*, XXVII(3), 755-774.
Shamasastry, R. (1929). *Translation of Kautilya's Arthasastra*. Mysore: Wesleyan Mission Press.
Shiva, V. (1989). *Staying alive: Women, ecology, and development*. London: Zed Press.
Shiva, V. (1991). *Violence of the green revolution: Third World agriculture, ecology, and politics*. London: Zed Press.
Shiva, V., & Bandyppadhyay, J. (1986). Environmental conflicts and public interest science. *Economic and Political Weekly*, 21(2), 84-90.
Smythies, E.A. (1925). *India's forest wealth*. London.
Thomas, A. (1998). The national support group. *Wastelands News*, XIII(4), 14-17.
Thomas, K.V. (1983). *Man and the natural world: Changing attitudes in England, 1500-1800*. London: Allen Lane.
Throgmorton, J. A. & Fisher, P. S. (1993). Institutional change and electric power in the city of Chicago. *Journal of Economic Issue*, XXVII (1): 117-153.
Tucker, R.P. (1988). The British Empire and India's forest resources: The timberlands of Assam and Kumaon, 1914-50. In J.F. Richards, R.P. Tucker (Eds.), *World deforestation in the twentieth century*. Durham: Duke University Press.
Upadhyaya, M.D. (1991). Historical background of forest management and environment degradation in India. In A.S. Rawat (Ed.), *History of forestry in India*. New Delhi: Indus Publishing Company.
Veblen, T. (1975). *The theory of business enterprise*. Clifton: Augustus M. Kelley.
Vira, B. (1995). *Institutional change in India's forest sector, 1976-1994 – reflections on state policy*. OCEES Research Paper No. 5. Oxford Centre for the Environment, Ethics, and Society, Mansfield College, Oxford.
Webber, T. (1902). *The Forests of Upper India and their inhabitants*. London.

Westoby, J. (1962). The role of forest industries in the attack on economic underdevelopment. In *The state of food and agriculture 1962* (pp. 88-128). Rome: FAO.

Williamson, O.E. (1985). *The economic institutions of capitalism*. New York: The Free Press.

World Bank. (1991). *Development strategies: World development report 1991*. London: Oxford University Press.

Young, O.R. (1982). *Resource regimes: Natural resources and social institutions*. Berkeley: University of California Press.

# CHAPTER 5

# VALUING FOREST ECOSYSTEMS – AN INSTITUTIONAL PERSPECTIVE

## ARILD VATN

*Department of Economics & Resource Management,*
*Agricultural University of Norway, Box 5033, Aas, Norway,*
*Email: arild.vatn@ior.nlh.no*

**Abstract.** This chapter is focused on the fundamental problems related to determining social choices in the realm of the environment – specifically choices with relevance to the preservation of forest biodiversity. The chapter is divided in three sections. First I will characterize the main features of biodiversity, emphasizing also the ethical implications of the common goods properties involved. This part concludes that the fundamental issue is choosing value-articulating institutions that are consistent with the underlying problem characteristics. The second part of the paper is thus devoted at clarifying the role of the institutional context in the valuation process. The final and main part concerns an evaluation of different value articulating institutions to be used when evaluating biodiversity. Both cost-benefit analysis/contingent valuation and various deliberative institutional structures are discussed. It is concluded that deliberative institutions are the only ones that can offer contexts being consistent with the type of cognitive and normative issues involved. A list of more or less unresolved challenges to the application of these methods, is, however, also emphasized.

## 1. INTRODUCTION

The decline in biodiversity is a major concern in contemporary society. The issue arises in a wide variety of ecosystems. Forest ecosystems warrant special attention since these are both hosting a large fraction of existing species and are under particular threats – e.g., conversion into arable land or construction, unsustainable logging practices, fragmentation.

Pimm, Russell, Gittleman, and Brooks (1995) estimate that across ecosystems, the present rates of extinction are 100 – 1.000 times higher than those observed in pre-human times. Still, why should we worry? As economists we know that goods are substitutable, and that technological development can even increase substitution potential over time. If we concern ourselves simply with managing extinction in an optimal way – i.e., at the optimal speed and order – we could concentrate our efforts

*Kant and Berry (Eds.), Institutions, Sustainability, and Natural Resources: Institutions for Sustainable Forest Management*, 115-134.

on the economic valuation of the various species. By and large, biodiversity loss is an externality, and there is a need to correct the involved market failures.

In this chapter I take up these issues from a distinct set of perspectives. First, I question the standard neoclassical economic perception of environmental goods as an array of substitutable commodities, a perception more likely to obscure rather than to enlighten the analysis. Second, I discuss both the sufficiency and the necessity of economic valuation and cost-benefit analysis as a way to inform choices about the preservation of biodiversity. Finally, I evaluate an alternative type of procedure built on the perspective of deliberative democracy.

While the standard economic model is a great endeavor, it is crucial that one understands its assumptions when applying it to real world issues. Especially the high level of abstraction and thus the seeming universality of the model makes it attractive to apply across a variety of issues. Still, this property may also in many cases lead us astray. Keeping some critical distance is thus important. It is also important to develop alternatives to monetary valuation. I have found that deliberative institutions offer a good basis for such a development. This is not so because they are capable of removing any of the fundamental problems involved. It is so because the challenges can be treated in a way that is consistent with their particular characteristics.

## 2. THE CHARACTERISTICS OF BIODIVERSITY

The standard economist vision of natural resources is well captured in the following passage from Robert Solow:

> ...history tells us an important fact, namely, that goods and services can be substituted for one another. If you don't eat one species of fish, you can eat another species of fish. Resources are, to use a favorite word of economists, fungible in a certain sense. They can take the place of each other. That is extremely important because it suggests that we do not owe to the future any particular thing. There is no specific object that the goal of sustainability, the obligation of sustainability, requires us to leave untouched.... Sustainability doesn't require that any *particular* species of fish or any *particular* tract of forest be preserved. (Solow, 1993, p.181, italics in the original)

There are three distinct characteristics of this quotation that deserve attention. First, no restriction on the substitution of goods is assumed. Second, environmental goods are perceived as items or commodities. Third, the perspective is welfarist in that no ethical concerns, for example human responsibility for other species, are emphasized. Species are mere instruments in the hands of humans.

The passage is not a reflection on the characteristics of natural systems as such and the ethical issues involved, but rather the superimposing of a distinct and very generalized perception on the physical world without really asking whether that perception fits. To say 'history tells us' begs the question: which history? History also illustrates the restriction to substitution and a pure means-end framework. Furthermore, natural goods like biodiversity are not foremost items or commodities. Biodiversity is not easily demarcatable. The various species are parts of nested systems of matter and energy transformation where both diversification and integration of tasks is a very important characteristic (Barbier, Burgess, & Folke, 1994; Wilson, 2001).

Thus expanding the idea of the market to encompass features of nature like biodiversity is problematic, indeed. Norgaard (1984, p.160) puts it this way: "It is ironic that environmental problems in economics are thought of as problems of market failure rather than evidence of the applicable limits of the market model". The aim of the first part of this chapter is thus to try to clarify why it is erroneous to treat environmental goods or services like biodiversity as commodities. Three characteristics will thus be highlighted: a) that biodiversity is not a set of items but a systems good; b) that biodiversity is a common good; and c) that there are ethical issues involved that cannot be handled well under the perspective of commodity exchange, substitution and trade-offs.

## 2.1 Biodiversity – A Systems Good

Wilson (2001) defines biodiversity as follows:

> The variety of organisms considered at all levels, from genetic variants belonging to the same species through arrays of species, to arrays of genera, families, and still higher taxonomic levels; includes the variety of ecosystems, which comprise both the communities of organisms with particular habitats and the physical conditions under which they live (p. 377).

Specifically within systems ecology, the structural and functional relationships between species are emphasized. Species are understood as integrated in webs of matter and energy cycles that reproduce or maintain the systems. As such they have developed over long time spans, where different trial and error mechanisms have shaped the internal relationships of matter and energy transformation. In this way they are structures that perform systems' functions. In this process the biota has furthermore greatly influenced the physical environments in which they live, not least the composition of the atmosphere (Graves & Reavy, 1996; Wilson, 2001). This way life has a certain meaning 'created its own conditions'.

The above perspective does not deny the fact that species composition changes due to natural forces. Wilson (2001) thus emphasizes that species go extinct all the time as part of the process of evolution in which those species that are more fit take over the resource niches. Thus, he suggests an average lifetime of a species of about 1 million years. In some periods more abrupt developments – i.e., mass extinctions – have been observed. There have been altogether five such periods over the last 450 million years. He concludes that it has on average taken approximately 10 million years to recover from these.

Empirical studies suggest that – from a systems view – there is a certain redundancy in ecosystems. Ecosystem functioning can in most situations be maintained by a smaller number of processes and a reduced number of species (e.g., Schindler, 1990). Thus some species can be termed 'keystone process species' (e.g., Walker, 1992). In evaluating this, Barbier et al. (1994) still comment:

> Although keystone process species are necessary for ecosystem functioning, they may not be sufficient for ecosystem sustainability. The remaining species that depend on the

niches formed by keystone process species are also important for maintaining the
resilience of the ecosystem (p. 28).

Resilience meant the system's ability to counteract perturbations from
external shocks – i.e., its capacity to return to its original state (Holling, Schindler,
Walker, & Roughgarden, 1995; Perrings, 1997). Ecosystems are viewed as complex
systems, characterized by self-organized and adaptive structures. The complexity of
the systems tends to make them fairly robust in the face of historically repeated
perturbations of state variables. This is a basic characteristic of self-organization. To
be able to develop this kind of complex resilience, the system does, however, always
have to produce some randomness, which serves to constantly develop and 'test'
better ways of adaptation. It creates 'searches' within the available state space
through processes like genetic mutations. Variation increases and the system
becomes better able to handle shifts in environmental conditions. Concerning this
duality of complex systems Nicolis and Prigogine (1989, p.218) write:

> .... complexity has been connected to the ability to switch between different modes of
> behavior as the environmental conditions are varied. The resulting flexibility and adapt-
> ability in turn introduces the notion of choice among the various possibilities offered. It
> has been stressed that choice is mediated by the dynamics of the fluctuations and that it
> requires the intervention of their two antagonistic manifestations: short scale
> randomness, providing the innovative element necessary to explore the state space; and
> long-range order, enabling the system to sustain a collective regime encompassing
> macroscopic spatial regions and macroscopic time intervals. A necessary prerequisite of
> all these phenomena is a nonlinear dynamics that gives rise, under suitable constraints,
> to instability of motion and to bifurcations[1].

The challenge is thus that it is impossible to predict what will happen if changes
which appear are either too large or too frequent – i.e., beyond levels not earlier
repetitively observed. They may change system performance in an essential way –
i.e., an attractor shift is observed.[2] We are, however, unable to predict where and
when such a shift may happen. The system is characterized by radical uncertainty
(Lemons, 1998).

The above description focusing on systems and functions demonstrates a
remarkable difference to the perception based on the commodity concept. The latter
would mainly give rise to a perception of species variation and richness as a set of
isolated items. Put bluntly, nature is viewed in a way equal to that of an animal or a
pet shop. At the same time, the quality of human life, as that of any other species,
depends on the quality of these natural systems. Humans may have the capacity to
transform the systems far beyond that of any other species, and to study the systems
and maybe single out key species, but they do not have the capacity to 'pick and
restructure' and at the same time count on the original level of resilience. The
system cannot be mastered in any such way; doing so would imply the capacity of
acquiring the information that is stored in the system as a product of trial and error
processes going on for millions, if not billions of years. This is a vast – i.e.,
impossible – endeavor.

In this sense, ecosystems are functionally opaque (Vatn & Bromley, 1994). The
exact and full contribution of a function or species in an ecosystem is not known,
indeed is probably unknowable, until it ceases to function. Furthermore, it will then

be very difficult to establish what has really happened. This is the essence of the perspective of complex systems and challenges the idea of substitution at its fundamentals. Ehrlich and Ehrlich (1992) use the 'rivet-popper' problem as an analogy to describe this. Airplanes are constructed with considerable redundancy. Thus the removal of a single rivet from its wing would most probably not cause a crash. However, continuing to remove rivets would certainly end in a disaster. Still, it would be impossible to foresee exactly when this might happen.

## 2.2 Biodiversity – A Common Good

The above perspective embeds another message. The environment is not external to the human being. The human is part of it, as it has become a part of the human. In physical and biological terms we are just like any other species. We utilize our niches. We depend on their capacity to produce goods and restructure waste. We depend on their 'health', as also the quality of our immune system is mutually dependent on various capacities of the environment. What distinguishes us is our capacity to shape our niches. This provides us an opportunity, but as is evident from the above, it is also the core of our problem.

The various relationships constituting the webs of the ecosystems link different human beings to each other in a very concrete sense. What everybody does influences everybody else's opportunities. If I pollute a lake, you cannot at the same time have it clean with its original life and capacities. If you drain a marsh to build a road through it, I cannot at the same time enjoy its cleaning capacities, its species richness etc. The environment, the ecosystems and their species, is a common good in all its dimensions. Here the commodity concept and the idea of simple trade-offs desert us fundamentally. If there is anything that correctly describes the functioning of an ecosystem, it is the idea of interlinked processes. Even if we could envision the impossible – i.e., to attach individual property rights to each molecule of the biosphere to make its basic operating unit become a demarcatable commodity – it would still not help, since the function of each molecule is first defined via its position or motion in the system it belongs to.

A commodity becomes just a commodity when it is taken out of the system it is shaped by – i.e., when a bird is put into a cage, a cow is fenced or made into meat, a tree is cut into planks. Still even this separation is fictitious. Any matter used – any commodity – becomes waste and through that process the fabricated and short lived status as a commodity vanishes. Though the institutionalization into a commodity is just partial, it is still important in simplifying many operations or transactions. This is the key point: the idea of a commodity is just a simplification, nothing more. In many situations this simplification deserts us, typically in matters that are most essential.

## 2.3 Biodiversity – An Ethical Good

In the world of common goods, what I do influences your opportunities. This simple observation holds a very strong implication. Choices made in the realm of the environment are fundamentally ethical in the sense that the preferences we hold, and the corresponding actions concerning the valuation and use of goods like biodiversity, influences which sort of environment is left for others to value and use. Through the linkages existing in nature, also a social interconnectedness is forced upon us.

In the case of common or 'public' goods, orthodoxy then tells us to make choices on the basis of what is evaluated the most by the collective measured as the sum of individual willingnesses to pay. This can, however, be a unanimous choice only in very rare occasions. Most probably some interests will have to suffer. Some favor the road, while others favor the wetland. In reality there can, however, only be road or wetland. This raises a long series of ethical problems that cannot be solved within a purely individualist, non-communicative model. The market is the wrong 'metaphor' for issues where common or public goods are involved.

Let us compare the market with another institutional structure – the family. In its idealized form, the focus of that institutional structure is developing the common good. Decisions about the rules that should govern its working, about which solutions to apply to specific challenges are not made on the basis of who pays the most. Instead as members of the family we are listening to the arguments, evaluating who has the greatest needs, what is considered a good solution for the sake of the family as a whole etc. Furthermore, the family is a place where preferences are developed, fostered, sanctioned or disapproved of. The dynamics are very much about developing norms that make the collective work better and solving conflicts by fostering a common appreciation of what is a good or defensible way of living.

All this is in principle abstracted from in the (idealized) market model. It fosters individuality, the reign of unquestioned and given individual preferences.[3] Historically, at least from Bentham onwards, the establishment of hedonist utilitarianism went hand in hand with the liberation of the individual from the oppressions of the 18th century state and church. In spite of its merits and importance for many aspects of life, this perspective is unhelpful when the issues faced involve developing the common good. In such a situation we need to talk to the others about what is reasonable to do, rather than to just offer a price in a market. We need to reason over which preferences are good or defensible. Concerning the future we need to reason with our contemporary fellows about how we should 'trade-off' our interests against those coming later, how we can best secure the interests of those yet not born. This perspective directs the focus towards the forum, not the market (Elster, 1986; Jacobs, 1997).

The above discussion, while focusing on the common good, is still anthropocentric in its orientation. Important criticisms of the economic perception of environmental goods, not least biodiversity, focus on the narrowness of just looking at nature in instrumental terms. Discussing the issue of nature's 'own right' or 'intrinsic value', Holland (1997, p. 130) emphasizes that "the natural world contains many items which undeniably in the case of sentient animals, or arguably in the case of other animals and plants, have moral claims on us." Being arguable, conclusions about the nature and extent of the moral claim may vary between cultures and over time. This does not, however, eliminate the challenge.[4]

Certainly, a definition of 'nature's right' has to be culturally or socially established. The fact that the issue is raised follows from our capacity to reflect about what constitutes life and the position of our species in relation to other species. Some have questioned why the utilitarian calculus has been defined only over human interests and needs, since this seems to be an arbitrary choice. Environmental economics has in a way tried to alleviate this problem by introducing the concept of 'existence value'. This way the rights of nature, as each individual perceives it though, becomes part of the calculus. However, as argued by Sagoff (1988) and Holland (1997) this reveals a serious misunderstanding of the character of moral claims. Such claims have to go beyond *individual* evaluations, since ethics and morality are social phenomena. They belong to another category from those to which ordinary trade-off calculations are appropriate. Commodity preferences and norms are incommensurable entities (Vatn & Bromley, 1994).

In many cultures, the sense of sacredness is attached to (part of) the natural environment, such as certain places and species. Even in more secular societies, the natural environment is of great importance in creating identity and defining belonging. It is further viewed as heritage by many – i.e. primarily as something we inherit with the responsibility to hand it on to later generations in good shape (Burgess, Clark, & Harrison, 1995). As such it becomes difficult to put it within the bounds of a trade-off calculation. Conceptualizing it as a commodity is a 'category mistake' (Sagoff, 1988).

## 3. THE INSTITUTIONAL POSITION

The above reasoning has turned our attention towards the role of the institutional setting under which valuation takes place. However, before focusing directly on that issue, we need to get a better understanding of the relationships between institutions and rationality, preferences, preference formation, and choices. This is not an easy task given the range of positions in the literature. Even within the specific branch of institutional economics, where I place myself, there exist several positions. I will simplify by distinguishing two of them.

First we have the so-called 'new institutional economics' – e.g., Coase (1984); North (1990); Eggertsson (1990). Here the influence from standard neoclassical economics is quite strong.   Second, we have the so-called classical school of institutional economics, originating with American institutionalists like Veblen and Commons who took a very different stand from the neoclassical. Their position has been 'modernized' by integrating ideas from sociology, anthropology and organizational science. It is this modern variant of the classic position that is of importance to us. To fully understand its characteristics, it is, however, helpful to first give a short overview of the 'new' position.

According to new institutional economics, institutions are 'the rules of the game' (North, 1990). Institutions thus define the context for transactions – e.g., measurement scales, accepted behavior when striking a deal, formalized rights and duties, etc. These formal and informal rules are important not least to establish necessary order and thus simplify transactions between individuals and firms. This school focuses dominantly on the consequences for economic theory of accepting positive transaction costs. The state exists, it is argued, not only to define and defend rights, but also to reduce transaction costs. Such reductions are also thought to be the motive behind establishing firms. Most authors of this school support the neoclassical position of individual rationality – i.e., rationality as maximizing individual gain. *The individual is self-contained*; implying that preferences are stable and thus independent of the institutions.

The classical school, on the other hand, looks at institutions as something more than external rules. They are also constitutive for the individuals and the communication between them. Veblen (1919, p. 239) defined institutions as "settled habits of thought common to the generality of man". Scott (1995, p.39) looks as institutions as "cognitive, normative, and regulative *structures* and activities that provide stability and meaning to social behavior".[5] Thus both these authors emphasize that *institutions have formative influence on individuals*. Scott is explicit on underlining that institutions are both external rules and structures shaping the individual. The latter capacity is related to the way we understand *what is* (the cognitive aspects) and *what should be* (the normative aspects). What is rational is not just a result of an individual calculation given external institutional constraints (the new institutionalists). Institutions also influence what we observe, which values we find it right to defend, which preferences we hold, etc. Choices, more precisely rational choices, are thus not only about what is *optimal* for the individual. They are also about what is *right* to do in a certain situation or institutional context.

There is thus a fundamental difference between the way the individual is perceived in the two approaches. On the one hand we have the 'new' position that sees the individual, in his/her origin, as fundamentally independent of institutions. Given this, institutions become external constraints.[6] On the other hand we have the view that the individual is very much a social creation. Unsurprisingly, there is a logical connection between the theory of individual choice and the understanding of what institutions are and do. Put differently: There is a necessary link between the *definition of what rational choice is and the perspective of what institutions are and do.* This has great impact on how we understand and institutionalize social choices.

According to the classic view there is a two-way interaction between the individual and the institutions. We produce institutions at the same time as these constructions influence what we become. In a complex world, societies use institutions to create necessary order and social cohesion. As we become socialized into an institutional structure, we also internalize the values and logic upon which it is based. The institutional view, as understood here, emphasizes the role both of choices/agents and of structures. The most important choices are those defining which institutional structures should exist to provide the context and rules for a specific area of decisions. The choice of institutions defines the (implicit) rationality of the arena within which 'secondary' choices like specific resource allocations are then to be made. Thus, while the market supports and fosters individual, calculative rationality, other institutional structures may support more cooperative types of rationalities.

## 4. BIODIVERSITY, VALUATION AND VALUE ARTICULATING INSTITUTIONS

From the above it follows that the way in which we institutionalize valuation or choices influences what become the preferred actions. Our perspective is that of plural rationalities and the fact that the institutional structure evokes certain ways of thinking about and treating an issue.

### 4.1 Value Articulating Institutions

From an institutional perspective, cost-benefit analyses (CBA) and contingent valuation (CV) are specific types of institutions – i.e., value articulating institutions.[7] They are based on a certain set of perspectives and choices concerning a) who participates in the valuation and in which capacity, and b) what is accepted as data and how data is to be produced and handled. In the case of CBA/CV people are asked to act as consumers, and data must take the form of prices or price bids. These bids are furthermore aggregated according to a specific set of rules where discounting may be the most prominent.

The aim here is not to go into any detail about what characterizes CBA/CV. The point I want to make is that the various types of existing value articulating institution can be characterized by the answers they give to the above questions. Thus *CBA/CV* (e.g., Boardman, Greenberg, Vining, & Weiner, 2001; Mitchell & Carson, 1989), *multi-criteria analysis* (e.g., Janssen, 1994; Keeney & Raiffa, 1976; Munda, 1995) and various *deliberative institutions* like focus groups, consensus conference and citizens' juries (e.g., Burgess, Limb, & Harrison, 1988; Joss, 1998; Lynn & Kartez, 1995; Armour, 1995; Smith & Wales 1999) all represent distinct responses to the above list. Basically, the different answers given relate to different assumptions concerning what kind of rationality is involved or should be fostered, how preferences are formed and the characteristics of the issues involved.

A value articulating institution is foremost a set of external rules. Through its structure it does, however, emphasize or evoke specific rationalities (Vatn, 2004). In the case of CBA/CV it is individual willingness to pay that is emphasized. Thus the basic rationale is individual gain, consumer sovereignty and calculative rationality. In the case of a consensus conference, on the other hand, the underpinning set of ideas is quite different. Here it is dialogue, the development of common perspectives and the evaluation of preferences and the scrutinizing of arguments – i.e., communicative rationality – that is fostered. It is the search for the best common solution based on an argumentative practice that is governing.

With reference to the above, it is important to emphasize that a value articulating institution may not have much capacity to change peoples' preferences. This is obvious in the case of a CV, which takes individual preferences as given, and involves the individual to responding (quickly) to a survey or a short interview. The consensus conference may have some capacity in the direction of change, since it focuses on exploring and defending specific views, on reasoning together. The basic issue here is to learn from each other and through deliberation to reach some kind of common view. Still, even in this case, the greatest influence is through the capacity the institutional structure has to evoke a certain type of rationality under which the involved individuals already have some developed capacity to act. The institutional setting defines for us whether we are supposed to behave more as 'consumers' or as 'citizens'.

Thus people may be able to handle the same issue, value forest biodiversity, under the influence of different value articulating institutions. They may, however, still have preferences concerning what is a reasonable, good or proper institutional structure to use – i.e., which structure conforms best to the kind of rationality they think the given issue should be treated under. I know of no study that has explicitly focused on peoples' views on this. We do, however, have some indications in the valuation literature about its importance. One indirect sign is the amount of 'protest bids'[8] observed in this literature. A more direct sign is found in a valuation study documented by Burgess, Clark, and Harrison (1998) and Clark, Burgess, and Harrison (2000). Here respondents revealed dissatisfaction with the CV method when they understood in which institutional context their statements (i.e., monetary bids) would be used. Furthermore these authors conclude "When deconstructed by the respondents themselves, their WTP figures proved to have little substance and they unequivocally rejected CV as an acceptable means of representing their values.... valuing nature in monetary terms was incommensurable with deeply held cultural values" (Clark et al., 2000, p. 60). Similar observations are found in Vadnjal and O'Connor (1994) and Schkade and Payne (1993).

Thus people may deliver bids, but still not be supportive of the institutional context into which such bids fit. Spash (2000) offers a similar observation. He documents that among people responding in a lexicographic way – i.e., people that find it wrong to trade-off environmental values against money – there were some that still offered a monetary bid when asked. An explanation to such behavior may be that some feel compelled to conform to the rules of the institutional setting they find themselves moved into. Since the setting is monetary valuation, they feel obliged to follow that logic even if it is against generally held values (Vatn, 2004). Still, not all respondents that value the good positively adapt this way. They react instead by protesting – e.g., by delivering so-called 'protest zero bids' (Spash, 2000; Stevens, Echeverria, Glass, Hager, & Moore, 1991).

Thus people may have ideas about what is a reasonable or better value articulating institution. Still they may be willing to comply with the one they are offered. Alternatively, they may choose to oppose the logic of that specific institutional structure by protesting. What is common to all these observations is that the institutional context influences the valuation process. Despite the fact that there is room for individual adaptation to any institutional context, the choice of value articulating institution is crucial.

*4.2 Building Institutions for Valuing Biodiversity*

So what would be an ideal value articulating institution for forest biodiversity? Though such an institution most probably does not exist, I believe it is very important to formulate this kind of question and take it seriously. It forces us to think systematically about the issues involved and try to define the ideal, rather than simply picking what might be in the toolbox we happen to be equipped with. The challenges we face when valuing e.g., biodiversity cannot be changed or removed by the choice of a value articulating institution. What we are looking for is a system that is well fit for treating the kind of problems that we are confronted with.

### 4.2.1 Responding to the Challenges

We noted earlier that biodiversity is characterized by high complexity, by including elements of radical uncertainty, and by its being a common good. The first implies that there are great challenges concerning how to define the good and how to understand its importance. The second and third implies, as we have seen, that several ethical issues are involved. There may also be conflicts concerning outcomes. There will most probably be no best solution in Paretian terms. We are not involved in a situation characterized by exchange, where everybody may gain. Instead we will have to consume the same good and thus accept that interests will be taken care of at a varying degree. The challenges are thus both cognitive and normative.

*The Cognitive Aspect – Treating Complexity*    There are two cognitive issues involved. First, the good must be defined: What is it that should be evaluated for protection? Second, it is necessary to distinguish the values involved. In the case of complex goods like forest biodiversity these issues are linked and demanding.

Defining the good has been a great challenge to handle for practitioners of contingent valuation. The NOAA panel[9] (Arrow et al., 1993) discussed a long list of issues related to improving the quality of CV estimates. They emphasized that "respondents must understand exactly what it is they are being asked to value" (p. 4605). Given Wilson's definition of biodiversity (Section 2.1), this is a tremendous task. It is both a question about who should make that definition and how the result of that process should be transferred to respondents.

In CV studies these issues are kept apart. Mainly it is the researcher who defines the good. The problem is envisioned as that of transferring an objective definition of the good to the respondents. Not least in the case of environmental issues like biodiversity protection it seems, however, quite problematic to agree on what it is that should be valued. Thus, the questions come: Is it some single, maybe 'red listed' species that should be valued? Is it (some of) the systems functions, or is it maybe the whole ecosystems we should focus on?

The social contingencies involved become visible not least if we look at the shifts in the way forests have been viewed over the last few decades While there are variations across societies, we observe a rather substantial shift from viewing forests mainly as a source of direct use values – e.g., timber, berries, hiking tracks – towards giving greater emphasis to them as the home for threatened species, as a carbon sink, as a great regulator of the hydrological cycle etc. This change is partly the result of observing the medium to long run effects of modern forestry and partly due to more general changes in the relationship between society and nature. Most fundamentally it concerns learning about qualities that have always been there, but which we have been able to recognize first when we have challenged one or more of the 'invisible' functions involved (Vatn & Bromley, 1994).

One might argue that 'what the good is about' is an issue for experts – e.g., biologists or ecologists. Certainly, given the complexities involved, experts will be very important participants in the process. They can systematize and organize perspectives. They provide necessary concepts and models. It does not, however, fall under their competence to draw conclusions about which perspective is the *right one* to use – that of the tree as timber or a threatened species. That is a normative issue and hence a matter for the citizen. The information from the experts should be screened and reinterpreted by ordinary members of a society.

The implication of this is furthermore that it should be the same group that ultimately defines the good as well as evaluating it. The practice of using a focus group of laymen to test and fine-tune the definition of the good, as done in some CV studies, represents progress compared to a situation where the researcher has exclusive control over that process. Use of focus groups is still no guarantee for securing good fit between the description made and the perception it next evokes by the final respondents. The more complex and novel the issue at stake is, the more serious the problem.[10]

To evaluate resources like forest ecosystems, insight into their functioning is warranted. To ask someone 'in the street' about their willingness to pay for protecting a certain species or forest area is problematic in two senses. First, according to the above, monetary valuation may not be the institutional structure under which they want to inform decisions. Second, if a person has not reflected over which values are involved, or just has a rudimentary knowledge of what the characteristics of a certain good are, valuation becomes rather random. This problem is certainly greater the more complex and unfamiliar the good is.

In situations where the respondents observe that they lack information about the qualities of the good, they will tend to look for different clues in the material presented to help them out (Vatn, 2004). They may search for information that helps them to link the actual good to already well known goods. They may believe that the offered bids in a CV study carry information – i.e., in the case of a closed bidding procedure. Information about protection costs may, if offered, influence the bids. More specifically it is observed that people, when such information is given has anchored their bid in what they consider to be their reasonable share of that cost, illustrating that they think more in obligation to the group rather than individual utility. Finally, there seems to be a tendency to favor visible goods as opposed to functional aspects when valuing less familiar goods (Vatn, 2004).

There is thus a need for an educational process through which citizens better inform themselves about the values involved. Expert knowledge seems important. Once again, however, there is the question of which perspectives and thus which value elements should get priority. Again it seems warranted to foster a dialogue between experts and lay people/the citizen. This can be achieved through deliberative/participatory value articulating institutions.

While knowledge development and transfer is an important element of any deliberative institution, they handle it differently. As an example, focus groups do not normally call upon expertise; the knowledge base is that of the citizens who are participating. The variety of information they have may still provide a considerable enlargement of that with which each one starts. In the case of citizens' juries, hearing various experts is an integral part of the process, adding to the knowledge base by both giving access to highly developed knowledges and by offering a structured way of obtaining it.

*The Normative Aspect – Valuing the Common Good* The above reasoning has illustrated that normative issues are involved already at the level of defining the good. This is inherent in the notion of complexity. There is no single, or neutral way to define it. While this favors the use of participatory/deliberative institutions, the normative issues underpinning that conclusion go deeper. As already emphasized, ecosystems are common goods, so the action or the valuation made by one citizen influences directly the opportunities to be experienced by others. In such a situation the issue of which preferences or values to hold can hardly be seen to be an issue for the individual alone. Instead societies tend to develop sets of common norms concerning what is important (Douglas, 1986). The idea is that since my choices influence your opportunities, you, as a fellow citizen, would like to reason with me over which preferences are defendable in the actual situation or problem area. Ultimately, this concerns who we want to become as members of a society (Page, 1997).

Neoclassical economics and CBA circumvents this problem by just claiming that preferences are not open to reason. Its response is to aggregate given individual preferences in the form of price bids and in this way calculate which option gives the highest value. The practice is plagued by a series of practical problems and ethical paradoxes (e.g., Hanley & Spash, 1993; Niemeyer & Spash, 2001). Other positions favor negotiations as a way to 'aggregate' or to handle the underlying conflicts.

While both CBA and the tradition of negotiation take preference structures as given, the deliberative model follow the classical institutionalist view of the human and emphasizes the aspect of preference learning and change. Thus this model has a structure that captures that there is also a social dimension of preference formation. It opens an opportunity to reason over which values are important and which interests should get protection. As the issue is formulated here, this is the core problem when deciding over the common good.

The ideas of deliberation run actually back as far as to Aristotle. Rousseau, and later Dewey and Arendt have delivered important contributions (see also Pellizzoni 2003). Finally, the more recent work of Habermas (e.g., Habermas, 1984), Dryzek (e.g., Dryzek, 1990), and Elster (e.g., Elster, 1986; Elster, 1998) is of significance. The idea behind deliberative institutions is that of communicative rationality (Habermas, 1984). It is about the creation of understanding through dialogue and the force of the better argument. It is a form of common reasoning where mutual learning, understanding and preference changes are all elements of the process. In the ideal Habermasian form communication is thought to be free of coercion, strategic action and manipulation. The idea is both recursive and reciprocal in that it opens up an ongoing communication both about what is of value and to whom it is of value. O'Neill emphasize that "... dialogue involves not just recording given views and attitudes, but ideally the transformation of ...actors' self-understandings ...through conversation" (O'Neill, 2001, p.488).

Thus the deliberative model is based on the presumption that preferences are to a large degree learned attributes and that the individual is capable of questioning own preferences in the light of *new knowledge* and *the needs of others*. Certainly, conflicts may be too deep to bridge gaps. No consensus is really available. There may thus be a problem involved in focusing too heavily on reaching consensus. Furthermore, deliberative institutions may become power instruments in the hands of the well articulated. Still, the appropriate response to that problem is hardly to return to institutions based on a pure individualistic understanding of the problem. It simply does not fit the problem structure.

As earlier emphasized it is a mistake to try to treat all values – e.g., both individual preferences and social norms – as if they were fully comparable. There are incommensurabilities and hence different non-reducible perspectives, even different rationalities, involved. Applying the above reasoning to the ethical issues concerned, it is no surprise that incommensurable values are often observed within the realm of the environment (O'Neill, 1993; Spash, 2000). Such values cannot be treated by the logic of CBA. They are, however, consistent with the logic under-pinning communicative rationality.

*The Issue of Radical Uncertainty – The Cognitive Becomes Normative* There are two aspects of complexity of interest to us. One is related to the kind of goods we are looking at – i.e., the fact that they are systems goods and as such are difficult to define and demarcate. This was the perspective of the previous discussion about the cognitive aspect. There is, however, one more aspect of importance. This is the fact that complex systems are characterized by radical uncertainty – as noted in Section 2.1.

What does radical uncertainty mean? The definition is best understood if we contrast it to that of risk and (normal) uncertainty. Risk is defined as characterizing a situation in which there are known outcomes with known probabilities. Uncertainty is in standard terms characterized by known outcomes, but unknown probabilities. Finally, radical uncertainty or ignorance describes a situation where even (some) outcomes are unknown. Typically the long run effect of (mass) species extinction is of that kind. The restricted resiliencies involved and the various bifurcation points make it almost impossible to determine the developments that might follow from such major condition changes.

In situations with radical uncertainty, so typical of modern societies with rapid and large changes in technology and resource use patterns etc., the role of expertise becomes rather unclear. The content of "knowledge" and the traditional distinction between expertise and lay-people's evaluation has thus become challenged, resulting in less lay trust in pure scientific advice, and in the development of concepts like post-normal science to capture this new relation between citizens and experts (e.g., Funtowicz & Ravetz, 1993; Ravetz,1999).

The tradition of 'normal' science has been based on our ability to determine, with high certainty, the relations between the variables making up a system. It is based on the assumption that ignorance can be reduced at least to risk. In line with this, emphasis has been given to the importance of not accepting a false statement as true, i.e. to avoiding so-called type I error. In the case of studying the effect of the loss of a certain plant species, a type I error would be to claim that it will influence e.g., the functioning of an ecosystem, when it in reality will not.

Given radical uncertainty, the role of science changes in two important ways. First, radical uncertainty changes the focus of the treatment of making errors. Second, the distinction between facts and values, between the cognitive and the normative, becomes fundamentally blurred. In the case of radical uncertainty it is the so-called type II error that becomes crucial. A type-II error occurs when one accepts a false negative result – i.e., that no harm will result from eliminating a species – where harm in the end still appears. Given that ignorance is irreducible, one will have to make type-II errors when working on the basis of standard practices that seek to avoid type-I errors (Lemons, 1998). It is almost impossible to prove the effect of extinction with the necessary 95 – 99 % certainty. Complex systems are very demanding in this respect. Changes operate at different scales, both in time and space. Short run positive effects of changing a system may very often turn into long run negative shifts of great significance. If irreversible (and large), very high stakes are involved. Type II errors are hard to avoid.

The point is that traditional science has little to offer in the case of radical uncertainty. It may not be demonstrable with any high degree of certainty that damage will occur, but the opposite cannot be proven either. Thus, the issue becomes a normative one: Who should be given the burden of proof? Given the potential existence of radical uncertainty, it is the way the proof is defined that determines the outcome. This tells us that the role of lay-people's evaluation, that of the ordinary citizen, is crucial. This fact is strongly reflected in the declining faith in science as a solution to the problems we face concerning the future health of eco-systems. The issues become foremost normative. They need a normative framing and a normative evaluation. This again points in the direction of deliberative institutions where not only the results from science are evaluated and discussed from various perspectives. It becomes also possible to formalize interaction between normative and cognitive competences in a systematic and informed way.

### 4.2.2 The Challenges

So we can conclude that both the cognitive and the normative aspects involved when valuing ecosystems point in the direction of using deliberative institutions. While I believe that the above arguments undermine the validity of standard economic valuation, both on the basis of cognitive and normative aspects, shifting to deliberative evaluations is no simple solution. It seems possible to develop deliberative institutions that are theoretically consistent with the perception of the problems raised in this paper. Still, I think there is a long list of potential fallacies, both theoretical and practical, that needs to be addressed. The format of this chapter only gives me the opportunity to briefly address the most important ones – i.e., the problem of representation, the problem of unequal argumentative resources, and finally the problem of institutional perversion.

The problem of representation has received a lot of attention in the literature – e.g., Dryzek, 1990; Goodin, 1996; O'Neill, 2001; Pellizzoni, 2003). It is typical to make a distinction between two 'ideal types' – stakeholder and citizen representation. The latter can furthermore be grouped into random or discourse based representation à la Dryzek (1990). In the case of stakeholder representation, people or groups that are directly involved in a case would be the natural participants in the deliberation. Concerning forest biodiversity, stakeholder representation would typically involve members of environmental activist groups, forest owners, local community representatives, etc. In the case of citizens' representation, the participants are randomly selected among ordinary citizens or selected on the basis of the discourses they represent. In any case there is no simple way to determine what universe or constituency to draw from. Thus according to Niemeyer and Spash (2001) we observe a variety of ways participants are recruited to the same kind of institution – e.g., a citizens' jury.

No single solution to the problem of representation seems obvious. Good deliberation depends furthermore, on a rather restricted number of participants. This makes the issue of representativeness even more crucial (O'Neill, 2001). Stakeholder representation offers insight and engagement. If an agreement is reached, it is furthermore one that has a good chance of being accepted when implemented. Pellizzoni (2003) argues, however, that a stakeholder representation may force the deliberative institution to disintegrate into mere negotiations.

The problem with citizens' representation à la Dryzek is that it may idealize and decontextualize the discourse. It may thus have low legitimacy in the practical situation of implementation. Finally, the random citizen representation may have the danger of 'maximizing' the representation of disinterest. Combinations of different representations at different stages of the process may counter some of these effects – e.g., Renn (1999).

While the willingness to pay established by contingent valuation studies are influenced by the ability to pay (distribution of income), the results of deliberative institutions depend on the 'willingness and ability to say'. The conditions for the Habermasian ideal of domination-free deliberation are hard to achieve in practice. There is still one argument in favor of some limited effect of the issue concerning the ability to say or speak. In deliberative institutions participants are forced to justify their position. This implies that it is less easy to argue one-sidedly out of purely private interests. Instead, appeal to the general interest is necessary. This places some restrictions on individual behavior that are of importance when building legitimacy. Still, there is certainly a need for building elements into the institutional structures that counter the effect of potential monopolization of perspectives and arguments.

As already hinted at, deliberative institutions run a final risk – that of disintegrating into confrontation instead of dialogue, strategic action instead of deliberation. This risk depends partly on the type of controversy and partly on the chosen system for representation. Again the fostering of dialogue does have the merit of discouraging strategic behavior. Basing the deliberation on citizen representation also reduces the chance of such disintegration. On the other hand, there are strong arguments in favor of letting those most closely engaged in the issues participate, both to utilize their insights and to foster greater trust and legitimacy of the conclusions ultimately drawn. Again we have a need to balance conflicting objectives.

The issue we face here is that of an infinite regress – of a chicken and the egg problem. The choice of representation is dependent on the issues involved. On the other hand, those representing should define these issues. There is no way to consistently cut this knot. The only option is to treat it as a second level issue through open dialogue in society at large concerning which stakes or issues are worthy of a more comprehensive focus and in which way. This opens up a kind of communication that at least offers some way through which the untying of the above knot becomes intelligible and visible.

## 5. CONCLUSION

This chapter has discussed a set of issues concerning the evaluation of biodiversity with reference to that of forest ecosystems. We have observed that biodiversity is foremost a systems good – a set of relations carried by functionally specialized species. Given its characteristics, biodiversity is best described as a common good. This raises immediate ethical issues since the behavior or priorities of one human influence the opportunities of others. The ethical issues go furthermore beyond the anthropocentric perspective, raising the question of how other species should be treated.

Evaluating choices in the realm of the natural environment demands a thorough analysis of the above characteristics. Choosing evaluative instruments implies choosing between different perceptions both of the good and the (potential) rationalities involved. Cost-benefit analysis with its focus on monetary evaluation/CV is one type of value articulating institution. It has here been argued that CBA/CV fails to treat the issues involved in a way that is consistent with the characteristics of the good and the ethical concerns involved.

The case is thus made that the kind of problems we face when evaluating biodiversity in general and forest biodiversity in particular, is best treated using deliberative value articulating institutions. While there are important differences across the set of deliberative institutions, they generally offer a better fit to the problems involved. This concerns both the cognitive challenges – where the potential for communication between citizens and experts is pivotal – and the normative issues – the process by which we develop an understanding of the ethical issues and dilemmas involved. Specifically, these institutions offer possibilities for learning about and for handling competing or incommensurable perspectives. They also offer ways to handle issues where radical uncertainty is involved, by providing the necessary opportunity to resolve the relevant cognitive and normative issues in a reasoned way. Certainly, no method can do away with radical uncertainty. The point is to construct an institutional context under which it can get appropriate recognition.

Still, using deliberative institutions is not an easy path to take. A lot of decisions to be made, not least about who should participate and how to secure the integrity of the process, are not straightforward and are hence open to debate. There is (as yet) no clear or agreed basis for making choices on these procedural issues. Some further development in the direction of defining common 'rules' is certainly possible. Still, the character of the problems involved seems to imply that several decisions need to be made with concrete reference to the actual issue and local contexts. The future challenge from a research point of view lies not least in sorting out the principal issues involved, but also in increasing the knowledge about effects of various choices to support the selection of procedures to be made in these local contexts.

**Acknowledgements.** The author would like to thank Albert Berry for thoughtful comments to an earlier version of this paper.

## NOTES

[1] Bifurcation: Beyond a certain value of a control variable, a state variable may take on either of two values – forcing the system to follow either of two trajectories. Which one is followed is indeterminable despite the fact that we are able to describe the movement of the system up until the bifurcation point.

[2] The idea is that events may have forced the system to shift attractor. If the system has stayed essentially within the same attractor after some shock, we know that it is able to retreat from such a change in state variables.

[3] Certainly, the position has gotten into problems in cases where individuals have preferences, which bring direct negative consequences to others like sadism. Should such preferences be accepted? Actually, the discussion about what are defensible preferences in the realm of the environment has structurally the same form as the one observed with the sadist preference – i.e. implementing a certain preference influences the opportunities or conditions for others.

[4] See O'Neill (1993) for a discussion of various inconsistencies and confusions in the literature concerning the concept of 'intrinsic value'. O'Neill, furthermore, develops an objectivist position of value, the idea that there are objective – like biological – reasons why something is of value to us, including the existence of other species. His position is clearly against strong versions of cultural relativism. Even though I emphasize that claims and norms will vary across cultures, I very much agree with O'Neill that there are important objective constituents of a good life, which individual and cultural differences cannot do away with. My emphasis on the importance of the functioning of natural ecosystems for humans to thrive rests precisely on this argument.

[5] Scott is a sociologist working mainly within the theory of organizations. I find his position quite representative also of more modern variants of 'classic institutional economics'.

[6] It must be emphasised that important new institutionalists like North and Williamsson over the years have accepted the idea of changing preferences. This is an interesting development. It has, however, not yet resulted in a changed view on what institutions are.

[7] Jacobs (1997) makes a distinction between value articulating institutions – e.g., CV – and decision supporting institutions – e.g., CBA. While semantically quite correct, it just complicates the presentation and discussion to emphasize this distinction here.

[8] The respondent refuses to participate or deliver a zero bid even in situations where it is clear that s/he values the good positively.

[9] The National Oceanic and Atmospheric Administration panel

[10] It may seem odd that representatives of a position that emphasizes subjectivity believe that it is possible to agree on what a good is. This is not so surprising. The neoclassical position is built on the view that nature is objective while preferences are subjective. The problem, as I see it, is hence the acceptance of a need to deliberate on the definition of something that should be objectively given. It implies an implicit acceptance of the fact that there is subjectivity also involved when defining the good.

Thus the demarcation between the object and the subject, so fundamental to neoclassical economics, is allowed to vanish at this stage. I think we here have a typical example of a situation where practitioners, to solve urgent problems, construct solutions that next challenges the theoretical consistency or underpinning of the model, still without reflecting on what it actually implies.

## REFERENCES

Armour, (1995). The citizens jury as model of public participation: A critical evaluation. In O., Renn, T. Webler, P. Wiedema (Eds.), *Fairness and competence in citizen participation (pp. 175-188).* Dordrecht: Kluwer.

Arrow, K., Solow, R., Portney, P.R., Leamer, E.E., Radner, R., & Schuman, H. (1993). Report of the NOAA Panel on contingent valuation. *Federal Register,* 58, 4601-4614.

Barbier, E.B., Burgess, J.C., & Folke, C. (1994). *Paradise lost? The ecological economics of biodiversity.* London: Earthscan Publ.

Boardman, A.E., Greenberg, D.H., Vining, A.R., & Weiner, D.L. (2001). *Cost-benefit analysis. Concepts and practice.* New Jersey: Prentice Hall Inc.

Burgess, J., Limb, M., & Harrison, C.M. (1988). Exploring environmental values through the medium of small groups: Theory and practice. *Environment and Planning A,* 20(3), 309-326.

Burgess, J., Clark, J., & Harrison, C. (1995). *Valuing nature: What lies behind responses to contingent valuation surveys?* Paper, Dept. of Geography, Univ. College London.

Burgess, J., Clark, J., & Harrison, C. (1998). Respondents' evaluation of a CV survey: a case study based on an economic valuation of the wildlife enhancement scheme, Pevensy levels in East Sussex. *Area,* 30, 19-27.

Clark, J., Burgess, J., & Harrison, C.M. (2000). 'I struggled with this money business': Respondents' perspectives on contingent valuation. *Ecological Economics,* 33, 45-62.

Coase, R.H. (1984). The new institutional economics. *Journal of Theoretical and Institutional Economics,* 140(1), 229-231.

Douglas, M. (1986). *How institutions think.* Syracuse: Syracuse University Press.

Dryzek, J. (1990). *Discursive democracy: Politics, policy and political science.* Cambridge: Cambridge University Press.

Eggertsson, T. (1990). *Economic behavior and institutions.* Cambridge: Cambridge University Press.

Ehrlich, P.R., & Ehrlich, A.H. (1992). The value of biodiversity. *Ambio,* 21(3), 219-226.

Elster, J. (1986). The market and the forum: three varieties of political theory. In J. Elster, Aa. Hylland (Eds.), *Foundations of social choice theory ( pp. 103-132).* Cambridge: Cambridge University Press.

Elster, J. (Ed.). (1998). *Deliberative democracy.* Cambridge: Cambridge University Press.

Funtowicz, S., & Ravetz, J.R. (1993). Science for the post-normal age. *Futures,* 25(Sept.), 739-755.

Graves, J., & Reavy, D. (1996). *Global environmental change.* London: Longman.

Goodin, R. (1996). Enfranchising the earth, and its alternatives. *Political Studies,* 44, 835-849.

Habermas, J. (1984). *The theory of communicative action. Volume one: Reason and the rationalization of society.* Boston: Beacon Press.

Hanley, N., & Spash, C. (1993). *Cost-benefit analysis and the environment.* Aldershot: Edward Elgar.

Holland, A. (1997). Substitutability, or why strong sustainability is weak and absurdly strong sustainability is not absurd. In J. Foster (Ed.), *Valuing nature? Economics, ethics and environment* (pp. 119-134). London: Routledge.

Holling, C.S., Schindler, D.W., Walker, B.W., & Roughgarden, J. (1995). Biodiversity in the functioning of ecosystems: An ecological synthesis. In C. Perrings, K.-G. Mäler, C. Folke, C.S. Holling, B.-O. Jansson (Eds.), *Biodiversity loss. Economic and ecological issues* (pp. 44-83). Cambridge, Cambridge University Press.

Jacobs, M. (1997). Environmental valuation, deliberative democracy and public decision-making. In J. Foster (Ed.), *Valuing nature? Economics, ethics and environment* (pp. 211-231). London: Routledge.

Janssen, R. (1994). *Multiobjective decision support for environmental management.* Dordrecht: Kluwer Academic Publishers.

Joss, S. (1998). Danish consensus conferences as a model for participatory technology assessment. *Science and Public Policy,* 25(1), 2-22.

Keeney, R.L., & Raiffa, H. (1976). *Decision with multiple objectives: Preferences and value tradeoffs.* New York: Wiley.

Lemons, J. (1998). Burden of proof requirements and environmental sustainability: Science, public policy, and ethics. In J. Lemons, L. Westra, R. Goodland (Eds.), *Ecological sustainability and integrity: Concepts and approaches (pp. 75-103).* Dordrecht/Boston/London: Kluwer Academic Publishers.

Lynn, F., & Kartez, J. (1995). The redemption of citizen advisory committees: A perspective from critical theory. In O. Renn, T. Webler, P. Wiedeman (Eds.), *Fairness and competence in citizen participation. Evaluating models for environmental discourse (pp.81-100).* Dordrecht: Kluwer Academic Publishers.

Mitchell, R.C., & Carson, R.T. (1989). *Using surveys to value public goods: The contingent valuation method.* Washington D.C.: Resources for the Future.

Munda, G. (1995). *Multicriteria evaluation in a fuzzy environment.* Heidelberg: Physica Verlag.

Nicolis, G., & Prigoigne, I. (1989). *Exploring complexity. An introduction.* New York: W.H. Freeman and Company.

Niemeyer, S., & Spash, C.L. (2001). Environmental valuation analysis, public deliberation, and their pragmatic syntheses: a critical appraisal. *Environment and Planning C: Government and Policy,* 19(4), 567-585.

Norgaard, R.B. (1984). Coevolutionary development potential. *Land Economics,* 60(2), 160-173.

North, D.C. (1990). *Institutions, institutional change and economic performance.* Cambridge: Cambridge University Press.

O'Neill, J. (1993). *Ecology, policy and politics. Human well-being and the natural world.* London: Routledge.

O'Neill, J. (2001). Representing people, representing nature, representing the world. *Environment and Planning C: Government and Policy,* 19(4), 483-500.

Page, T. (1997). On the problem of achieving efficiency and equity, intergenerationally. *Land Economics,* 73(4), 580-596.

Pellizzoni, L. (2003). Uncertainty and participatory democracy. *Environmental Values,* 12(2), 195-224.

Perrings, C. (1997). Ecological resilience in the sustainability of economic development. In C. Perrings (Ed.), *Economics of Ecological Resources. Selected Essays* (pp. 45-63). Edward Elgar.

Pimm, S., Russell, G., Gittleman, J., & Brooks, T. (1995). The future of biodiversity. *Science,* 269, 247-350.

Ravetz, J. (1999). What is Post-Normal Science, *Futures,* 31(7):647-653.

Renn, O. (1999). A model for analytic-deliberative process in risk management. *Environmental Science and Technology,* 33(18), 3049-3055.

Sagoff, M. (1988). *The Economy of the earth: Philosophy, law and environment.* Cambridge: Cambridge University Press.

Schkade, D.A., & Payne, J.W. (1993). Where do the numbers come from? How people respond to contingent valuation questions. In J.A. Hausman (Ed.), *Contingent valuation: A critical assessment.* (pp. 271-293) Amsterdam: North Holland.

Schindler, D.W. (1990). Experimental perturbation of whole lakes as tests of hypotheses concerning ecosystem structure and function. *Oikos,* 57, 25-41.

Scott, W.R. (1995). *Institutions and organizations.* Thousand Oaks, California: Sage Publications.

Smith, G., & Wales, C. (1999). The theory and practice of citizens' juries. *Policy and Politics,* 27, 295-308.

Solow, R. (1993). Sustainability: An economist's perspective. In R. Dorfman, N. Dorfman (Eds.), *Economics of the environment (pp.179-187).* Selected readings. 3$^{rd}$ edn. New York: Norton.

Spash, C.L. (2000). Multiple value expression in contingent valuation: Economics and ethics. *Environmental Science & Technology,* 34(8), 1433-1438.

Stevens T.H., Echeverria, J., Glass, R.J., Hager, T., & Moore, T.A. (1991). Measuring the existence value of wildlife: What do CVM estimates really show? *Land Economics,* 67(Nov.), 390-400.

Vadnjal, D., & O'Connor, M. (1994). What is the value of Rangitoto Island? *Environmental Values,* 3(4), 369-380.

Vatn, A., 2004. Valuation and Rationality. *Land Economics,* 80(Feb.):1-18.

Vatn, A., & Bromley, D. (1994). Choices without prices without apologies. *Journal of Environmental Economics and Management,* 26, 129-148.

Veblen, T. (1919). *The place of science in modern civilisation and other essays.* New York: Huebsch.

Walker, B.H./1992). Biodiversity and ecological redundancy. *Conservation Biology,* 6, 18-23.

Wilson, E.O. (2001). *The diversity of life.* London: Penguin Books.

# CHAPTER 6

# THE GREAT TRAGEDY OF SCIENCE: SUSTAINABLE FOREST MANAGEMENT AND MARKETS FOR ENVIRONMENTAL SERVICES

## CLARK S. BINKLEY

*Hancock Timber Resources Group*
*99 High St. Boston, MA 02110-2320*
*Email:cbinkley@hnrg.com*

**Abstract.** Sustainable forest management is one of the most capital-intensive activities imaginable. Governments are unlikely to provide the capital needed to emend landscape degradation. On the other hand, private capital will flow into the sector to fund the necessary solutions, but only if investors are rewarded for the environmental services provided by healthy landscapes. Creating markets in environmental services is a necessary, but not sufficient, condition for sustainable forestry.

## 1. INTRODUCTION

This chapter reports on a personal odyssey in the quest for sustainable forest management, an odyssey from being an academic economist to joining "dark side" to work in a private equity firm specializing in forestry investments. My company, the Hancock Timber Resource Group (a subsidiary of the Canadian financial services company Manulife Financial), manages about 1.2 million hectares of forest in the US, Canada, Australia and New Zealand. Most of these forests are plantations, and all are managed to a high standard of environmental stewardship. Because our ultimate investor base is very sensitive to environmental issues, sustainable forestry is a key element of our business strategy. But, what does that mean, and how do we achieve it?

During my academic days, my research focused on forestry investments, including both timber and environmental services. In 1998 I thought that the time was propitious to move to the private sector and see if it was possible to make some of this theory work in practice. Four hypotheses underpinned my decision to jump into these murky waters:

*Kant and Berry (Eds.), Institutions, Sustainability, and Natural Resources: Institutions for Sustainable Forest Management, 135-139.*

i.   that forestry is a very capital intensive enterprise, so attracting capital to the
     sector and ensuring that it is efficiently allocated is critical to sustainable
     forest management;

ii.  that environmental services comprise an material value stream from most
     forestry investments;

iii. that national and international policies increasingly seek to use markets as a
     means to secure these values for society;

iv.  and finally, that as a result of the first three factors, it would be possible to
     capitalize the value of environmental services into investment decisions, and
     to get more trees grown on a larger fraction of the world's land.

The evidence seems to be consistent with the first three of these, but, regrettably,
not the fourth.

The capital intensity of forestry is now widely recognized by economists and
financial analysts alike. Because of the long production period and large amount of
standing inventory in a sustained-yield forest, forestry is among the most capital-
intensive activities humans pursue. For example, the capital-output ratio for a
sustained-yield loblolly pine forest regulated on a 23-year rotation is about 10, and
would be much higher for the kinds of long-rotation forests many proponents of
sustainable forest management prefer (Binkley, 1993). The forest sector is evolving
so the capital represented by forests is moving into the hands of investors with long
investment horizons and comparatively low capital costs. We figure that institutional
investors of this type—pension funds, insurance companies, university endowments,
foundations and families seeking multi-generation wealth creation—now have about
$US 15 billion invested in timberland. One investor—Harvard University—has
about 9% of its $US 20+ billion endowment invested in timberland. While the
allocation of institutional capital to timberland is not large when compared with the
overall quantum of institutional investment, it does comprise a material fraction of
the industrial forest land in the US, and is growing at about 20%/year.

Any doubt about the importance of environmental services that might have
existed at one time surely has faded. Both population and per-capita income
continue to grow, both globally and in most individual countries. As a result, the
demand for most environmental services is also increasing. At the same time, the
supply of these services is shrinking as habitat is lost, species are extirpated, and the
water, air and soil are polluted. The interaction of rising demand and shrinking
supply means that the implicit "price" for these services is rising, if only markets
existed to record the development.

The lack of markets for environmental services means that they are
systematically under produced and over consumed. Governments have attempted to
bridge the gap between supply and demand through regulations requiring pollution
abatement and habitat maintenance, but such regulations have turned out to be
costly, cumbersome and blunt tools to achieve society's objectives. As a result, even
the early proponents of regulatory approaches now recognize the value of markets in
achieving environmental objectives. Markets feature prominently in the flexibility

mechanisms of the Kyoto Protocol. An ENGO, Forest Trends is dedicated to the idea of establishing markets for environmental services as a means of achieving sustainable development—they just celebrated their fifth anniversary.

Regrettably, this enthusiasm for markets has not been matched by the reality. Gus Speth (who *The Economist* called "one of the grand old men of greenery") commented in his recent book *Red Sky in the Morning,*

> The flowering of market-based approaches is an important and encouraging development. But, paradoxically, we do not yet seem to be in the midst of a shift to full-cost, environmentally honest prices. Economic instruments...are increasingly being used, but mostly to achieve greater efficiency in environmental protection and not rigorously to get the prices right. (Speth, 2004; 163)

All of this brings us to the title of this paper. In his 1870 presidential address to the British Association for the Advancement of Science, Thomas Huxley described "the great tragedy of Science" as "the slaying of a beautiful hypothesis by an ugly fact" (Huxley, 1870). The rosy expectations related to the establishment of markets for environmental services have, in my view, been slain by some ugly economic facts—facts that economists concerned about sustainable forest management can help to change.

## 2. MARKETS FOR ENVIRONMENTAL SERVICES AND TRANSACTIONS COSTS

To understand how economists can help save the "beautiful hypothesis" of "environmentally honest prices", it is useful to go back to the theory of property rights. Society creates property rights when the benefit of doing so exceeds the costs. The absence of property rights creates losses to society through the misallocation of resources. The cost of creating property rights are transactions costs. So, it makes sense for society to create property rights when the costs of the misallocations exceed the transactions costs. In espousing markets for environmental services, economists commonly focus on the misallocations associated with the absence of markets—external costs and benefits that lead to over consumption and under production of environmental services. In the face of the obvious problems and clear economic prescriptions for solving them, we economists commonly imagine policy makers to be stupid and venal because they do not jump to adopt market-based mechanisms.

While the attention on one side of the equation—the misallocations—is useful and important, I think more focus needs to be placed on reducing transactions costs. The remainder of this chapter touches on three kinds of these costs.

The first is political—establishing property rights inevitably produces winners and losers. Despite the fact that a policy might create Pareto improvements, the losers may be powerful enough to block the necessary changes. Economists could usefully help policy makers craft strategies to mollify the losers. Policy makers may be venal, but, in my experience, they are rarely stupid.

The second relates to measuring environmental services. One cannot trade in a product or service unless one can measure the quantum being bought or sold. For example, to sell a carbon sequestration credit from a forest, one needs to measure the amount of carbon that is actually sequestered. For plantation forests this is comparatively straight forward, but there is a need to reduce the costs and increase the precision of the measurements. For other, more complex kinds of forests— precisely the kinds of forests that many proponents of sustainable forest management prefer—the task is daunting.

The third relates to actual financial transactions costs associated with the market. Let me provide a personal example. In 2000 my company established its "New Forests" program. This program was designed to take advantage of emerging markets for environmental services from forests as a means of providing another, uncorrelated income stream for forestry investors. The better returns would attract more capital to the sector, or so our reasoning went. We were and are especially interested in carbon sequestration credits. It turns out that to sell carbon credits we actually needed to be licensed to deal in financial derivatives. We can sell trees without a license, but—amazingly—we need a complicated financial services license to sell the carbon embodied in the trees! Fortunately, we are part of a large and sophisticated financial services company, so it only took us about six months at a cost of perhaps several hundred thousand dollars to obtain the necessary licensing. But, for many who want to enter this market this will be a prohibitive expenditure of time, effort and money. Political economists or our colleagues from Law Schools could surely help figure out ways to characterize carbon credits such that they were more easily traded.

## 3. CONCLUSIONS

In conclusion, establishing markets for the environmental services that flow from forests appears to be a necessary (if not sufficient) condition for sustainable forest management over the long run. Enormous amounts of capital are required for sustainable forestry, and governments are unlikely to provide this capital. Private investors have shown a substantial willingness to invest in forestry, and, I believe, will make investments that increase the supply of forest-related environmental services as long as there are (i) reliable markets for these services, and (ii) adequate income streams to compensate investors for their investments.

There is reason for optimism. Australia is a good example. That country has put in place the institutional framework to make carbon markets a reality—carbon rights can be registered on land deeds, and the Australia Greenhouse Office has done extensive work on measuring carbon embodied in trees. One state, New South Wales, has established a "cap and trade" regime so there is a solid market in "NGACs"—NSW Greenhouse Abatement Certificates (which currently trade at about \$A 10/ tonne of $CO_2$-e). Our New Forests program there has completed two large transactions where carbon credits are a material aspect of the investment proposition.

In short, a broad consensus seems to be emerging around the desirability of market-based approaches to sustainability. But, much work remains to be done on designing the details of these markets. Until those details are in place, markets for environmental services, and, indeed sustainability itself, may tragically remain beautify hypotheses slain by ugly facts.

## REFERENCES

Binkley, C. S. (1993). Long-run timber supply: price elasticity, inventory elasticity and the use of capital in timber production. *Natural Resource Modeling*, 7, 163-181.
Huxley, T. H. H. (1870). Address delivered to the British Association for the Advancement of Science. : Red Lion Court: Taylor and Francis Printers.
Speth, J. G. (2004). *Red Sky at Morning.* .New Haven: Yale University Press.

# CHAPTER 7

# THE KYOTO PROTOCOL: PROPERTY RIGHTS AND EFFICIENCY OF MARKETS

## GRACIELA CHICHILNISKY

Department of Economics, Columbia University,
New York, USA
Email: gc9@columbia.edu

**Abstract.** The origin of today's global environmental problems is the historic difference in property rights regimes between industrial and developing countries, the North and the South. In industrial countries resources such as forests and oil deposits are often under well defined and enforceable property rights, mostly private property regimes. In developing countries they are generally under ill-defined and weakly enforced property rights, mostly community or state property regimes but de-facto open-access regimes. Ill-defined and weakly enforced property rights lead to the over-extraction of natural resources in the South. They are exported at low prices to the North that over-consumes them. The international market amplifies the tragedy of the commons, leading to inferior solutions for the world economy. In developing countries, the conversion of natural resources regimes from community or state property regimes or common access to private property regimes faces formidable opposition due to international economic interests, or to heavy dependence of local and poor people on these resources. The weakness of property rights in *inputs* to production, such as timber and oil, could be compensated by assigning well defined and enforceable property rights to products or *outputs*. The 1997 Kyoto Protocol provides an example as it limits countries' rights to emit carbon, a by - product of burning fossil fuels, but the atmosphere's carbon concentration is a public good, which makes trading tricky. Similarly, trading rights to forests' carbon sequestration services involve public goods. Markets that trade public goods require a measure of equity to ensure efficiency, a requirement different than the markets for private goods

## 1. INTRODUCTION

Human beings—or their close genetic relatives—have lived on Earth for several million years. Yet only recently has human activity reached levels at which it can affect fundamental natural processes—such as the concentration of gases in the atmosphere, the planet's water mass, and the complex web of species that constitute life on earth. Scientists find that the most environmental damage has occurred in the last 50 years, the period in which the human species has consolidated its dominance of the planet and has embarked in an unprecedented rapid phase of industrialization. Our current global environmental problems originate from the industrial growth in the world economy since World War II. Fueled by abundant and inexpensive raw materials, most of which were exported by poor countries and imported by industrial countries, this industrialization has been voracious in the use of natural resources. In the last 50 years international trade grew three times faster than the countries

*Kant and Berry (Eds.), Sustainability, Institutions, and Natural Resources: Institutions for Sustainable Forest Management,* 141-154.

themselves—and with them grew the international demand for energy derived from fossil fuels, and the demand for other natural resources such as wood, which are extracted from developing countries' forests.

International trade in natural resources is directly implicated in the current global environmental problems. Most of the natural resources we use worldwide are extracted from developing countries, where they are usually held under conditions of ill-defined and/or weakly enforceable property rights such as state and community regimes which are in many cases de-facto open-access regimes, and end up being consumed in the rich industrial countries. In a divided world economy in which poor countries trade with rich nations, ill-defined and weakly enforceable property regimes of natural resources distort the market behavior, and these natural resources from developing nations are sold internationally at low prices, often under replacement costs. Low resource prices leads to poverty at home, and to over-consumption in the rich nations that import them. Most of the planet's carbon emissions come from oil that is burned in rich nations. The US, for example, imports most of its oil from developing nations—and it is the largest oil consumer in the world, using about 26% of the world's oil production and generating about 26% of the planet's carbon emissions even though it has less than 4% of the world's population. Now we know that carbon emissions could change the global climate and become catastrophic for the survival of the human species.

Even though international markets are at the root of the problem, this chapter suggests that they could also be instrumental in providing solutions. Resource markets play a key role in the problem—and a solution may be found in markets involving global public goods, such as markets for trading the rights to emit. A word of caution is needed here. Emission markets that trade 'rights to use the planet's atmosphere' are in reality trading global public goods, and as such very different from the markets that economists have known for centuries. Following our recommendations, global emission markets appeared in the United Nations Kyoto Protocol created in 1997 by 166 nations at the United Nations Framework Convention for Climate Change UNFCCC (Chichilnisky 1996c, 1997; Chichilnisky and Heal, 2000). In 2005, the Kyoto Protocol became International Law. Markets for emission trading are key to the global environment—and global equity issues are important for the efficient functioning of these global markets. A resolution of the global environmental problems that concern us today depends therefore upon achieving a measure of equity in the global economy (Chichilnisky 1996c, 1997; Chichilnisky & Heal, 2000). While often competing objectives, the notions of equity and efficiency now converge in a world economy that is increasingly dominated by goods and services based on **environmental resources** and on **knowledge**—both of which are global public goods.

This chapter analyses the economic issues underlying the origins of today's global environmental problems and seeks solutions. Markets are implicated in the problem, and are part of the solution. But economics needs to be developed further to understand and foster the functioning of markets involving privately produced public goods, such as the global emissions markets. To ensure the proper functioning of these markets new institutions are needed, as discussed at the end of this chapter.

## 2. MULTIPLE PERSPECTIVES ON GLOBAL ENVIRONMENTAL PROBLEMS

Global environmental problems include the impact of Chlorofluorocarbons CFC's on the ozone's layer of the atmosphere, the loss of the planet's biodiversity, the problem of acid rain, and the international transport of $SO_2$. Ozone depletion was successfully tackled by the international community through the Montreal Protocol of 1987, which restricted the use of CFC's in industrial products. With respect to greenhouse gas emissions, in 1996 the Intergovernmental Panel on Climate Change (IPCC) reported that human-induced emissions of carbon and other greenhouse gases have a 'discernible effect on climate'. While some uncertainty still surrounds the scientific evidence on climate change, the risk of climate change is potentially catastrophic. The greenhouse effect is a typical example of a problem of global commons, where no single country can tackle this problem on its own and an international cooperation is necessary. The concentration of $CO_2$ in the planet's atmosphere is uniform, and the whole world is subject to the same concentration.

### 2.1 Global Environmental Problems and Economic Incentives

Global environmental problems are driven by economic incentives. Human energy use contributes almost half (49%) of the green house gases while industrial processes contribute almost a quarter (24%). The two other sources of green house gases are deforestation (14%) and agriculture (13%) (WRI, 1990, p.24). Hence, the threat of climate change is driven by the use of energy that increases with industrialization. Across the world, energy is produced mostly by burning fossil fuels—leading directly to higher emissions of greenhouse gases. Biodiversity destruction is led by the destruction of habitat in forests—for economic purposes. Forests, where most known biodiversity resides, are cleared for the extraction of natural resources (such as oil and wood products) or for growing cash crops and grazing livestock. These products go mostly to export markets. CFC emissions that damage the ozone layer originate from industrial products.

While the causes of global environmental problems are economic, the initial effects are physical or biological. Because the effects are physical, economists underestimate them. Since the causes are economic, physical and biological scientists cannot find solutions. Hence, global environmental problems, such as climate change, require thinking and acting across social and physical disciplines, which is a major challenge in an era of compartmentalized approaches across disciplines.

### 2.2 Global Environmental Problems and Population

Many believe that global environmental problems emanate from the enormous growth of human population on the planet. The term "population bomb," created by Paul Ehlrich more than twenty-five years ago, symbolizes this perspective. The view has been erroneously used to imply that the developing countries—whose populations grow on the whole faster than industrial nations—are the main source of danger to the global environment. The view is not without merit but misses the main point. Yes, global environmental issues are related to the human dominance of the planet. Indeed if there were no humans, the problem would cease to exist; this is what I call the 'ultimate solution'. However the regions in the world with fewer humans and with lowest population growth are the ones responsible for most of the problems.

For example, developing nations have higher rates of population growth on the whole. However, it is widely known that developing nations and the regions of the world with the highest population growth are not the main cause of global environmental damage; they contribute far less to the global environmental problems than countries with lower population growth. This is because it is industrialization that causes the environmental problems we have today and not population pressures by themselves. The most industrialized regions have lower population growth, but are the main cause of biodiversity loss, carbon and CFC emissions. Data regarding the share of world carbon dioxide emissions, population, and GDP, for Industrial and Less Developed countries, is given in Table 7.1.

**Table 7.1.** *Share of the Total World Carbon Dioxide Emissions, Population, and GDP (in terms of purchasing power parity) for Industrial and Less Developed Countries*

| Countries | Cumulative $CO_2$ emissions | Current $CO_2$ emissions | Population | GDP |
|---|---|---|---|---|
| Industrial | 70% | 60% | 24.5% | 68.5% |
| Less Developed | 30% | 40% | 75.5% | 31.5% |

The data in Table 7.1 clearly indicates that the usually drawn connection between global environment problems and the population is incorrect. Historically and currently, economic output is the major determinant of carbon emissions. Indeed, industrial countries account for 68.5 of world GDP and emit 60 to 70% of $CO_2$ emissions, though having only 24.5% of the population. Reciprocally, developing countries have 75.5% of the world's population, 31.5% of GDP and account for only 30 to 40% of $CO_2$ emissions. Hence, there is a direct positive relation between GDP and $CO_2$ emissions, but a direct negative relationship between $CO_2$ emissions and population. If the currently less developed countries eventually become the major polluters, it will be because of their industrialization, not their further population growth. Ehlrich's predictions of run-away population growth in the planet have in any case proven incorrect.

*2.3 Global Environmental Problems, the Concept of Basic Needs, and Sustainable Development*

To address global environmental problems, in 1974 I introduced a way to measure economic progress that is different from GDP—the concept of development based on the 'satisfaction of basic needs' as presented in Chichilnisky (1977). The basic needs approach does not assume GDP to be the defining feature of economic progress, but rather measures such progress by the satisfaction of the population's basic needs. This concept was introduced to make economic development patterns more consistent with environmental constraints and was developed in empirical and mathematical studies undertaken in 5 continents, within the Bariloche World Model (1974, 1976). This led directly in 1987 to the Brundtland Report, which introduced the concept of Sustainable Development in the Earth Summit in Rio de Janeiro, Brazil. Sustainable development is based on the satisfaction of Basic Needs. But the Brundtland report links the basic needs of the present and those of the future: the definition proposed here for Sustainable Development is "development that meets the needs of the present without compromising the ability of future generations to meet their own needs" (WCED, 1987). A formal operational definition of sustainable development is presented in Chichilnisky (1997). In addition, the UN's Millenium Development Targets (2005) are based on the satisfaction of Basic Needs

## 2.4 Global Environmental Problems and North–South Issues

How does the current situation in the global environement arise, about 50-60 years ago on the whole? What happened 50-60 years ago? After World War II, the US economy accounted for 40% of world output, following the destruction of Germany and Japan. Today the US share is back to 25%, as it was before the war. Following World War II, the US pattern of economic development became a global benchmark. Based on rapid industrialization, it was fuelled by a deep and extensive use of natural resources, in this sense being a "frontier" type of growth. Important global institutions were created at this time (the World Bank, the International Monetary Fund, the General Agreement on Trade and Tariffs, the United Nations as a whole, and the current system of National Accounts) whose metrics for economic progress reinforced this vision of resource-intensive economic development. Thus the Bretton Woods Institutions created by Lord Keynes played an important role in taking the "American Dream" global. Keynes saw the role of these institutions as replacing wars by trade—using the differences among nations as a source of gains from trade. His dream succeeded beyond anyone's expectations and in the 50 years since the end of World War II, international trade grew three to four times faster than the world economy.

The rapid increase in emissions of carbon dioxide of the last fifty years has been due to the burning of fossil fuels linked to intensive energy use for production of goods and services in industrial nations. The globalization of the world economy since World War II has intensified a pattern of resource use by which developing nations extract most natural resources, exporting them to industrialized nations at prices that are often below replacement costs. Through the international market, industrial nations, which house less than a quarter of the world's population consume most forest products (pulp, wood); consume most products produced through the clearing of forests (cash crops such as cotton, livestock including beef and veal); and consume most mineral products (copper, aluminum, and fossil fuels such as petroleum) (Table 7.2). Hence, the North's[1] economy represents the main driving force in global environmental problems, producing 60-70% of the world's $CO_2$ emissions (Table 7.1) and emitting most CFC's, responsible for the damage to the earth's ozone layer. Most emissions of greenhouse gases originate in energy use and production (including the production of electricity)—and a major share of the world production is located today in industrial nations. The South emits fewer greenhouse gases into the planet's atmosphere, roughly 30% of the world's total, even though it has more than three-quarters of the world's population. At the same time, in the developing countries—which are geographically located on the whole in the Southern hemisphere of the planet—there is currently an intensive and extensive destruction of ecosystems for agricultural production and for mineral extraction, mostly directed towards export markets. Because the industrial countries have already exhausted most of their own forests in their own process of industrialization, it follows that most environmental resources, such as forests and biodiversity are now found in the developing countries, where tropical deforestation is occurring most rapidly today.

*Table 7.2.* Consumption of Natural Resources by Industrialized and Developing Countries

| Resource | Country | 1961-1965 | 1966-1970 | 1971-1975 | 1976-1980 | 1981-1985 | 1986-1990 |
|---|---|---|---|---|---|---|---|
| Fossil Fuel (giga joules/ person) | Indust. | 115.82 | 142.53 | 165.7 | 169.52 | 153.81 | 160.06 |
| | Develop. | 7.37 | 8.26 | 10.34 | 12.91 | 14.53 | 17.28 |
| Alumi-nium (metric tons/ 100 people) | Indust. | 5.99 | 9.00 | 11.89 | 13.50 | 12.56 | 14.13 |
| | Develop. | 0.13 | 0.23 | 0.37 | 0.51 | 0.58 | 0.69 |
| Copper (metric tons/ 1,000 people) | Indust. | 6.17 | 7.00 | 7.46 | 7.90 | 7.50 | 8.06 |
| | Develop. | 0.17 | 0.17 | 0.26 | 0.34 | 0.38 | 0.48 |
| Beef and Veal (kilograms/ person) | Indust. | 24.53 | 27.37 | 28.59 | 29.65 | 27.69 | 27.17 |
| | Develop. | 3.98 | 4.06 | 3.84 | 4.21 | 4.05 | 4.29 |
| Cotton (kilograms/ person) | Indust. | 6.91 | 5.32 | 5.30 | 4.70 | 4.77 | 5.35 |
| | Develop. | 1.93 | 2.29 | 2.40 | 2.29 | 2.76 | 2.60 |

*Source:* World Resource Institute (1993)

Even though the South has most of the remaining forests and biodiversity, and has produced less damage to the global environment, it is more vulnerable to the ill effects of environmental damage, such as climate change, on its food production, living conditions, and rising of the sea level. The North therefore creates the most risks, but the South will bear the brunt of the resulting damage. The origins of today's environmental dilemmas involve the historical coupling of two different worlds through the international market.

## 3. PROPERTY REGIMES, MARKETS, AND GLOBAL ENVIRONMENTAL PROBLEMS

*3.1 North–South Trade and the Property Regimes of Natural Resources*

What explains the pattern of North-South trade in which the developing countries are the main exporters, of fuels and natural resources, to the OECD countries? One possibility is that there is a geographic coincidence, in which the developing nations are rich in natural resources. This explanation would view the pattern of world trade simply as a manifestation of countries' respective comparative advantages—as traditional theory of international trade would predict. Countries like Mexico and Ecuador exporting oil to the USA contradict this view. Playing a substantial role in this pattern of North–South trade is a historical difference between agricultural and industrial societies, a difference in the property rights regimes of natural resources which prevail in these two types of nations. In developing nations natural resources are typically held as state or communal property, for example oil deposits and forests in many countries, such as Mexico, Nigeria, and the Arab States, are often government property. In many cases, due to ill-definition of property rights, high-transaction costs of enforcement of well-defined property rights, the lack of financial resources to enforce property rights, physical properties of natural resources, high dependence of local communities on natural resources, and non-complementarity between social norms and property rights, the state regimes become de-facto closer to open-access regimes (Kant, 2000). On other hand, well defined and enforceable property regimes, which are generally private regimes but also state or communal regimes such as forest property regimes in Canada, Italy, and even the USA (about one-third forests are owned by the government), are the dominant category of natural resource property regimes in the developed countries. This difference in property regimes has been shown to lead, through the international market, to a pattern of trade such as the one we observe between the North and the South, in which the latter export resources to the North even though they may not be resource-rich—the industrial countries may be richer in resources themselves (Chichilnisky, 1994).

In a world where agricultural societies trade with industrial societies, international markets can magnify the 'tragedy of ill-defined and non-enforceable rights'—the over-extraction of natural resources that typically occurs under open access regimes. The resulting outputs (wood, cash crops, and livestock) are mostly sold in international markets (Barbier in Chichinisky, 1994). Both these natural resource exports and world's use of natural resources exceed what would be optimal or would occur if property rights were well-defined and were enforceable, the conditions which are generally equated with private property rights but can also exist under state or community regimes, and the attendant prices in the global markets are also below what would prevail with well-defined and enforceable rights (private property rights). International markets—even if they work competitively— fail to produce an optimal solution. International trade is therefore skewed, leading to resource exports from countries that do not have a true comparative advantage in resources—and resource imports in countries that do. The historical coupling of the North and the South through the international market leads directly to over extraction of resources in the world, to resources prices that are lower than replacement costs, and to over-consumption of these resources in the industrial countries that import them.

Figure 7.1 contrasts two different supply curves for resources in a domestic economy of the South and illustrates the problem of over-extraction and under-pricing of resources. The steeper supply curve is based on efficient supply behavior in (well defined and enforceable property regimes) private property economies. The price corresponding to each quantity supplied equals the marginal cost of extraction, ensuring a Pareto efficient solution. However, when property rights are not well defined and enforceable, some elements of open-access regime are present, the

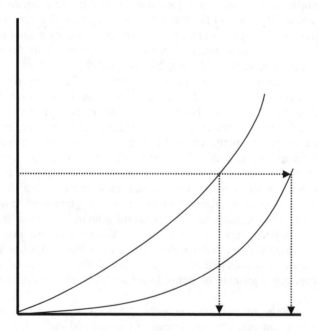

***Figure 7.1.*** *Supply Curves for Private Property Regime and a Property Regime with Ill-defined and Weakly-Enforceable Property Rights*

supply curve of the resource is 'flatter' than the private property supply curve (see Chichilnisky, 1994, 1996a)[2]; at each price the country will supply more resources than it would under private property regimes, with the result that more will be exported than is optimal and that exports will be sold at lower than appropriate prices. Through this process, resource intensive trade leads to an increasingly divided North-South world.

## 3.2 Property Rights and the Atmosphere as a Sink

The problematic North-South trade patterns just discussed could be improved by improving the property regimes of natural resources in developing countries. History suggests that in any case, this would probably occur naturally in those countries that are undergoing a transformation from agricultural to industrial societies. However, privatizing of natural resources in developing countries may be impractical, due to various social and cultural factors, in a reasonable time frame. The world is trying to find a short-term solution to the overuse of natural resources now in order to prevent biodiversity destruction and climate change—both of which are potentially catastrophic and irreversible events.

Rather than privatizing on the input side, it would be possible to privatize on the output side, i.e. to privatize the use of the global atmosphere rather than privatizing the developing countries' use of natural resources. The lack of well defined and enforceable property rights (private property regimes) in natural resources, which are inputs to production, leads to the overuse of the planet's atmosphere, that is the "sink" in which the outputs are deposited. Over-consumption of petroleum as an input leads, for example, to the overuse of the atmosphere as a "sink" for the greenhouse gases that are part of the output. The planet's atmosphere is held as "open access regime" in the entire world. One would expect somewhat less conflict in the process of allocating property rights to the use of the atmosphere, simply because these are property rights that have not yet been defined so the problem is still in a more fluid state. This is in fact what happened in the 1997 Kyoto Protocol, which limited the rights to emit green house gases of Annex B countries—the industrial nations. The Kyoto Protocol is an international attempt to determine various countries' property rights to the use of the atmosphere as a 'sink' for greenhouse gases associated with burning of fossil fuels and other industrial activity. In September 2004, Russia expressed its decision to ratify the Kyoto Protocol which, under the provisions agreed in 1997 by 166 nations, has now, in 2005, become international law.

## 3.3 Global Emissions Markets, Efficiency, and Equity

Assigning property rights in the use of the planet's atmosphere was a first step. The Kyoto Protocol goes further, offering also a first step in the creation of global markets for trading such rights, following our suggestion for a global market for emissions trading in Chichilnisky (1996b). These 'global emission markets,' are a historical first. Emission markets by themselves are not new though they are still unusual—they have a short but successful history. In the US where the Chicago Board of Trade introduced tradable permits to emit $SO_2$ they have been deemed very successful and cost effective in the reduction in emissions of sulphur dioxide by power plants in the US.[3] But the Kyoto Protocol offered the first opportunity to trade a global public good—the use of the planet's atmosphere—by trading the rights to emit greenhouse gases.

Once global emissions markets are created, the next challenge is to ensure that they be efficient. Successful markets require good regulation—the best markets in the world are regulated, not to restrict trade but to ensure healthy competitive conditions. For example, the Securities Exchange Commission (SEC) in the US is active in promoting the sharing of information in securities markets and penalizes 'insider trading' in which asymmetric information exists. Efficiency of emissions

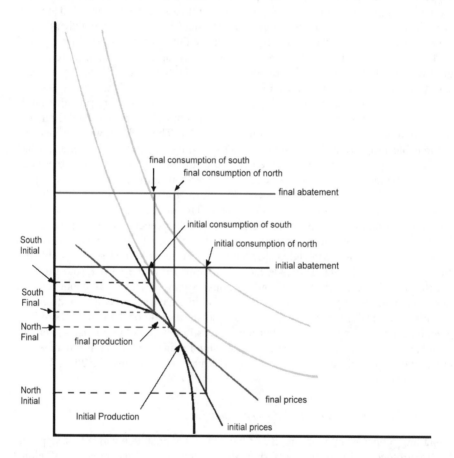

***Figure 7.2.*** *Redistributing initial property rights on emissions yields a net Pareto improvement for both the North and the South because the vertical axis represents a public good*

markets however requires different conditions than efficiency of standard markets for private goods. New economic findings establish that there is a deep connection between the distribution of property rights (rights to emit) and the efficient performance of markets with privately produced public goods—such as the use of the planet's atmosphere (cf. G. Chichilnisky and G. Heal 2000).

Efficiency in trading permits requires that more emission rights be given to the developing countries (i.e. than they would have under an auction system where such rights had to be purchased in the market, Figure7.2)—just as provided in the Kyoto Protocol. Indeed, the Protocol places no constraints on the emissions right of developing nations—all its restrictions are on Annex B countries which are industrialized. Therefore it implicitly provides more emission rights to the developing countries. But what is the connection between efficiency in emissions markets and the emission rights given to developing countries? Conventional wisdom has been that the distribution of property rights does not affect the efficiency of markets. Standard economic thinking is that equity and efficiency are independent of each other in competitive markets and indeed often orthogonal to each other as well. What makes this situation different?

The so called "Coase theorem" has shown that equity and efficiency are unrelated in those markets with private goods that have the capacity to internalize externalities. This means that where externalities exist, assigning property rights and allowing them to be traded leads to Pareto efficient solutions no matter what is the initial distribution of property rights. The textbook case is the property rights to pollute that are assigned to a factory producing 'soot' that interferes with a laundry's capacity to produce clean clothes. The externality here is the 'soot'. Soot is a private bad – in the sense that it is 'rival in consumption—the soot deposited in one site does not deposit in others. One compares the rights of the factory to emit soot to the rights to clean air of the laundry itself. Coase showed that at the end of the day it does not matter who gets the rights to pollute or to breathe clean air; as long as property rights are assigned clearly and the parties are allowed to trade them, the market solution will be Pareto efficient. Of course, the assignation of rights does affect the welfare of each of the traders and therefore the equity of the situation, but it does not affect the efficiency of the market solution. Why does this widely accepted result not apply to our case? Why is equity in the assignment of carbon emission rights connected with the efficiency of markets?

Global emission markets for $CO_2$ are different, because they involve a global public good, namely the quality of the atmosphere of the planet as measured by its concentration of $CO_2$, as shown Figure 7.2 and in Chichilnisky and Heal (2000). In Coase's case, the initial distribution of rights does not matter because he considers markets involving private goods, goods that are 'rival' in consumption, such as soot. The soot that the factory deposits on the laundry's clean clothes is 'rival'—whatever soot is deposited on one shirt, it is not deposited on another shirt. Different people receive different amounts of soot and can bargain for their rights. The situation however is different in the case of $CO_2$, which spreads very evenly and stably throughout the entire planet's atmosphere, requiring 60 years to decay. These are physical properties of carbon dioxide, which do not depend on social organization. They make carbon dioxide concentration a global public good: the result is that everyone in the planet is exposed to the same concentration of $CO_2$, in China as well as in South America, Europe or Australia The concentration of carbon in the atmosphere is the same for all and therefore it is a global public good. And markets that trade the rights to emit carbon are therefore markets trading a global public good. Carbon dioxide is in addition a different kind of public good (or "public bad") from the traditional category of public goods produced by governments, such as roads and law. Carbon dioxide is produced by individuals as a by-product of private activities such as heating one's home or driving one's car. Trading of carbon emission rights exemplifies a market for privately produced public goods, an unusual market, of a type that economists are not used to. Such markets are increasingly important, however, because they include also the trading of knowledge rights. Like carbon concentration, knowledge is a privately produced public good and one that is fast becoming the most important input of production in advanced societies. Markets with privately produced public goods are new and different but should not be considered exotic. They are possibly the most important type of markets in the new century.

Market efficiency in the case of privately produced public goods requires an additional condition which alters fundamentally Coase's conclusions; this is the Lindahl, Bowen, and Samuelson condition whereby the marginal rate of transformation equals the sum of the marginal rates of substitution among the traders. In effect this means that the marginal private benefit to the firm or individual from the activity which causes an emission is equal to the sum of the marginal private costs (damage) to all those who are negatively affected by it[4]. This latter condition derives from the 'non rivalness in consumption' characteristic of public goods—at the end everyone consumes the same amount of the public good. In the present case, everyone in the world is exposed to the same concentration of carbon dioxide in the atmosphere.

This additional condition required for efficiency 'over-determines.' the market equilibrium. Therefore while market solutions exist, they are not efficient in general. New policy tools are required to reach and implement efficient market solutions. It turns out that the distribution of property rights across nations on the global public good, is the right tool and has the right dimensionality to solve this problem. Distributing properly these initial rights to emit allows one to reach solutions that clear the markets and are, simultaneously, efficient in the use of the global public good. This, in a nutshell, explains the tight relation between efficiency and equity in markets for global public goods.

Equity is an important consideration for developing nations in the climate negotiations. Industrial countries have emphasized, instead, market mechanisms and economic efficiency as their own priority. The unexpected connection that we discussed between equity and efficiency may therefore provide a way to reconcile the priorities and interests of the North and the South. Since North-South conflicts of interests have led to debate and delays in ratifying and implementing the Protocol, an overlap in interests of the North and of the South is welcome. However, the connection between equity and efficiency that emerges here is new in economic terms, and it is not still completely understood. More economic work remains to be done, academic as well as diplomatic and political. Properly interpreted and implemented, however, the Kyoto Protocol may signal the way to a sustainable future.

## 4. CONCLUSIONS

The origin of today's global environmental problems is the historic difference in property rights of natural resources in industrial and developing countries, the North and the South. The lack of well defined and strongly enforceable property rights leads to the over-extraction of natural resources in the South. These resources are exported at low prices to the North that over-consumes them. The international market amplifies the problem of "the global commons", leading to inferior solutions for the world economy (Chichilnisky 1994). Updating property rights on natural resources faces formidable opposition. However, the lack of property rights in *inputs* to production could be compensated by assigning property rights on *outputs*. The 1997 Kyoto Protocol is a right step in this direction. Yet trading emissions rights is tricky, because the quality of the atmosphere is a *public good*. Global emissions markets are therefore different from the market of private goods. A measure of equity is needed to ensure efficient trading in these types of markets (Chichilnisky 1996b, 2000), and is fortunately built into the Kyoto Protocol.

In addition to carbon sequestration, biodiversity and watershed services are among the other most valuable services that forests provide. Here again, assigning property rights to localities or nations on the use of genetic blueprints that are obtained from their forests would be a step in the right direction. Biodiversity, as any other form of knowledge, is a public good and therefore the observations made above about the properties of markets with privately produced public goods will apply (Chichilnisky & Heal, 2000).

New institutions are needed at the global level to implement these solutions. In Chichilnisky (1996b), I proposed the creation of an *International Bank for Environmental Settlements*, a self-funding institution that would help administer the rights to the global public goods such as the assignment of emissions rights on a global scale. This institution would have as mandate to derive economic value from the environment—such as economic value carbon sequestration services, and genetic blueprints from developing nations' forests—without destroying them.

## NOTES

[1] Most of the industrial countries are located in the Northern hemisphere, and are therefore often referred to as "the North".

[2] Using a game-theoretic approach introduced in Dasgupta and Heal (1979) to explain the Nash equilibrium.

[3] Other examples of environmental markets are mentioned in Chichilnisky (1996) and in Chichilnisky and Heal (2000).

[4] Generally, marginal private benefits will decrease with increasing emissions, at the firm level or country level, and therefore, marginal private benefits from incremental emissions will be higher to the producers in developing countries (low emission countries). Hence, equity in emission rights (increase in emission rights to developing countries (low emission countries at present) and decrease in emission rights to developed countries (high emission countries)) will move the global emission markets towards efficiency.

## REFERENCES

Chichilnisky, G. (1977). Development patterns and the international order. *Journal of International Affairs*, 31(2), 275-304.

Chichilnisky, G. (1994). North-south trade and the global environment. *American Economic Review*, 84(4), 851-874.

Chichilnisky, G. (1996). Property rights and the dyna mics of north south trade. In M. Bredahl, N. Ballenger, J. Dunmore, & T. Roe (Eds.) *Agriculture, trade, and the environment: Discovering and measuring the critical linkages* (pp.97-110). Colorado and Oxford: Harper Collins Publishers.

Chichilnisky, G. (1996). *Development and global finance: the case for an International Bank of Environmental Settlements (IBES)*. NDP and NESCO, Office of Development Studies, Discussion Paper No. 10, September 1996Available at http://www.chichilnisky.com

Chichilnisky, G. (1996). *The greening of the Bretton Woods* . Financial Times, January 10, 1996 p. 8. Available at http://www.chichilnisky.com

Chichilnisky, G. (1997). What is sustainable development? *Land Economics*, 73(4), 46-491.

Chichilnisky, G. (1999). *The Kyoto Protocol and the global carbon cycle* Pegram Lectures, Brookhaven National Laboratories, available at http://www.chichilnisky.com.

Chichilnisky, G. (2000). Eqity and efficiency in globa l emissions markets. In R. ReverszP. Sands, & R. Steward (Eds.), *Environmental law, the economy, and sustainable development* (pp. 26-279). Cambridge: Cambridge hiversity Press.

Chichilnisky, G. and G.M. Heal. (Eds.). (2000) *Environmental markets: Equity and efficiency.* New York: Columbia hiversity Press.

Dasgupta, P. & Heal, G. M. (1979). *Economic theory and exhaustible resources.* Cambridge: Cambridge hiversity Press.

Kant, S. (2000). A dynamic approach to forest regimes in developing countries. *Ecological Economics*, 32, 287-300.

World Commission on Environment and Development (WCED). (1987). *Our common future.* Oxford: Oxford hiversity Press.

World Resources Institute (WRI). (1990). *World Resources, 1990-1991.* (WRI in Collaboration with the NEP and NDP). New York: Oxford hiversity Press.

World Resource Institute (WRI). (1993).*The 1993 information please environmental almanac.* Boston, MA: Houghton Mifflin.

# CHAPTER 8

# DEFORESTATION AND POPULATION INCREASE

JOHN M. HARTWICK

*Department of Economics, Queen's University*
*Kingston, Ontario K7L 3N6, Canada*
*Email: hartwick@qed.econ.queensu.ca*

**Abstract.** We reflect on the reciprocal relationship between population growth and deforestation. In human history there must have been long intervals when, in contrast to a Malthusian scenario, land clearing was low-cost and led subsequent population growth. Trade and migration have taken the bite out of local land scarcity. We explore in theory and in simulations. An extended Hartwick-Long-Tian model relates deforestation to per capita income and relative prices for land in agriculture and in forestry. We report on land-use change since 1700 with the recent HYDE database.

> "... humankind resembles an acute epidemic disease, whose occasional lapses into less virulent forms of behavior have never yet sufficed to permit any really stable, chronic relationship to establish itself" with other organisms. (McNeill, 1976, p. 20)

## 1. INTRODUCTION

There is a reciprocal relationship between population and forest cover, in the large. Even if one accepts the hypothesis that deforestation is a consequence of population growth - more people, more food required, more land for cultivation and grazing[1]- there are still many subtle questions concerning the pace of deforestation relative to population growth, particularly about how land clearing for food production feeds back into population growth.[2] In some cases, it may make more sense to say that deforestation has been a principle ingredient of population growth than to say the reverse.[3] After all, migration and deforestation provide relief from local Malthusian pressures.[4] Far from being costly, deforestation often pays for itself in fuel, construction material, and timber sales[5] besides providing new land for agriculture.[6]

We are interested in the question of whether land-clearing might precede, in some sense, population growth. Implicit in the Malthusian view of the world is the idea that deforestation cannot be done rapidly enough to forestall a decline in arable land per capita: the creation of arable land is very costly and population growth will drive per capita consumption to subsistence, even if deforestation is an option. One senses the opposite to have been the case in early periods; arable land could be

*Kant and Berry (Eds.), Institutions, Sustainability, and Natural Resources: Institutions for Sustainable Forest Management*, 155-191.

obtained very cheaply by slashing and burning local forest cover, as we have seen in many parts of the world in recent decades.

Presumably the earliest large-scale clearing of forests occurred in the Golden Triangle in the mid-east some eight to twelve thousand years ago. Little is known about this phenomenon. And then there would have been large-scale clearing around major rivers in China followed by pressure on forested land around the Mediterranean shores. Again data on population and cleared acreage are few and are unreliable. Erosion, following deforestation, is written about by observers in ancient Greece. Deforestation in what is modern New Zealand about one thousand years ago is written about with some authority. Similarly for the case of the tiny Easter Island, some two thousand miles off the coast of Chile. Evidence of large-scale burning has turned up for these two cases. Then we have fragmentary evidence for large-scale deforestation in England before Elizabeth I, about 1600. Laws were enacted to conserve standing timber during Elizabeth's reign. And there is evidence of excessive clearing in France some decades later. Poland and Russia became timber suppliers to western Europe after about 1790. Hence considerable deforestation in eastern Europe is post 1800. Similarly for Canada and the United States. Haiti gets devastated by deforestation during the nineteenth and twentieth centuries. Its striking poverty is not unconnected to the deplorable state of its rural land, denuded of healthy forests. We touch on these cases below. Not surprisingly we would like to do economics with good data and we attempt to move to where solid information is available. We report below on population and land-use changes since 1750 with the recently developed database in the Netherlands. This is a case of "good news", here.

Our view, derived from Pomerantz (2000), is that England, France, Denmark and eastern China were close to a Malthusian state in the early eighteenth century.[7] There was large-scale deforestation and very slow population growth, with poor life expectancy. Some combination of the industrial revolution (technical and organizational innovation), urbanization, migrations abroad, and improved international trade resulted in rises in per capita incomes by 1850 and a noticeable decline in fertility about this time. The modern era arrived. Between 10,000 BC and 1750 AD a largely farming-based system of living swept over much of the earth and yielded Malthusian states in many places. Obviously there was population growth in many regions, more rapid than the clearing of land for new farms. Population growth at 0.1% per year will double population in about 700 years. 10,000/700 allows for some 14 doublings up to say 1750. One estimate has the world with a population of 300 million about 1AD. At again 0.1% net increase, we could expect world population to be well over a billion by 1750. It seems reasonable to refer to such slow population growth as essentially Malthusian. And temperate valleys and plains did get filled up while large-scale land-clearing took place. We have to consider the agricultural revolution as curiously retrograde, supportive of relatively large populations, some in towns and cities, but unable to improve life expectancy or per capita income (Steckel & Rose, 2002; Robson, 2003).

North America provides an interesting contrast. Here European immigrants virtually over-ran the land within about a hundred years. It is not hard to argue that the windfall of marketable timber in North America in the nineteenth century contributed significantly to the high per capita incomes enjoyed by the many immigrants. Free timber and farmland was to nineteenth century North Americans what free oil has been to many in the twentieth century. We take up a model of forest clearing in North America below. We note that in most situations, the possibility of trade and migration has taken the bite out of local resource scarcity. In western nations, the last centuries have been an era of population growth and deforestation caused by massive growth in per capita incomes, growth fuelled by factors other than timber sales and land clearing. Demand for timber presumably turns more on income currently than on population. We will argue that sustainable forestry is possible even in a situation of growing per capita income, though sustained population growth is not. In isolated areas, where trade and migration are constrained, the dependence of population maintenance and growth on deforestation is more clearly read. Models of simple forest-based economies show that social cohesion, property rights, and costs of harvesting are critical in determining the possibility of sustainable forestry.

## 2. DEFORESTATION WITH LAND RELATIVELY ABUNDANT

The claim that our "primitive" ancestors were prudent harvesters seems not to be supported by the archeological evidence. Anthropologists (and Vernon Smith, 1975) claim that inhabitants of the North American Plains "over-harvested" the mega-fauna to extinction. Low costs of harvesting seem here to have elicited over-shooting of appropriate harvest levels. Regarding deforestation, Malin (1956), cited by Clark (2000; p. 187), Krech (1999) and others contend that by the time Europeans arrived in North America,[8] plains Indians had perpetrated much deforestation with burning and were not in harmonious balance with the natural resource base.[9] The Maori are said to have stripped the islands of New Zealand of indigenous pines for agricultural land when they arrived in New Zealand about one thousand years ago (Clark, 2000, p. 208).

If abundant agricultural land can be obtained at very low cost by clearing forest, it makes sense for natives to acquire it and to pursue farming with its high land-to-labor ratio.[10] This is not necessarily a case of population pressure leading to deforestation, but one of specialization in production using the relatively more abundant factor.

In the early post-neolithic period, clearing forest areas would have provided people with wood for construction and fuel, and more, perhaps more fertile, land to cultivate.[11] With no land scarcity, the living standard, including family size and survival rates, would be determined by how hard people were willing to work the land. New families would appear "at the periphery" as people migrated to new places of lower density relative to the local carrying capacity in order to achieve the living standards their parents were used to. One might say that there was continual pressure by parents on some of the children to migrate to the hinterland.

This raises the important question of what determined the standard of living which was being replicated in new places as children moved to the frontier to establish themselves. Did consumption have to fall to subsistence levels before young people were forced to migrate to new, probably less desirable and less fertile, areas? A Malthusian equilibrium would be one with all arable land "used up" and only two children on average living to an age for reproduction. The calorie content of the harvest per family plot would be just sufficient to sustain two people on average. This leaves open the question of how parents survived while raising their surviving children.

It is fair to say that a Malthusian steady state was never reached over large areas. For hundreds of years, population growth was almost imperceptible but it was not restrained by a calorie constraint.[12] Rather, a combination of malnutrition, disease and warfare of various kinds kept population virtually constant over many hundreds of years. (In place of the term Malthusian equilibrium, we should perhaps use the term a malwardi (malnutrition, warfare, and disease) equilibrium.) One would expect a saw-tooth pattern in the long-run time profile of population, with population rising regularly above the trend line and being trend-reverted by the effects of war and disease.[13] Whether or not there was extensive deforestation in prehistory, the pressure of population is not likely to have been the reason.

## 3. THE DEVELOPMENT OF CITIES

The historical outline we have sketched so far begins with the migration of humans out of Africa about 60,000 years ago. Despite conflict, natural hazards, and disease, population increased fast enough to fill all the habitable niches on earth, presumably to a malwardi steady state.[14] Fertility must have been at a maximum. Settled crop-cultivation is radically different from hunting and gathering, and the Neolithic Revolution (circa 10,000 BC) represents a huge innovation.[15] Agricultural surpluses from farms allowed an urban class of persons to emerge[16] and this change must have resulted in dramatic shifts in political, social and economic organization. A population jump around 8000 BC might be attributed to the regularization of family life, brought on by the spread of farming in place of ubiquitous hunting and gathering. Still, the rate of increase in the overall population would have remained very, very slow on average.

There was probably no great surge in population because, although citizens higher in the new social hierarchy would be better fed[17] and live somewhat longer lives on average, the higher density of settlement in towns and cities should have allowed disease to spread rapidly locally. Town dwellers could expect more disease on average over a lifetime than rural folk and hunter gatherers, and in the long run, townsfolk would end up with better immune systems as the weaker among them succumbed.[18] Similarly, when Europeans arrived in the New World in 1492, they were at an advantage in terms of immunity because, they came from a more urban and disease-intensive lifestyle, so to speak.[19]

As humankind spread to every continent and island, those peoples that found themselves isolated in relatively small areas would be the first to feel the bite of Malthus.

## 4. WHEN LAND SCARCITY BITES: EASTER ISLAND

When trade and migration are not possible, the full force of resource-scarcity is brought to bear on consumption levels and population growth. This is often presented as a possibility for the world economy sometime in the future. No doubt it has played out several times in human history, most graphically in the case of Easter Island.

Easter Island is of particular interest because it experienced a population boom after 700 AD and then a crash following extensive deforestation and steady ecological deterioration. Archeological probes suggest that warfare and social disintegration coincided with the population's decline. That a crisis became so acute over a relatively short period of time is not surprising when we recognize that the Island is only about eight miles long. There was no frontier to migrate to when the natural base (the carrying capacity) for the population collapsed. The Island is also relatively isolated, with its nearest potential trading partner 2000 miles away.

A revised model of Brander and Taylor, discussed next, is able to predict the collapse of Easter Island, but only if the cost of harvesting or the yield is sufficiently sensitive to stock size per capita. The addition of an extra equation meant to describe increasing social disorder is not enough to seal the fate of the model economy, but points the way to its demise. Consideration of property rights and fertility also contributes to an explanation of how the population of Easter Island might have collapsed.

### 4.1 The Model of Brander and Taylor

Brander and Taylor (1997) have constructed a dynamic model[20] with two states, population and the "carrying capacity", to simulate the history of Easter Island. The central idea is that population increase is a response to current nutritional standards; there is a lack of foresight or anticipation of future bad times in "family planning". Population collapse ultimately occurs because "people growth" outstrips "environment re-growth" as in timber renewal over long periods of time. The link between" people growth" and the state of the environment is "mediated by" the harvest function. Our first critique of the Brander-Taylor model turns on an alternative harvest technology, one associated with higher harvesting costs with a denser population on the island. Recall that Easter Island is less than 66 square miles in area; less than 9 miles in diameter. Polynesian settlers arrived about 400-700 AD and the population grew steadily to about 7000 until apparent environmental stress caused the level to decline. Slash and burn agriculture resulted in soil erosion and the depletion of timber led to a cessation of fishing activity. A primary activity was statue carving and erection. Estimates range as high as 1000 units, the last lying in the quarry, too large to move.[21] Timbers would have been essential to moving the statues to their pedestals around the island. There may have been inter-clan warfare at various times but when the first European arrived in 1722, he and others after him found a well-functioning, farming society of more than 4500 people. At 1% per year, the population decline from 7000 would occur in 45 years. Presumably warfare would be needed to effect this dramatic change. At 0.1% the decline would take 450 years. Such demographic change would surely not stress the social fabric of the islanders. The final destruction of the society appears to have resulted from diseases brought by European whalers and slavers in the nineteenth century. Brander and Taylor's demographic, island-ecology model is intended to capture the original run up in population to 7000 and the subsequent decline to about 50 people in the 1880's. What they should be modeling is the run-up to 7000 and the decline to 4500. The subsequent decline appears to be due to factors noticeably distinct from the earlier decline.

The Brander-Taylor model of the history of Easter Island has a materials balance

$$\dot{S} = \phi(S) - hN$$

where $S$ is "the environment" as in the stock of harvestable fish, fruit and nuts, and trees, $h$ is per capita harvest and $N$ the population. The cost of obtaining $h$ is specified as $\lambda S$, making current total harvest, $\lambda SN$. Per capita harvest is linearly related to the size of the environment, $S$. A convenient form for natural growth of the environment is the S-shaped logistic function which reduces to $\phi(S)$ as $\dot{S} = S\alpha[1 - S/K]$ for $K$ the carrying capacity for "the environment". Here then "the environment" which is harvested grows on a base or substrate or carrying capacity. $\alpha$ is the growth rate of the environment when the substrate is hugely abundant relative to the current size, $S$.

The natural companion equation to the environment equation, $\dot{S} = \phi(S) - hN$ is the well-known population growth equation

$$\frac{\dot{N}}{N} = \gamma[h - \bar{c}]$$

where $\bar{c}$ is the subsistence consumption level. The assumption is being made that population growth depends on living standards. Brander and Taylor go with this specification and end up with $\dot{N} = \gamma\lambda NS - \gamma\bar{c}N$ and $\dot{S} = \phi(S) - \lambda SN$. Their convergent spiral of $S$ and $N$ over "history" mimics the predator-prey model of Lotka and Volterra, except the Brander-Taylor cycle converges. In Easter Island, the people are the Lotka-Volterra foxes and the environment $S$ is a stock of hares, predated upon by the foxes.

Though Brander and Taylor only extract a branch for a time period of about 1000 years from their spiral, their model will eventually converge, in cycles, to a point with a steady population and environment, a sustainable outcome. The model does not exhibit collapse to zero population and an exhausted environment. We turn to some critiques of the Brander-Taylor formulation.

More plausible is a per capita "cost of harvesting" equation, like

$$h = \chi(S/N)$$

with $\chi(.)$ increasing in $S/N$. A person does not harvest much when the current environment in under stress by a large population. Hence it becomes harder for a single harvester to "deplete" the environment when population is large relative to the current "size" of the environment. The analogy to fishing is as follows: it is difficult for a fisher to find any catch when stocks are low, measured by $S/N$. Per capita harvest varies with stock size $S$ and with competition from fellow "citizens", $N$, presumably via search and extraction costs.

We select the specific form

$$h = \beta \left( \frac{S}{N} \right)^{\eta} \text{ for } 0 < \eta < 1 \text{ and } \beta > 0.$$

In this model, population will grow when $h$ is "large" or when the harvestable environment, S, is large relative to the population. In contrast, a population that is large relative to resource stocks means less harvest, more costly harvesting for each individual and, ultimately, negative population growth once *ex post* harvests per capita fall below subsistence.

When we did numerical simulations of this revised Bander-Taylor model, we observed the dynamics to be sensitive to both $\alpha$ in $\dot{S} = S\alpha[1 - S/K] - hN$ and to $\eta$ in the harvest equation. We used initial conditions $K = 50, \alpha = 0.2, \gamma = 1.2, \eta = 0.7, \beta = 0.2$, and $\bar{c} = 1.2$. With a small initial population, ($N(0) = 0.001$) and a large initial stock for the environment, ($S(0) = 49.5$), we observed per capita harvest rise rapidly to 3.5 and then slowly decline to about 2.2 as $S(t)$ declined to a steady state value, close to zero. When "the productiveness" of the environment was less ($\alpha$ down to 0.02 from 0.2) we observed the same rapid rise in per capita harvest but then a fairly sharp decline as the environment headed toward zero. This is the classic case of population growing too rapidly relative to the harvestable stock. The environment is mined while population grows rapidly because harvesting is relatively easy. Trees (the environment) are not able to replenish themselves fast enough to keep up with demand for wood from a larger population. This might be termed a drastic Brander-Taylor scenario.

However, when the harvest parameter, $\eta$, was decreased from 0.7 to 0.3, we observed per capita harvest to rise rapidly to a low value (about 0.16 compared with 3.5 above) and then fall slowly to a steady state value at about 0.14. **With per capita harvest activity less productive, less responsive to increases in $\frac{S}{N}$, a high "cost of harvesting" saves the environment from collapse.** The dynamics in this case depart considerably from those in Brander and Taylor (1997). This case might be labeled the case of the very productive environment. For millenia, hunter-gatherers in say the Amazon region showed no sign of "depleting" the environment. No technology at their disposal would result in excessive harvesting. It has been pointed out that the climate at Easter Island is not tropical and slash and burn agriculture effectively exhausted the place of timber.

### 4.2 Social Cohesion

A useful addition to our model and the original Brander-Taylor formulation is *a social response to environmental stress*, such as theft and violence that might occur when canoes and other essentials cannot be acquired by harvesting trees.[22] The modern view of social decay is that peaceful people enjoying prosperity see increased crimes against property as well as muggings and are obliged to install more locks, alarms, and protective fences and also support more police and guards. A fraction of formerly "productive workers" are retrained as policemen and women, and as fabricators of "defensive" devices such as locks, etc. Formally we model this as a decline in the productiveness of harvesting activity because one is always on guard against crime. "Unproductive" environments can lead to declines in per capita harvesting.

We define environmental stress in terms of the ratio $\frac{S}{N}$. When stress is severe as in a low value of $\frac{S}{N}$, we posit that social disintegration, such as widespread banditry, occurs. This shows up in declining per capita harvest.[23] Formally, we assume

$$\dot{h} = \lambda[\beta\left(\frac{S}{N}\right)^{\eta} - \bar{s}]$$

for $\lambda$, $\beta$, and $\eta$ positive. $\bar{s}$ becomes an index number which relates environmental stress to social disintegration. An abundant environment (large $\frac{S}{N}$) is associated with low cost per capita harvesting, and per capita harvest is increasing. This upside should be non-linear as for example, small increases in $h$ when the environment is more abundant. A stressed environment (low stock to population) has high per capita costs of harvesting and per capita harvesting is declining. Violence on Easter Island increases with a low value of $\frac{S}{N}$, with say "over-population", and this violence shows up in a **declining** per capita harvest value.[24]

We now make the traditional link between low per capita consumption levels and a declining population. Inadequate caloric intake per person shows up in a decline in the population. That is

$$\dot{N}/N = \gamma[h - \bar{c}].$$

The model has three dynamic equations: one for stocks, $S$, for per capita harvest, $h$, and for population, $N$. Now many and more complicated histories that can emerge, given initial conditions, than was the case with only two dynamic equations.

A benchmark case for this model is a perpetual, unchanging cycle.

$\dot{h} = 0.007 * h * [0.9 * \{\frac{S}{N}\}^\wedge .06 - 1.1]$, and $\dot{N} = N * 0.2 * [h - 3.58]$. We were able to get our cycle to converge or diverge by decreasing or increasing our value of $\alpha = 1.29$. We were able to locate this cycle by trial and error on our computer. In this case, declining $h$ precedes the peaking of $N$ , that is, the rise of banditry precedes population decline. Abundant $S$ is associated with a rising $h$, and *vice versa*. The dynamics of the $N$ and $S$ pairing is a highly complementary cycle with a "large" value of $S$ matched by a "small" value of $N$ at each date, with no leading or lagging.

Corresponding to the arrival of a very small population on Easter Island, one has a large environmental stock, $S$ and a rising per capita harvest, $h$. The rising population precipitates a decline in $S$ and later, a switch to a negative $\dot{h}$ which in turn precedes the collapse in the population. Thus it is not only scarcity of the environment that leads to food shortages and population decline, but food shortages precipitate banditry which makes acquiring food "even more" difficult. Over the cycle, the population bottoms out at its initial low value, and this permits $S$ to renew itself. There is then not only plain food scarcity causing population decline here, but food scarcity contributing to more food scarcity via higher costs of "harvesting". It is the arrival of banditry which makes harvesting activity become less productive. One might say that social disintegration amplifies the effect of environmental scarcity on population decline.

The coincidence of poverty and social disorder has been observed but the coincidence of poverty and "environmental scarcity" is less well documented. Homer-Dixon (1999) for example exhibits noticeable circumspection in linking known episodes of social dislocation to strict natural resource scarcity. Newly impoverished people strive to migrate in order to fend off further decline. Haiti is an example of a somewhat isolated nation with pervasive poverty and conspicuous deforestation and soil erosion. Governments there have maintained order by force and intimidation but have not been able to build up physical and social infrastructure.

## 4.3 Property Rights

Hartwick and Yeung (1997) have a model in which lack of property rights to natural resources leads families or tribes to have "too many" children in the steady state, which leads in turn to population pressures and lower per capita consumption than when rights to resources are secure and tied to each family. Individuals want more children of their type as well as more food per person. On a given territory, more children can lead to less food per person. But one can contemplate having more children than the next family and then establishing a right to more land at the margin. In that model, when generalized squatting is the means of securing land, a family claims farmland by essentially "sending out" more children at the prevailing standard of consumption, and at the margin, each family "sends out" "excess" children and ends up with a "low" standard of consumption in equilibrium. It is the common property aspect of land that leads to an equilibrium with each clan having "excess" offspring at the margin and "deficient" per capita consumption at the margin. If clan ownership of land were secure, then the mutual excess population equilibrium would not occur. Alternatively if all clans agreed together on population policy and abided by the agreed policy, then the excess offspring equilibrium would not occur. This is a somewhat roundabout way of saying that population policy is myopic but fundamentally in this case it is the common property in land aspect that causes the myopic population policy to work its way out in "excess competition" for land at the margin, via "excess" offspring.

There is no possibility of migration in this model. With regard to Easter Island, there seems to have been competition among clans for resources and as the population grew any "extra" growth by a clan may have been a mechanism for appropriating resources, at the margin. This could of course lead to general "over-population" and, with explicit environmental decay, to subsequent population retrenchment. The run-up in population to 7000 from the original founders and its subsequent decline may have been quite gradual and seeds for any violence could have been set down from other causes. Early European visitors reported on the good health of the citizens, when they made brief stops. One is of course left wondering about the effects of inbreeding on the health of the people.

### 4.4 Social Disorder among the Maya

The collapse of the Mayan system of cities resembles an Easter Island. Some scholars argue that excessive irrigation salinized the soil and led to a productivity collapse. But Burroughs (1997, pp. 21-22) points out that Maya country was hit with a dry spell in about 750 AD which lasted 150 years, and sediments from Lake Chichancanab suggest that the dryness was more severe than any other spell in the last 8000 years.[25] This must have de-stabilized the social order leading possibly local revolts of peasants against the oppressive rulers and certainly to inter-city warfare. Why did the Maya not migrate? Strangely, the forests of the Maya have returned in full glory, unlike those eliminated on Easter Island and elsewhere.

### 5. THE MAORI

The Maori case is an interesting variant of the Easter Island scenario. The Maori arrived in New Zealand at about 1000 AD and proceeded with major deforestation. They were interested in harvesting a large ostrich-like fowl and burning forests allowed them to concentrate the numbers of the bird for a convenient slaughter.[26] At least thirteen species of Moas (ostrich like birds) were hunted to extinction (Krech, 1999; p. 42).[27] Given the size of New Zealand, the Maoris could have enjoyed relatively high consumption levels for many hundreds of years. Relative to Easter Island one might have expected the Maori to reach a Malthusian state in the year 3000 AD or later. Europeans intervened and this Easter Island-type experiment never got played out. The Maori case differs from the Easter Island case in that population pressures were greatly lower in New Zealand.

## 5.1 The Spread of People over Land with costly Clearing

We take a materials balance relation

$$\dot{A} = \zeta[F(A,N) - C - \delta A]$$

where $A$ is hectares of agricultural land, $N$ is labor (population), $C$ is aggregate consumption, and $\zeta$ is a parameter indicating how much new land is cleared with a given input to clearing, namely, $F(A,N) - C - \delta A$. $\delta A$ can be thought of as maintenance of cleared land, as with for example fertilizing activity. There is no produced capital, $K$, here. Current gross product from cleared land, namely, $F(A,N)$ gets used up in land-clearing, consumption or labor maintenance, and fertilizing or land maintenance. Constant returns in $F(.)$ suggests unlimited virgin forested land; we prefer to deal with a case of aggregate scarcity of forested land, and to this end we posit decreasing returns to scale in $F(.)$. The scale diseconomy captures the notion of land clearing, at the margin, being more expensive, given $A$ hectares currently cleared.

This two dimensional model allows for the spreading of families into all available niches of usable land. This is appropriate because we are not dealing with a captive population on a small island. Ecological stress in one region should induce local migration out of that region sooner than an outbreak of violence. In this sense this model is more Malthusian than was the one for Easter Island. To show the limited migration possibilities we used this version of the above equation

$$\dot{A} = [A^{0.6} N^{0.3} - \delta A - cN]\zeta$$

and the traditional population dynamics

$$\frac{\dot{N}}{N} = \gamma[c - \bar{c}]$$

with $\bar{c}$ subsistence consumption and $\gamma$ a parameter linking the speed of population response to deviation in $c$ from $\bar{c}$. We selected per capita consumption or harvest to be more costly with a higher labor to ag-land ratio. That is, $c = \beta\left(\frac{A}{N}\right)^{\eta}$. Abundant farmed land relative to population implies a high per capita consumption of farmed produce. The special case of $c = \xi A$ is not intuitively compelling but yields interesting simulated histories. For a "large" initial stock of ag-land and a small initial population,[28] we observe a convergent cycle with early rapid population increase and reforestation and then a sharp drop in population with some small clearing of forested land. The early arrivals lived off the abundant initial ag-land and then suffered a population decline when the population to ag-land got high. Of interest is that the economic "logic" of our model is very different from the Brander-Taylor model but the dynamics are similar, given our per capita consumption function.

With the more complicated per capita consumption function, $c = \beta\left(\frac{A}{N}\right)^{\eta}$ , we obtain a saddle point for the dynamics rather than a convergent spiral.[29] Our central case, featuring a small initial population and "medium" initial agriculture land, exhibited a simultaneous "rush" of land clearing and an increase in population. As the maximum population level was approached, reforestation commenced and population declined to a stationary "large" value. This resembles the case of the Maoris arriving in New Zealand around 1000 AD. A "rush" of land clearing was followed by some population increase, but the population did not grow rapidly enough to drive per capita living to subsistence. It would have taken some hundreds of years more. This non-Malthusian outcome seems plausible. Low-cost land clearing leads to a good standard of living ("high" caloric intake) and steady but not dramatic population increase. Per capita subsistence is approached only in the very long run. China and Europe were both able to avoid the Malthusian outcome inevitable in the very long run, by "buying into" the accumulation of produced capital, "harvesting" technical change, and somehow getting fertility reduced. There was a regime switch – an industrial revolution and a demographic break with the past.

The Maori case reminds us that there are indirect payoffs to deforestation whatever the primary motive may be. For the Maori, deforestation meant easier hunting as well as fuel, construction material, and agricultural land. In a different situation such as the Roman Empire, the indirect payoff to clearing farmland is trading timber for other products useful in the development process. One thinks of timber exported for metals or foodstuffs. The American and Canadian economies experienced much development in the nineteenth century funded by the investment income earned by selling wood and timber abroad, while at the same time, much land was opened up for crops. No one worried that a Malthusian outcome was imminent because growth in per capita income kept ahead of population growth.

We turn now to a discussion of deforestation and population growth in areas such as the Roman Empire, where geographic, political, and economic integration with other regions kept Malthusian pressures at bay.

## 6. THE MEDITERRANEAN AND ROME

Like the ancient Egyptians before them, the Romans were city builders and serious accumulators, producing prodigious amounts of capital, largely structures: buildings, colliseums, roads, aqaducts, and sewers.  Rome provided its many provinces with law and order, solid and secure transportation networks, reasonable transactions and credit arrangements, and a tax system that did not press incomes in the hinterland all the way down to subsistence levels. Most historians argue that the Romans were able to accomplish these things by maintaining a well-functioning military linked to a competent and committed central administration. Many other "civilizations" demonstrated remarkable skills in construction and organization but none tied cities together in a huge trading system like "Rome", except perhaps China. Efficient trade among regions and specialization among workers were pushed to new levels. Growth without technical change is referred to as Smithian as in "complete" exploitation of worker specialization and regional specialization and trade. The success of the Roman empire may have been a simple consequence of the spread of law and order over such an extensive geographical area. No "nation" had been as successful in promoting the integration of so many productive regions and thus the world had not witnessed such extensive regional specialization and trade. One might also argue that the Romans exhibited a genius for getting useful labor from slaves. Skillful management involves setting incentives for productive effort from slaves and not tying up large amounts of labor in supervision. Estimates of the size of the slave population range as high as thirty per cent. Since machines for building roads and structures were very primitive, most of the building was carried out with human and animal power. One might say that the Roman empire was unusually successful because it was able to develop administrative skills to an unprecedented extent. Such skills resulted in law and order over a huge area and contributed to the extraction of a large fund of labor power from slaves.

But Roman efficiency did not extend to mechanization and mass production. Critically, Rome failed to develop modern energy generators such as steam engines; perhaps abundant slave-power was the substitute (Principe, 2002, p. 37). Why the large-scale production of consumer goods failed to take root in Rome has always been a mystery. What in fact did the city of Rome trade for the treasures it did not simply appropriate: textiles and exotica from the Far East, grain from Egypt, and ores from deposits far and wide? When silk clothing became fashionable in Rome, the Emperor Tiberius forbade the wearing of silk because too much gold bullion was being shipped to China in return for silk (Clark, 2000, p. 148). Much of this gold could have been tribute and appeasement money. Prosperity was widespread in the Roman empire. No one has ever suggested that Rome was tending to a Malthusian steady state, say with a small elite taxing peasant farmers of their surpluses. At about year 1 AD, the Roman empire, including slaves and non-citizens has been estimated to have had a population of about 120 million in a world of about 300 million (Primuspilus, 2003).[30] Towns had a large number of shopkeepers. The Mediterranean economy appears to have been keeping ahead of a crush from population overload quite satisfactorily. Growth historians would argue that the gains from regional specialization and trade had been developed sufficiently so that per capita income was respectably high. There was obviously good specialization by workers as well. Open is how much of the "high" income could be attributed to the exploitation of "advanced technology" relative to other "nations" such as India, China, and the empire in what is present-day Nigeria. Certainly the Roman economy had grown fast enough relative to population that per capita income was not at a subsistence level over much of the empire.

We must ask how much deforestation was feeding the growth of per capita incomes in the Roman Empire. Williams (2003, p. 100) writes that there had already been considerable clearing of Mediterranean forest before the Greeks and Romans, and expresses skepticism that the Romans were responsible for large-scale deforestation. Manning (1991) indicates that the Greeks did substantial deforestation in the Mediterranean, including southern Italy, before the rise of Rome. Plato spoke of Attica as a "mere relic of the original country", a place denuded of forests and subject to serious soil erosion.

Could the Greeks have taken all the wood? For each excavated house from Roman times, surely fifty or a hundred wooden houses have rotted to earthen smudges. All the stone homes, temples, etc. were capped with wooden roofs. For Nero's Rome to burn, there had to be an abundance of wood. A reasonable hypothesis is that the rise of the Roman economy coincided with the deforestation of the remaining large areas of timber around the Mediterranean basin[31] and that this "free" injection of building material and fuel was a major factor in the relatively high output per person in the Roman economy.[32] Of considerable interest would be detail on the trade between Rome and other nations. Wood may have been brought from abroad, as when the King of Tyre supplied cedars of Lebanon for Solomon's temple. Rising per capita incomes emerging because of rising regional specialization and trade would have induced a demand for more agricultural products and wood, and thus more deforestation. This would not be population growth pressing on food supply and agricultural land so much as rising incomes pressing on food supply and agricultural land. Rising incomes would to some extent show up in larger families as well as in in-migration from the edges of the empire but the tendency to a Malthusian equilibrium with subsistence consumption was not apparent. "Development" must have proceeded faster than population increase.

Manning (1991; p.15) is less sanguine: "The Greeks eventually reached to southern Italy, and Spain "thick with woods and gigantic trees"... Then Rome reached. The Greeks already had denuded the south half of Italy, so the Romans logged Spain and North Africa of naval timber, fuel and decorative hardwoods. Then Rome collapsed." It is true that rapidly rising wood prices could have damaged the Roman economy. One is inclined to follow historians and look for larger causes of the decline of Rome,[33] but serious shortages of low-cost inputs to the Roman economy may have triggered or reinforced problems as Rome slid downward. This fits into the dynastic cycle view (Usher, 1989, and others) that the rulers at the center lose control when their appropriated surplus shrinks. They then have difficulty supporting their soldiers. A period of chaos and decay is followed by the emergence of a new and vigorous ruling group which is able to maintain law and order until it "decays".

### 6.1 The Dynastic Cycle Model

The dynastic cycle model has the rulers controlling heavily taxed peasants and dealing with a perpetual fringe group of bandits which prey on rulers and peasants. One version has the elite leaving some small surplus for the peasant families so as to discourage peasants from turning to banditry in desperation. Small surpluses for peasants would of course connect to some positive population growth. Collapse at the center coincides with a large-scale rise in the bandit class and a general decline in per capita product as law and order fade.[34] A collapsing ruling class sounds much like the scenario for Rome favored by historians. The collapse of Rome was followed by a very long interlude of bandit dominance. No successor dynasty was able to re-create anything nearly as expansive as the Roman Empire. And certainly the scale and grandeur of the Roman public buildings has not been seen since. China also had to recover from periods of banditry and balkanization between dynasties. The new dynasty not only had to reassert central control over a large hinterland, it was obliged to re-integrate regional economies which had splintered off the former empire. It is curious that in China, re-integration of the disparate regions seemed to recur under each new regime whereas in Europe, nothing as extensive as the Roman Empire re-appeared.

Many factors, including the plague that began in 165 AD,[35] may have contributed to the fall of Rome. Perhaps interregional trade was choked by a stressed financial system.[36] While it is possible that Rome rose on supplies of low-cost wood passed over by the Greeks, it is a bit of a stretch to suggest that Rome declined when the low-cost sources of wood were depleted.

### 7. EUROPE AND CHINA: TECHNOLOGY AND TRADE INSTEAD OF MIGRATION

Pomeranz (2000) suggests that eighteenth century Europe and China resembled the Roman Empire in the sense that the gains from "Smithian growth" had been exhausted.[37] Both areas were approaching a Malthusian state, with serious depletion of wood supplies a feature of the later decades. Both were facing new highs in population densities as 1800 approached. Early in the eighteenth century, Europe and China had experienced a turning point in population trends. The population growth rate rose above its long run glacial rate, and in Europe at least, fertility was gradually reduced in families.[38] Figure 8.1 shows world population growing faster after 1750 and accelerating again after 1900. European deforestation (increase in the areas of cropland and pastureland[39]) never matched the rate of population increase but deforestation was taking place after 1700. One might argue for land-clearing first in western and eastern Europe before 1750 and then a population increase. (The population figures are for the world and we acknowledge the great advantage of good population figures by region. We were not able to obtain the disaggregated figures for population. And some crop and pasture land could have been brought on stream without deforestation but we are not considering this possibility in our data.)

Why was population able to grow so quickly relative to the historical trend? Three causes stand out: food raising in a more benign climate, the rapid diffusion of new crop varieties, and a "respite" from war and disease. These seem exogenous to the population growth-deforestation dynamics. In Figure 8.1, we do not have an explanation for the curious case of Oceania exhibiting rapid deforestation. However, rapid world population increase after 1900 coincides with more rapid growth in farmland in all regions except for Europe. In Figure 8.2, we consider land-use in Asia and then observe rapid clearing up to 1850 in East Asia with a slight slow-down beyond 1850.

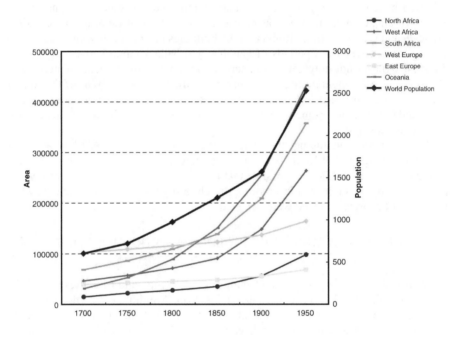

*Figure 8.1. Pasture and Crop Land by Region of the World*

### 7.1 The Role of Climate

The influence of climate on population growth is unclear. Kelly (2003) has recently surveyed the mechanics of population growth in pre-industrial England and, with the aid of careful econometric tests, and arrived at the conclusion that the positive population growth rates correlate well with benign climatic conditions.[40] Common sense would suggest that better weather improves crop yields and the standard of living. Improved yields can translate into population growth at the extensive margin (no nutritional improvement on average) or less population growth with some average nutritional improvement. In Europe, average nutrition does not seem to have improved, though there is evidence of declining fertility, possibly due to a rapid increase in the standard of living.[41] The agricultural surplus was "soaked up" in population increase. Massimo Livi-Bacci (1991)[42] believes that ordinary people achieved high caloric intake when meat consumption rose following the increase in acreage devoted to pasturing. But he discounts nutritional improvement as the principal spur to population increase. Instead, Livi-Bacci emphasizes family formation which is helped by the bringing of new land into cultivation, the spread of agricultural improvements, and improvements in transportation.[43] Lee (1973) does not believe that population growth is correlated with an increase in the standard of living,[44] but he and Anderson (2002) mention climate as a factor in changing mortality and fertility rates. A more benign climate should reduce disease in crops, cattle, and man. While other authors find little connection between climate and the timing of disease, Scott and Duncan (2001, p. 105) link "famine, sometimes in conjunction with bad weather conditions" to nutritional stress in the population and then to susceptibility to disease.[45] Warmer conditions in the eighteenth century do seem to have been accompanied by reductions in epidemic disease.

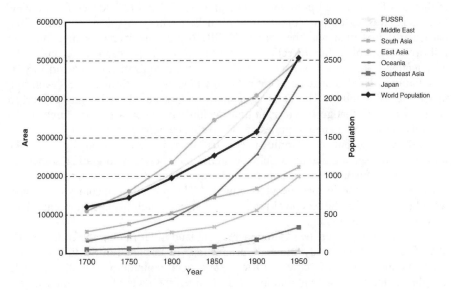

*Figure 8.2. Pasture and Crop Land by Regions of Asia and Oceania*

The eighteenth century also witnessed a reduction in armed conflict among the nations of Europe. Peace and widespread law and order can contribute to regional specialization and trade which of course can contribute mightily to increases in average living standards. It may not be a coincidence that Smith's Wealth of Nations (1776) appeared after a lengthy interlude of relative peace in Europe, a period in which the advantages of regional specialization and trade could play themselves out.

Doctrinal differences which fuelled religious wars abated, perhaps because of the spread of the new physics associated with Isaac Newton in England and Euler and the Bernoullis on the continent.

## 7.2 New Food Crops

Perhaps the most important technical change in the eighteenth century was the introduction to Europe and Asia of maize, potatoes, and buckwheat from the Americas. These crops increased the productivity of agricultural land in terms of calories produced per hectare, and made previously marginal land suddenly valuable in agriculture.[46] The arrival of turnips in England (1730) meant that livestock could now be fed through the winter. A popular view is that large areas of new land were brought into pasture and crop-growing because of technical change. This includes land within Europe and Asia as well as lands in the Americas.[47] Migration of "homesteaders" to the north and southwest of China followed. Figure 8.2 indicates large increases in pastureland and cropland in East Asia after 1700.[48]

The numerical data for South Asia and the former USSR display large increases in the use of land for crop-growing alone after 1700. This is evidence for the view that there was extensive colonization of new territory by farmers after 1700, particularly in China and the former USSR. Noticeable increases in population followed. The plausible direction of causation is from technical change (the introduction of potatoes, corn and buckwheat) to the colonization of new territory by new farmers. This colonization continued in China and the former USSR well into the nineteenth century. The extra agricultural output could feed a growing urban class. Technical change in transportation would have been a driving force in the growth of cities, given new supplies of food from new territories.

The technical change in farming meant that, from an economic point of view, the farmer at the spatial margin would be indifferent between migrating somewhat further out or remaining on his existing plot. The colonization of new territories does not imply a decline in average productivity.[49] The prevailing incomes of farmers were buoyed up by the technical progress, an effect which made submarginal land suddenly worth colonizing.

In Figure 8.3, we see the striking increase in land in farming in the United States after 1850 as well as steady and fast "production" of farmland in South America.[50] Worth noting is the fact that Costa Rica experienced rapid deforestation after 1945 and this is not really showing up in our data, given the level of aggregation we are dealing with.

## 7.3 Population and Deforestation: Moving in Tandem

Painting with a broad brush, we can make a rough link between population growth and new land drawn into agriculture in Europe and China. The supply of agricultural land increased faster than population up to about 1850 as huge areas were deforested. Rather than population increases causing deforestation, or deforestation leading to population growth, it would seem that both population growth and deforestation and deforestation were driven mainly by technical change in agriculture, principally the introduction of new crops, under favorable climatic conditions. Technical change in agriculture has continued to the present day, but probably was not the most important force driving farmers into new territories in the nineteenth century.

Technical change helped provide an agricultural surplus to feed cities; technical change developing within cities could have raised per capita income and demand for food, with a resulting increase in food prices that would feed back into new rounds of deforestation and colonization of agricultural land. In Britain, food and fuel prices rose rapidly at the end of the eighteenth century; Pomeranz (2000) argues that Britain was facing a crisis around 1800 (during the Napoleonic war) and could have failed to experience an industrial revolution had it not been able to trade industrial goods for food and fuel (Britain became a permanent importer of grain in 1795). This brings us to the subject of importation as a proxy for migration and deforestation.

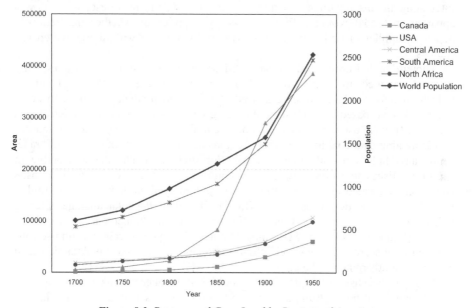

*Figure 8.3. Pasture and Crop Land by Regions of America*

## 8. TRADE AS A BUFFER

Trade for timber and food can ease population pressures on local supplies of wood. The evidence that the world's distribution of population within continents has not changed since pre-industrial times (see Diamond, 1997)[51] suggests that trade in food and timber did not induce people to settle away from relatively fertile areas. Instead, the needed food and timber that could not be squeezed out of the local soil must have been acquired from less prosperous areas via international trade. Consider the different histories of Europe and China in the period 1700 to 1850. Europe was rescued from serious fuel and food scarcity by new technologies and new lands across the Atlantic ocean and down-under.

Pomeranz argues that both Europe and China were facing serious energy scarcity because of deforestation. In China, "the primary cause of the denudation was, of course, agricultural clearing to feed a population that grew from about 300 million in the early eighteenth century to reach 430 million by 1850 and 500 million by 1900. Some observers suggest that the ruling elite did not take deforestation seriously, unlike the government of Japan, where there was a shift from land to sea in search of a supply of protein (Williams, 2003, p. 326). Deforestation in Europe was also serious." The rise in fuel prices in eighteenth-century Europe generally seems to have greatly outpaced other price increases....In Britain, firewood prices had already risen 700 percent between 1500 and 1630 and three times as fast as the general price level between 1540 and 1630; for much of the country the seventeenth century was a period of energy crisis. After 1750, the country was perpetually short of wood, charcoal, naval stores, and bar iron (made with charcoal)" (Pomeranz,2000). Interesting in this regard is whether Britain "liquidated" its forests when it was apparent that coal would suffice for energy supplies or whether the seriously depleted supplies of timber induced innovators to come up with alternative energy sources.

Though both Europe and China faced severe food and fuel shortages, only Europe managed to switch to coal and avoid an energy bottleneck in its industrialization.[52] China, despite a head start in iron production, failed to get a large-scale coal-based iron and steel industry going once convenient supplies of iron ore and coal were used up. China was not able to develop a transportation system for accessing abundant coal in its northern regions despite efficient inter-regional trade. China also failed to trade internationally for timber, though she was active in exporting dishes to Borneo. Europe, in contrast, had iron ore and coal deposits conveniently located near rivers and coasts, and enjoyed an active trade in timber from Norway, Sweden, Prussia, and Russia. Over half the total tonnage entering British ports in the 1750's was timber. Fir imports grew 700 percent between 1752 to 1792 (Pomeranz, 2000; p. 221, citing Thomas). Europe was rescued from serious fuel and food scarcity by new technologies and new lands across the Atlantic ocean and in the southern hemisphere.

Not only because of "impressive technological advances in manufacturing" but also because of trade, Europe's economy 1700-1850 sailed ahead of China's. Because of trade, "the limits imposed by its finite supply of land suddenly became...less important. This was partly because its own institutional blockages had left significant unexploited agricultural resources...,partly because far more extreme institutional blockages (above all serfdom) in eastern Europe ... had left lots of slack there; and partly because new land management techniques were brought home from the empire in the early nineteenth century....Even so, Europe's transformation also required the peculiar paths by which depopulation, the slave trade, Asian demand for silver, and colonial legislation and mercantilist capitalism shaped the New World into an almost inexhaustible source of land-intensive products and an outlet for Europe's relatively abundant capital and labor." (Pomeranz, 2000; pp. 22-23).

Some historians argue that nations in Europe had developed a culture that could innovate successfully in the face of social, political and economic transformations. Europeans did respond to their potential "ecological impasse" with inventiveness in energy production and in trading arrangements broadly defined, including the colonization of the Americas and the mass production of new exportables like cotton.[53]

## 8.1 A Note on Venice

No one was more dependent on trade than the maritime merchants of the lagoon city, Venice. The Venetian elite was well aware of the importance of maintaining a supply of low cost oak and fir for shipbuilding. Ultimately Venice's timber supplies were depleted and in 1606 fully one half of Venice's fleet was constructed "abroad". Lane (1968) is quite explicit about timber scarcity ending the maritime supremacy of the Venetians. He sees the torch being passed to the Dutch who, themselves having little forest stock, "controlled the lumber resources of Baltic". It is not clear, however, why Venice could not have continued its maritime commerce with purchased ships. The price of buying and building should not be much different, unless local tree-felling is used to subsidize ship production. A separate question is self-sufficiency in time of emergency. Britain was perpetually anxious about maintaining her navy for her defense.[54] She did not want to be caught relying on distant places for timbers.

It seems that Venice could have remained a great maritime power with bought-vessels, but saw her maritime hegemony decline for reasons other than the high cost of new vessels. Rapp (1976) describes the economy of Venice in the sixteenth century as modern and diversified. She led in "industrial" activities such as ship-building, dying and cloth production, soap-making and sugar refining, metallurgical activities and printing and book manufacture. Her decline involved being out-competed in these activities by the Dutch, English and French. One naturally thinks of lower labor costs being the central factor in a competitive race and Cipolla (1968) puts emphasis on a rigid guild system preventing labor markets working effectively in Venice. Rapp is less persuaded of the contribution of the guild system to decline.

## 9. EXPORTING TIMBER TO FINANCE DEVELOPMENT

Hartwick, Long and Tian (2001) developed a simple dynamic model of a country like Canada in the eighteenth century which exported much timber for importables (consumption and investment goods) while clearing land for new farms. World prices of traded goods were fixed, but internally, forested land (and timber) was in excess supply because farmland was scarce, scarce given the current world prices for agricultural products and products from forested land (e.g., sap and nuts). This then is land-use response in a small open economy to the opening up of trade with a large, rest of the world. It is a model of stock adjustment in the face of "new" initial conditions. The model exhibits a gradual "replacement" of timbered land with farms while the local prices of forested land and land in farms moved gradually to equilibrium. The price of farmland fell to the rising price of forested land. Formally the model operated as if forested land was being mined (cleared) until a balance was achieved between the competing uses. Timber, forest products from sustainable foresting and farm products were exported and funded imports of consumption and investment goods. This was a case in which the local prices for land responded to world prices for the products of the land with a process of deforestation (timber "mining"). Property rights were secure here and there was no explicit new demand entering the picture arising from immigration or local population increase. Latent rest of world demand was present, awaiting the opening up of a new small country. Forests got cleared in this model because the initial distribution of land uses was out of equilibrium, out of equilibrium relative to land prices "dictated by" the world prices for the products of the land. Land use change brought the areas in different uses into a land price equilibrium. In the end the land uses brought prices in competing uses into an equilibrium that was indirectly set by the world prices for the products of the land. And land use adjustment was not a jump to equilibrium for the same reason that with discounting, oil stocks for example, do not get dumped instantaneously on the market. In general the oil stocks in depletion models get run down smoothly. The same with the model of land use change.

This view of funding of imports with primary product exports during development was looked at again in Hartwick and Long (2001) where bits of an explicit stock of productive machine capital were built up via imports. Local capital was used to produce local consumption goods. And consumption goods were potentially importable as well and were thus at times competing with investment goods for the foreign exchange earned by exporting agricultural goods. This model was fully analyzed "with pencil and paper" and in fact was "tested" with computer simulations. Factor intensities in agricultural production and in local consumption goods production affected local land rent changes during development in notable ways. In some cases development occurred with land rent increasing and in other cases with land rent decreasing. This model was extended by us to include timber exports from land currently being cleared for agricultural use.[55] That is, the two models immediately above were merged into a quite complicated new model, with endogenous deforestation. Once again it is not population increase that presses on forested land. Land uses respond to the exogenous world prices of goods producible on land of various types and deforestation is one of the salient features of development.

We proceed to set out the model and its steady state. The economy produces four goods, $Q_a$ of agricultural goods, $Q_f$ of goods form sustainable forestry, $Q_c$ of consumer goods in the city, and $Q_h$ of timber when land is cleared. The prices of these goods are $P_a$, $P_f$, $P_c$, and $P_h$, all exogenously given in the world market. Subscripts are implicitly defined by the $Q's$ above. The investment good is not produced in this economy, it is imported at the price $P_I = 1$. The production functions for these goods are

$$Q_a = N_a^{a_1} L_a^{1-a_1}$$

$$Q_f = N_f^{f_1} L_f^{1-f_1}$$

$$Q_c = N_c^{c_1} K^{1-c_1}$$

$$Q_h = 2\alpha N_h^{1/2}$$

At any time $t$, the allocation of the fixed supply of labor, $N$ must satisfy the constraint

$$N_a + N_f + N_h + N_c = N$$

(these are control variables.) The state variables are capital level $K$ and land in sustainable forestry, $L_f$, and we have $L_a = L - L_f$ for $L$ the fixed supply of total land. At time $t$, the stock $L_f$ is given. The rate at which $L_f$ changes is determined by $N_h$:

$$\dot{L}_f = -2N_h^{1/2}$$

Thus we may think of labor $N_h$ in land-use change as performing two functions: trees are cuts, and forested land is transformed into agricultural land, (a kind of "joint product" from labor, $N_h$). Thus, the marginal value product of $N_h$ is not just $\alpha P_h N_h$, it is $(\alpha P_h + \eta)N_h$ where $\eta$ is the "worth" of the activity of transforming forest land to agricultural land. (We will see how $\eta$ is determined below.) We note that while the allocation of land satisfies $L_a + L_f = L$, at any given time we cannot choose $L_a$ and $L_f$ in the Heckscher-Ohlin way because the value of marginal product of agricultural land is not equal to the marginal product of forest land.

Let $C$ be the economy's consumption of the consumption good. Net import of the consumption good is $C - Q_c$. Each of $C$ or $K$ can be viewed as an indicator of the current level of development. Each rises in our development scenarios. Let $I$ be the amount of investment good currently imported by the economy and added to current stock, $K$. Then trade balance implies

$$P_c(C - Q_c) + I = P_f Q_f + P_a Q_a + P_h Q_h$$

The economy seeks to maximize

$$\int_0^\infty U(C)e^{-\rho t}dt$$

subject to

$$\dot{K} = I - \delta K$$

$$\dot{L}_f = -2N_h^{1/2}$$

and the trade balance constraint and production and employment constraint. Further details of the model's solution are in the appendix. This model has been numerically solved[56] for specified parameter values and initial conditions. Given its relatively complicated structure, one gets a variety of solutions (e.g., the economy as consumer goods importer or exporter over various intervals of development, depending on factor intensities). This then is a complicated model of a small open economy importing consumption and investment goods in return for its exports of timber, products from forested land, and products from agricultural land. Land use change occurs because the activities on land respond to the world prices of the products on the land. Part of the complexity of this more detailed model of a staple-exporting nation lies in the possibility that at some points in development the nation may be importing consumer goods or exporting them. Endowments, world commodity prices, and factor intensities all affect the nature of the development path. Novel here is the constant population or labor force. This is a departure of the standard model of development of a small, new economy but allows us to focus on other forces driving deforestation. The summary would be that it is the "arrival" of world demand for primary products that drives deforestation and land use change.

## 10. POPULATION PRESSURES IN THE TWENTIETH CENTURY

Between 1900 and 1990, world population tripled. Recall that "things" double in seventy years at 1% growth, compounded. Hence world population was growing at about 1.2%. Advances in public sanitation were the most important factors. Massive die-offs from epidemics have been reduced and may well be eliminated in the next few decades. It seems that very modest supplies of medical care from nurses and clinics can eliminate vast amounts of death and disease in the slums of the world (Dugger, 2004). While population tripled, per capita food production only doubled, but this was for lack of demand (millions do not participate in the market due to poverty) and not lack of possible supply: output per hectare has mushroomed and continues to expand (at about 3% per year in the 1980's). Because of increased agricultural productivity, declining fertility, and a generous initial stock of natural resources, the Malthusian outcome has not yet come to pass on a global level. Nor have global forest stocks been destroyed.

One third of the world remains forested (Clark, 2000, p. 226). This remaining woodland is not well-suited to agriculture; most or all of the low-cost agricultural land has already come into production. Nevertheless, continued deforestation can be expected. Currently, in Africa, about 44,000 km$^2$ is logged each year and severely degraded while about 110,000 km$^2$ is cleared for agriculture. (Williams, 2003, p. 493). Deforestation in the rainforest of South America is also a concern. Because of poverty and the lack of property rights, new agricultural land continues to be sought, even if it is substandard and cannot, without fertilization, provide more than one or two years of output. In addition to obvious deforestation as in loss of land cover, there is forest degradation taking place without noticeable forest-cover depletion. India for example registers no loss of forest-cover over the last two decades in satellite photographs, though degradation of forests caused in part by animal foraging is apparent.[57]

The United States provides a strong counter-example to the notion that continued deforestation is necessary to provide enough food for Earth's inhabitants. It is true that by 1900, only half of the original forest cover of the United States remained and much lumber and paper is imported. But gains in agriculture since then are largely due to productivity gains, for which the United States set the pace. See Figure 8.4 for a comparison of the "production" of farmland in the USA and in the former USSR after 1900.

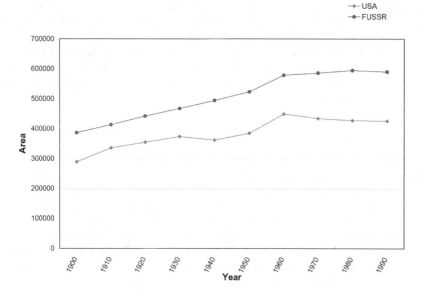

**Figure 8.4.** *Pasture and Crop Land of USA and FUSSR*

Gardner (2003) points out that eighty-five per cent of current U.S. agricultural output is produced from just 150,000 large farms. Labor productivity has increased fifty-fold in cotton and corn production. The 20 million horses and mules used in farming in 1900 were down to 3 million in 1960, given large-scale tractorization of U.S. agriculture. New hybrid seeds raised corn production per unit of land seven or eight-fold. And dairy output has increased dramatically with the exploitation of cross-breeding and artificial insemination. Further large increases in milk production can be expected as genetic-engineering, embryo transplantation, and cloning are exploited more and more (cited in *The Economist*, January 11, 2003). Most of the gains in food production in the future will come not from the colonization of new land but from the poorer regions of the world adopting leading capital- and research-intensive techniques. Better irrigation, increased application of fertilizer, and the development of better strains for cultivation should provide huge increases in output per hectare.

That being said, agricultural land scarcity and pressures to deforest continue to bite in various regions of the world. Homer-Dixon (1999) cites two contemporary cases of deforestation tied somewhat indirectly to violence. Haiti was once abundantly forested and is now 90 percent denuded of tree cover (pp. 119, and 135-36). There was large-scale looting of mahogany stands in decades past. This has led to much soil erosion and an impoverished peasant class with farm families working tiny plots of land. Efforts at deforestation have reportedly been interfered with by the military. The other case is Pakistan (pp. 119-120) where a so-called timber mafia ravages the forests. The harvesters keep themselves beyond the reach of government regulation.

But Homer-Dixon emphasizes that few or no wars between nations appear to have been fought over renewable resources. He also concludes that there is relatively little violence within or between regions that is associated with renewable resource scarcity. We infer that he fails to link migration as a response to local environmental stress. Our work so far points to migration and trade as providing critical relief to resource-constrained populations. Homer-Dixon documents very large growth rates for cities, often in poorer nations, but fails to see what appears to be the strong link between these high rates, each with its large migration dimension, and environmental scarcity.

## 11. MIGRATION AS A BUFFER. THE CITY AS FRONTIER

According to Malthus, the effect of land scarcity on per capita consumption will only bind when there is no additional land to farm. This means that local Malthusian tendencies can be mitigated by migration. In our examination of Rome, Europe, and China, migration seems to have largely negated any genuine Malthusian outcomes in history. One rarely sees a Malthusian equilibrium because groups of people migrate away from the area of environmental stress before per capita subsistence is reached. This observation suggests a general Mathusian detour in operation: as long as agricultural land can be added to the stock at low cost, humans will add it in such a way as to keep their per capita consumption at a "comfortable" level - certainly above subsistence, which of course implies further population increase, in general. The full Malthusian outcome will only occur when further migration (filling in of people on more agricultural land) is impossible. Ironically, much Malthusian-driven migration is to cities, not arable land. Slums of third world mega-cities are not Malthusian sinks, since some medical care, education, and subsistence income is most often available. Life expectancy in the slums of third world cities is higher than in eighteenth century times in Europe because there is lower mortality in the early years of life. Children do get vaccinations and mothers do know how to hydrate sick babies. Still, subsistence for many people involves picking through fresh garbage dumping each day.

## 12. CONCLUDING REMARKS

Throughout human history, people have migrated away from areas of subsistence income to new lands. Deforestation has provided fuel, materials, and tradable goods as well as land for agriculture. Income decline to subsistence levels was forestalled. But population growth was on average glacial, at less than 0.1% on average because disease and violence were uncontrolled. Then about 1700 population growth became noticeable in Europe and east Asia. Malthus saw a population bomb ticking in Europe and wrote about it in his famous 1790 monograph. But even then France had turned a corner and was in fertility decline. Britain followed in about 1850. Growth in per capita income became linked to fertility decline.[58]

In a situation of geographic isolation and a small resource stock, deforestation becomes critical as forests shrink and livelihood is imperiled. Models by Brander-Taylor, Hartwick, and Hartwick-Yeung suggest that, even then, sustainable forestry is possible, though population growth is not. The conditions most conducive to sustainable population and forestry in an Easter Island scenario are property rights, social order, and a not trivial cost of harvesting. In contrast to the cases of geographically isolated areas like Easter Island, nations in Europe and cities in China have benefited from trading networks and an extensive hinterland providing resources and migration possibilities. Deforestation has abetted population growth for centuries. As for the effect of population on deforestation, increases in population mean an increased demand for forest products and, more fundamentally, an increased demand for food grown on cleared land. Wherever population has grown, forests have been cleared. Yet demand for food and fuel is a function not only of market size (population) but of market value (per capita income). Many increases in deforestation associated with increased population are actually the result of independent causes that boost per capita income, changes such as improved weather conditions and improved technology. At the same time, exogenous improvements in agricultural productivity provide hope that income growth can continue without further extensive deforestation. However, even while many parts of the United States, for example, are experiencing re-forestation, the US continues to be a large importer of forest products. High consumption levels as well as high population growth rates threaten world forests. Our models of staple-exporting development were basically those of deforestation at home caused by robust demand abroad.

Somewhat odd was the measured decline in timber prices in the twentieth century up to 1950, even while population growth in the world was at "high" rates (Christy and Potter, 1962; Manthy, 1978). Timber prices jumped in the 1970's along with other primary resource prices, but have not displayed a noticeable upward trend since the 1950's (Sedjo & Lyon, 1996). Markets are not signaling a basic timber scarcity in spite of past aggressive deforestations, and others that people draw attention to today. Markets may well be defective at signaling impending dire scarcities or dire scarcity of forests and forest products may be a thing of the remote future. The best we can do is think deeply and carefully about the past and its links to the present and exhort prudence to those who are aggressively harvesting timber in various places around this not particularly expansive planet.

## APPENDIX

## SOLVING FOR PATHS OF DEVELOPMENT FOR A PRIMARY PRODUCT EXPORTING NATION

Let $\lambda$ and $\mu$ be the shadow prices of $K$ and $L_f$. We define the Hamiltonian

$$H = U(C) - 2\mu N_h^{1/2}$$

$$\lambda \{ P_f N_f^{f_1} L_f^{1-f_1} + P_a N_a^{a_1} L_a^{1-a_1} + 2\alpha P_h N_h^{1/2} - \delta K$$
$$+ P_c (N - N_f - N_a - N_h)^{c_1} K^{1-c_1} - P_c C \}$$

Differentiating $H$ with respect to the control variables $C$, $N_f$, $N_a$, $N_h$ respectively, and equating the derivatives to zero, we get

$$U'(C) = \lambda P_c \text{ or } C = (P_c \lambda)^{-1/\sigma}$$

$$N_f = L_f \Pi_f \left[ \frac{N_c}{K} \right]^{(1-c_1)/(1-f_1)} \text{ where } \Pi_f \equiv \left[ \frac{f_1 P_f}{c_1 P_c} \right]^{1/(1-f_1)} \tag{1}$$

$$N_a = \left( L - L_f \right) \Pi_a \left[ \frac{N_c}{K} \right]^{(1-c_1)/(1-a_1)} \text{ where } \Pi_a \equiv \left[ \frac{a_1 P_a}{c_1 P_c} \right]^{1/(1-a_1)}$$

$$N_h = \Pi_h \left[ \frac{N_c}{K} \right]^{2(1-c_1)} \text{ where } \Pi_h \equiv \left[ \frac{\alpha P_h - (\mu/\lambda)}{c_1 P_c} \right]^2 \tag{2}$$

Since $N - N_f - N_a - N_h = N_c$, we have

$$N - L_f \Pi_f \left[ \frac{N_c}{K} \right]^{(1-c_1)/(1-f_1)} - \left( L - L_f \right) \Pi_a \left[ \frac{N_c}{K} \right]^{(1-c_1)/(1-a_1)} - \Pi_h \left[ \frac{N_c}{K} \right]^{2(1-c_1)} = N_c \tag{3}$$

This equation determines $N_c$ as a function of the state variables $K$ and $L_f$, and the shadow prices $\mu$ and $\lambda$.

(Example: If $c_1 = f_1 = a_1 = 1/2$, then we get from (3):

$$N_c = \frac{NK}{K + \Pi_h + \Pi_a(L - L_f) + \Pi_f L_f}$$

assuming $K + \Pi_h + \Pi_a(L - L_f) + \Pi_f L_f > 0$.)

In general, equation (3) yields

$$N_c = N_c(K, L_f, \mu/\lambda) \tag{4}$$

Substituting this into (1), we get

$$N_f = L_f \Pi_f \left[ \frac{N_c(K, L_f, \mu/\lambda)}{K} \right]^{(1-c_1)/(1-f_1)} = N_f(K, L_f, \mu/\lambda)$$

and similarly for $N_h$ and $N_a$.

$$\dot{\lambda} = (\delta + \rho)\lambda - (1-c_1)P_c \left[ \frac{N_c(K, L_f, \mu/\lambda)}{K} \right]^{c_1}$$

Next, we turn to the differential equations and

$$+\lambda(1-a_1)P_a \Pi_c^{a_1} \left[ \frac{N_c}{K} \right]^{a_1(1-c_1)/(1-a_1)} \qquad \dot{\mu} = \rho\mu - \lambda(1-f_1)P_f \left[ \frac{N_f}{L_f} \right]^{f_1}$$

$$+\lambda(1-a_1)P_a \left[ \frac{N_a}{L - L_f} \right]^{a_1}$$

or

$$\dot{\mu} = \rho\mu - \lambda(1-f_1)P_f \Pi_f^{f_1} \left[ \frac{N_c}{K} \right]^{f_1(1-c_1)/(1-f_1)} \tag{5}$$

$$\dot{L}_f = -2N_h^{1/2} = -2\Pi_h^{1/2} \left[ \frac{N_c(K, L_f, \mu/\lambda)}{K} \right]^{(1-c_1)}$$

$$\dot{K} = -\delta K - P_c (P_c \lambda)^{-1/\sigma} + P_f Q_f + P_c Q_c + P_h Q_h$$

where

$$Q_f = L_f \left[ \frac{N_f}{L_f} \right]^{f_1} = L_f \Pi_f^{f_1} \left[ \frac{N_c(K, L_f, \mu/\lambda)}{K} \right]^{f_1(1-c_1)/(1-f_1)}$$

$$Q_a = (L - L_f) \Pi_a^{a_1} \left[ \frac{N_c(K, L_f, \mu/\lambda)}{K} \right]^{a_1(1-c_1)/(1-a_1)}$$

$$Q_c = K^{1-c_1} \left[ \frac{N_c(K, L_f, \mu/\lambda)}{K} \right]^{c_1}$$

$$Q_h = 2\alpha N_h^{1/2} = 2\alpha \Pi_h^{1/2} \left[ \frac{N_c(K, L_f, \mu/\lambda)}{K} \right]^{(1-c_1)}$$

Thus we have four differential equations of the form

$$\dot{K} = F_1(K, L_f, \mu, \lambda)$$

$$\dot{L}_f = F_2(K, L_f, \mu, \lambda)$$

$$\dot{\lambda} = F_3(K, L_f, \mu, \lambda)$$

$$\dot{\mu} = F_4(K, L_f, \mu, \lambda)$$

At the **steady state**, $\dot{L}_f = 0$ hence $N_h = 0$, implying, via (2)

$$\overline{\mu} / \hat{\lambda} = \alpha P_h \tag{6}$$

where the hat denotes steady state values. Then using (4), we get

$$\overline{N}_c = N_c(\overline{K}, \hat{L}_f, \alpha P_h)$$

Setting (5) to zero, and using (6), we get

$$\alpha P_h = \frac{1}{\rho}(1-f_1)P_f \Pi_f^{f_1}\left[\frac{\bar{N}_c}{\bar{K}}\right]^{f_1(1-c_1)/(1-f_1)}$$

$$-\frac{1}{\rho}(1-a_1)P_a \Pi_c^{a_1}\left[\frac{\bar{N}_c}{\bar{K}}\right]^{a_1(1-c_1)/(1-a_1)} \tag{7}$$

And setting $\dot{\lambda} = 0$, we get

$$\hat{\lambda} = (1-c_1)P_c\left[\frac{N_c(\bar{K},\hat{L}_f,\alpha P_h)}{\bar{K}}\right]^{c_1} \tag{8}$$

Setting $\dot{K} = 0$, we get

$$\delta\bar{K} = -P_c(P_c\hat{\lambda})^{-1/\sigma} + P_h\bar{Q}_h(\bar{K},\hat{L}_f,\alpha P_h)$$

$$+P_a\bar{Q}_a(\bar{K},\hat{L}_f,\alpha P_h) + P_c\bar{Q}_c(\bar{K},\hat{L}_f,\alpha P_h) + P_h\bar{Q}_h(\bar{K},\hat{L}_f,\alpha P_h) \tag{9}$$

The three equations (7), (8) and (9) determine $\hat{\lambda}$, $\bar{K}, \hat{L}_f$. Then $\bar{\mu}$ can be calculated from (6).

## NOTES

[1] "Simply, one of the prime driving forces of deforestation – the sheer pressure of people – will continue unabated, and as cultivation has always been the greatest devourer of the forests, many more millions of hectares will be destroyed. Similarly, the demand for fuel wood will remain immense for the poor of the world." (Williams, 2003; p. 495).

[2] And we should not slight the direct demand for wood for building, including ships in earlier days, and for fuel. Both increasing population and income create this direct demand. Williams cites as revolutionary for timber demand: the introduction of iron and steel ships, about 1860.

[3] United Nations Development Program (2002, p. 5) indicates that 50% of forest cover has been lost since "pre-agricultural times".

[4] The idea that per capita "income" will fall to subsistence because population growth is more rapid than the increase in arable land is a useful abstract reference case. In fact global population growth was very slow until about 1750 in part because of food scarcity but also because disease could attack without any systematic checking mechanism. It would also appear that economic growth had difficulty taking root because law and order were seldom maintained over wide areas and over reasonably long periods.

[5] Marketable timber as a by-product of clearing land for farming was central to the analysis of Hartwick, Long and Tian (1999). Clearing stopped short of the "corner" in that model because forested land yielded marketable products and land for these activities was ultimately bid up to the value of another hectare for

farming. This was a model of an initial stock of forested land getting adjusted to a sustainable configuration. A general equilibrium model of moving a renewable resource to its sustainable configuration is contained in Tian and Cairns (2002).

[6] There is a sense that land clearing is like disposing of waste in a river. It is a costless activity locally in time but has negative effects, seemingly unanticipated, in the long run. Costless in the short run becomes costly in the long run. Cumulative "costless" actions turn out to yield large costs down the road.

[7] Pomerantz explicitly dismisses these places being in Malthusian equilibria. He sketches a convincing case of significant fuel scarcities around 1700. By Malthusian equilibrium, we mean zero population growth with subsistence living standards. History confounds such outcomes with the arrival of pervasive diseases at various times. Such epidemics can be sparked off and abetted in spreading by the poor nutrition of people but can also take hold independently of the general health condition of the mass of citizens.

[8] In 1621, the Pilgrim ship *Fortune* returned to England loaded almost exclusively with clapboard (Cronon, 2003, p. 109). This testifies to timber scarcity in England.

[9] Williams (2003, p. 65) observes that the North American Indians practiced settlement development and farming. One settlement near present-day St. Louis had between ten and twenty-five thousand residents. The land was cultivated around the site up to 15 km away. The town collapsed around 1100 AD, apparently from the exhaustion of timber resources in the surrounding area.

[10] In Hartwick, Long and Tian (2000) forested land gets cleared in a place like Canada in the nineteenth century because of robust demand abroad for timber as well as local demand for a high ratio of agricultural land per farm family. Again it is not the pressure of subsistence consumption and population increase that is driving forest clearing. Rather it is the anticipation of good family incomes, following clearing and farming, which drove the deforestation.

[11] "Almost everywhere in the world the colonization of the forest has been the means of social advancement and improvement for the landless peasant – the "little" man and his family. It was true, for example, on the frontiers of Rome during the first centuries BC and AD, of medieval Europe, eighteenth and nineteenth century America, nineteenth century New Zealand and Australia, large parts of colonial India and Burma, and even, from what little we know, of Ming China, especially in the southern part of the country." (Williams, 2003, p. 474). "But in many parts of the world the peasants have taken the initiative themselves and have gone ahead and cleared the land in the time-honored fashion, bit by bit, year by year, making enough new ground to establish themselves and feed a family. In toto this massive, undocumented movement is thought to be one of the greatest impacts on the forests of the tropical world." (Williams, 2002, p. 485).

[12] 13% of Canadians' disposable income goes to food compared with 60 to 70 % in the middle ages and 50% in India. Stephen Strauss, p. D13, *The Globe and Mail*, May 31, 2003. "Shankar Subramanian and Deaton calculate that in rural Maharashtra in 1983, 2000 calories (in the form of standard coarse cereals) could be purchased for less than 5 percent of the day wage, a finding that is consistent with the observation that poor agricultural workers in India typically eat their fill of cheap calories at he end of the work day." (Deaton, 2003, p. 131).

[13] Scott and Duncan (2001, p. 98) refer to the post plague situation for one English town: "Here again, the ecological niches were quickly filled and the community returned to pre-crisis population dynamics." The addition of 60 million people to Europe between 1500 and 1750 made agricultural expansion the primary cause of deforestation. England's population doubled between 1550 and 1700 making it the densest nation, at 7 persons per acre, after Holland and China. (Williams, 2003, p. 173).

[14] Recent analysis of skeletel remains of ancient folk suggest that hunter-gatherers were relatively healthy when Europeans invaded the Americas (Steckel & Rose, 2002).

[15] Diamond (1997, Chapter 6) argues for the spontaneous development of farming on different continents but is agnostic on whether higher population density led to farming or vice versa. Cohen (1977) also finds a diffusion process improbable for explaining the arrival of farming in different regions of the world between 12,000 BC and 2,000 BC.

[16] Wrigley (1986) presents an interesting calculation of the technical progress in agriculture needed to provide a surplus to feed the increasing numbers of town and city folk in England, 1550 to 1800. I think he neglects the large payoff from specialization of workers. For example if a farmer and his family spent half their time farming and half say weaving and sewing in 1550 and full time farming in 1800 there could be huge gains from specialization without any technical progress. The gains from specialization are

referred to today as "Smithian growth". If one adds modest investment in internal navigation and roads, one could expect additional large gains in farm output reaching towns. These are gains from town-country trade. Regional specialization and trade is another aspect of productivity enhancement. Technical progress could be extremely modest while a complicated system of towns and cities, with their specialized production activity, emerged. One thinks of China with its five cities with over a million people in 1200 or Rome with over a million people in 150 AD.

[17] Tentative new material (Steckel & Rose, 2002) suggests that the pre-Columbus farming and town people of the Americas were on average poorly fed because their economies were unable to distribute income evenly enough. One wonders about a ruling class exploiting a working class, and of the fact the tradeoff, even in modern times, between the productivity of the city and the relatively free and autonomous life possible without the restrictions and regulations of paid employment. There may be a general rule that societies with relatively equal "incomes" are the exception in "modern" history. Hunter-gatherer groups lived better because "income" was (a) abundant and (b) relatively equally accessible or equally distributed. However, Roosevelt (1984, p. 577) suggests that productivity increased with settled agriculture and that birth rates must have increased if population could continue to grow in "dense, sedentary settlements" more prone to disease.

[18] It is argued that for many decades London was unable to maintain its population by natural increase and relied on in-migration in order to grow.

[19] It is common to argue that the people of Asia and Europe lived in closer contact with livestock, including horses, dogs and cats, than did the natives of North and South America and thus, that the people of Asia and Europe had built up stronger immune systems in so living. We have no quarrel with this view.

[20] People base their family size of current "income", not anticipated future incomes and this results in an "excess" of population relative to the carrying capacity of the island, "in the future". In a dyspeptic aside, McNeill (1976, p.277) comments: "If pleistocene extinctions were the work of human hunters, that catastrophic ancient overkill closely parallels our modern industrial squandering of fossil fuels... moderns will probably require fewer centuries to destroy the principal energy base of their existence..."

[21] www.islandheritage.org/eihistory.html

[22] When trees were abundant, acquiring a new dugout would have been relatively free of social friction. One would replace one's worn-out canoe by felling a suitable tree and hollowing it out, likely with the help of neighbors.

[23] The Malthusian view appears to be that environmental stress (a low ratio of arable land to population) gets manifested first in migration and then in the somewhat passive die-off of the weak. Malthus detoured around the delicate question of violence breaking out during the process of increasing environmental stress.

[24] Homer-Dixon (1999, p. 153) cites data for the Peruvian southern highlands in the 1980's indicating that per capita caloric intake was 70 percent of the requirements of the FAO. This was an area of Shining Path insurgency.

[25] Williams (2003, p. 63) endorses the view that the collapse of the Mayan cities need not have been due to nutritional and environmental degradation. .

[26] Williams (2003, p. 21) explains the burning of forests in New Zealand by the Maori in order to access large groups of the huge bird, Maos, which they hunted for food and skins for clothing. "A mere 8,000 to 12,000 people in South Island destroyed not less than 8 million acres of forest by the mid-thirteenth century" in their pursuit of the Maos. The North Island became cleared also, in part in response to the demand for land for agriculture and sheep raising.

[27] And Polynesians drove over half of the endemic spcies of birds in the Hawaiian archipelago to extinction (Krech, 1999, p. 42).

[28] Our two equations were $\dot{A} = A^3 N^{-3} - .3AN$ and $\dot{N}/N = 0.2[0.3A - 1.1]$. The initial values were $A = 15$ and $N = 1.3$.

[29] The first equation is the same as that above but with $c = [A/N]^3$ and 1.2 in the second equation changed to 0.2. Initial values were $N = 20.28$ and $A = 1542$.

[30] At 0.1% growth rate population would have reached 1.5 billion in 1600 years. It seems reasonable to infer that such rate of increase would be indistinguishable from zero to the living in any era.

[31] Some argue that the Romans desertified North Africa by deforestation and poor farming practice.

[32] Meanwhile (between 120 and 240 AD), in southeast England, the six largest Roman bloomery furnaces

of the Weald were producing about 550 tons of iron annually, according to estimates based on slag remains. This rate of production would have required the clearing of 2 square kilometers of dense forest per year.
[33] Williams (2003, p. 95) on the decline of Rome. Williams discounts deforestation as a principal cause. "While soil erosion certainly occurred, and devastatingly so in places, it is more likely that constant war, ravaging epidemics, rebellion, invasion from outside, a declining population, and an excessive degree of urbanization, separately or in combination, operated in the complexity of an empire that had extended beyond its means. In particular, the slender margin of surplus agricultural production needed to sustain city life could have been a crucial factor, given that over ten people were required to support one city dweller, even in a prosperous region."
[34] Though per capita production might decline to subsistence levels, for this to be a Malthusian state the decline would have to be linked to past population increases, not simply to collapse of control at the center.
[35] McNeill (1976, p. 103) reports that this plague spread through the Roman army from Mesopatamia and kept population growth down for the next five hundred years.
[36] Were transactions reduced because specie was scarce? Mediterranean trade flourished again under Venice and Genoa in the fifteenth century when ample quantities of gold flowed up from west Africa.
[37] Smithian growth represents gains from increased specialization in production by people in tasks and regions in lines of activity with a comparative advantage. Smithian growth is to be distinguished from growth due to technical change.
[38] Scott and Duncan (2001, p. 382) report: "Plague quickly fizzled out in England after 1666 because it could not be maintained through winter and there were no further introductions from continental Europe."
[39] The date in Figures 1 to 4 are taken from the HYDE database (http://www.rivm.nl.env/int/hyde/). Population figures are for the world in each case. Land use areal units are taken from HYDE.
[40] Kelly is obliged to exclude 1300-1450, "the famine and plague period" from his analysis. His central conclusion is that "periods of contraction were not due to low living standards" and real wages have no "explanatory power" for population growth rates. Between 1541-1800, the period of most detailed population records, Kelly finds that climate has a large effect on fertility, with a ten year lag. Here one would expect marriage rates to play the central role.
[41] Livi-Bacci (1992, p. 45) presents a table of average numbers of children per woman for a selection of currently advanced countries, since 1750. The series for England and Wales, and Sweden are most complete. Each series indicates a halving between 1800 and 1900 (from 5.54 to 1.96 for England and Wales and from 4.68 to 1.90 for Sweden). Other countries exhibit large declines from 1850 to 1900 also. Figure 4.8 relates children per woman for a cross section of countries in 1980 to real GDP per capita. Here we observe a dramatic decline in numbers relative to increased incomes. These data display a striking regularity: family size declines greatly with increases in per capita incomes. The limiting family size appears to be "replacement". Offspring are an inferior good: more family income, less children. The link to urbanization makes sense. Children can be valuable labor for a farmer but a burden to a struggling city family. Children also can be viewed as potential support for parents in old age when the accumulation of savings for old age is hard to carry out. The trend to small families certainly ante-dated modern birth control technologies. Women or women and their partners were clearly making substantial effort to control family size as the industrial revolutions of various kinds took root. Smaller family size feeds back into higher per capita incomes. This makes the isolation of causality difficult. The broad phenomenon is one of substituting material goods and costly leisure activities for children.
[42] Livi-Bacci (1991, p. 15) argues, in view of the extensive research of Wrigley and Schofield, that England's population increased with increases in household formation (by nuptiality). In good times, high wages relative to food prices, couples married younger and marriages rates increased, and population growth increased. This mechanism "prevented the repressive check of high mortality from coming into operation". "In contrast to the English example there is that of France, at least during the seventeenth and eighteenth centuries, which might be considered one of high demographic pressure. Here it was primarily mortality that checked demographic growth, keeping the system in a state of unhappy equilibrium." (p. 15) For England and many other places, Livi-Bacci argues for epidemics limiting population growth, not food scarcity. And Livi-Bacci argues that the timing of epidemics had "little or nothing to do with standards of living." (p. 18)
[43] Livi-Bacci (1991, p. 120-21) sums up with an argument that population growth resulted from increased

family formation, not from increased fertility resulting from improved nutrition standards. "The availability of virgin land accompanied medieval demographic growth and the colonization of eastern Europe. A similar role was played by the tilling and reclaiming of land in many parts of sixteenth-century Europe, drainage and the creation of polders reached a zenith at the beginning of the seventeenth century was achieved as farming pushed back the frontiers of uncultivated land. In France, when the ancien regime was nearing its close, cultivated territory amounted to almost twenty-four million hectares as against nineteen million of thirty years before. In England, whereas enclosures at the beginning of the eighteenth century amounted to a few hundred acres, by the second half of the century they were increasing by 70,000 acres per year. The swamps and marshes of the Maremma and Prussia were drained as were the bogs and fens of Ireland. It is a process which permeated a large part of eighteenth-century Europe."

[44] According to Lee, high levels of population are correlated with low standards of living, with no implication for population growth.

[45] Scott and Duncan (2001, p. 14) link at least three episodes of high mortality in northwest England in the seventeenth century to a coincidenc of high wheat prices and low wool prices. This fits well with they theory that suggests that famines generally occur from dislocation and a lack of purchasing power rather than from a direct shortage of food. Following three years of wet harvests, 1438-39 is identified as a year of famine and high mortality as well as "a year of pestilence" (p. 114). "There were frequent and virulent outbreaks of plague in France during 1520-1600 that were accompanied by food shortages, famines, flooding, peasant uprisings and religious wars." (p. 291). "1628-29 were years of widespread famine in northern Italy with unusually high prices of grain. This period immediately preceded the greatest outbreak of plague in continental Europe." (p. 384).

[46] The example of the New Guinea highlands is reported by Diamond (1997). "A population explosion" followed the arrival of the sweet potato, via the Philippines. "Even though people had been farming in the New Guinea highlands for many thousands of years before sweet potatoes were introduced, the available local crops had limited them in the population densities they could attain, and in the elevations they could occupy."

[47] "In Europe, maize and potatoes became significant only after 1650; in China, maize and sweet potatoes seem to have spread more rapidly, perhaps because the intensive labor characteristic of Chinese farming easily allowed experimentation with a new crop, whereas the rigidities of collective "open field" cultivation, which prevailed in most of northern Europe until the eighteenth century or later, powerfully inhibited any departure from custom." (McNeill, 1976, p. 317). See also the many references here to primary sources. Flinn (1981, p. 96) notes that "the settlement and colonization of new lands" in eastern Europe was important to population increase as well as the introduction of the potato, maize (corn) and buckwheat. Flinn suggests that the new crops provided more reliable harvests than some they displaced.

[48] Again the case of Oceania is an anomaly for which we have no explanation at this time.

[49] I am indebted to Frank Lewis for this point. See for example Olmstead and Rhode (2002).

[50] Livi-Bacci (1991, p. 19) argues that much population increase occurred without apparent improvement in average nutritional standards. More land and food supported larger populations but not apparently improved living standards. For example he cites opinion in support of the view that the Neolithic revolution (the arrival or crop-raising as distinct from hunting and gathering) exhibited no significant increase in average nutritional standards or in life expectancy. In the eighteenth century, Livi-Bacci argues for some small improvement in life expectancy but this turned on transportation improvements, new food stuffs (maize and potatoes from the new world), and a "decline in mortality". It seems that epidemics were less frequent and wide-spread. However, population growth from 1750 to 1815 outstripped food increase and resulted in "deteriorating standards of living both for farming and for factory or city populations." (p. 98). England became a food importer around 1795 and wheat consumption per head actually dropped right up to 1850.

[51] Compared to other continents, the population of North America has grown absolutely and relatively since colonization by Europeans. Although the Mississippi Basin was relatively densely settled in pre-industrial times, some areas of relatively abundant fertile land had yet to be effectively exploited and settled in 1500. This still leaves a large role for trade in food in allowing relatively densely settled areas in 1500 to remain relatively dense in 2000.

[52] Williams (2003, p. 140) The large iron industry of the Sung dynasty (around 1000 AD) required large quantities of fuel, mostly coal. Production in 1075 was achieved in England and Wales only in 1795. By

1300 production in China had halved. The Sung dynasty faded. By 1100 at least five cities had populations in excess of one million.

[53] In Hartwick (2003), I argued that the English were energy-conscious since at least the sixteenth century and it was not accidental that coal was turned to and that innovations in mine-draining machines (Savery and Newcomen pumps) came along when they did.

[54] Albion (1965) documents the struggle which the British navy had in maintaining a supply of timber for its ships. The early nineteenth century was a period of particular tension. "The ample heritage of excellent native oak which England had possessed when Henry VIII came to the throne was wasted during the next century partly by royal policy and partly by rival economics demands until the groves were so depleted by the time of the Restoration that the Navy felt the effects during the Dutch wars. Wise measures were drawn up to ensure an adequate supply from the royal forests, which might have met the entire oak demands of the Navy; but these were so neglected that the dockyards became dependent on the precarious contributions from private groves." (p. 412). Timber importers ran blockades during the Napoleonic wars, and supplies from Canada assisted greatly. Indian teak provided much needed wood for ships after local supplies were depleted. By 1860 it was clear that iron warships were superior and the perennial crisis of timber supply melted away. Williams (2003, p. 295) The success of iron-clad ships in the American Civil War signaled an end to the era of wooden ships, "the drain on the world's forests that had gone on for thousands of years stopped suddenly and dramatically."

[55] This family of models of development, via primary product exporting, illustrates deforestation as market driven, rather than being population driven or driven by property right failures. Many observers have emphasized that excessive deforestation occurs quite generally when property rights are not enforced, as with chop and grab.

[56] By Alexei Cheviakov, Queen's mathematics department.

[57] I am indebted to Alix Zwane for this information.

[58] Livi-Bacci (1991, p. 45) reports on declining family size in advanced countries since the eighteenth century.

## REFERENCES

Albion, R.G. (1965) *Forests and sea power: The timber problem of the Royal Navy.* Hamden, Connecticut: Archon Books.

Brander, J., & Taylor, S. (1997) The simple economics of Easter Island: a Ricardo-Malthus model of renewable resource use. *American Economic Review*, 88(1), 119-138.

Burroughs, W.J. (1997). *Does weather really matter?* Cambridge: Cambridge University Press.

Christy, F.T., & Potter, N. (1962). Natural resource commodities. *Baltimore: Johns Hopkins Press for Resources for the Future.*

Cohen, M.N. (1977). *The food crisis in prehistory.* New Haven: Yale University Press.

Cipolla, C.M. (1968). *The economic decline of Italy.* In Brian, S. Pullan (Ed.), Crisis and change in the Venetian economy in the sixteenth and seventeenth centuries (pp. 127-145.). London: Methuen.

Clark, R.P. (2000). *Global life systems: Population, food, and disease in the process of globalization.* Lanham: Rowman & Littlefield.

Cronon, W. (2003). *Changes in the land: Indians, colonists and the ecology of New England.* New York: Hill and Wang.

Deaton, A. (2003). Health, inequality, and economic development. *Journal of Economic Literature,* 41(1), 113-158.

Diamond, J. (1997). *Guns, germs and steel.* New York: Norton.

Dugger, C.A. (2004). Brazilian slums are seen as pawns in political games. *New York Times,* Jan. 18, p. 3.

Flinn, M.W. (1981). *The European demographic system.* Baltimore: Johns Hopkins University Press.

Gardner, B.L. (2003). *American agriculture in the twentieth century: How it flourished and what it cost.* Cambridge, Massachusetts: Harvard University Press.

Hartwick, J.M. (1978). Intergeneration equity and investing rents from renewable resources. *Economics Letters,1, 85-88.*

Hartwick, J.M., & Yeung, D. (1997). The tragedy of the commons revisited. *Pacific Economics Review,* 2(1), 45-62.

Hartwick, J.M., Long, N. V., & Tian, H. (2001). Deforestation and development in a small open economy. *Journal of Environmental Economics and Management*, 41, 235-51.

Hartwick, J.M., & Long, N. V.. (2001*). Land rent and staple exporting development*. typescript.

Homer-Dixon, T.F. (1999). *Environment, scarcity and violence*. Princeton: Princeton University Press.

HYDE Database, arch.rivm.nl/env/int/hyde/.

Krech, S. (1999). *The ecological Indian: Myth and history*. New York: Norton.

Kelly, M. (2003). *Climate and pre-industrial growth*. typescript.

Lane, F.C. (1968). Venetian shipping during the commercial revolution. In Brian, S. Pullan (Ed.), *Crisis and change (pp.22-46.)*. London: Methuen.

Lee, R. (1973). Population in pre-industrial England: An econometric analysis. *Quarterly Journal of Economics*, 87(4), 581-605.

Livi-Bacci, M. (1991). *Population and nutrition: An essay in European demographic history*. Translated by T. Croft-Murray. Cambridge: Cambridge University Press.

Livi-Bacci, M. (1992). *A concise history of world population*. Translated by C. Ipsen. Oxford: Blackwell.

Malin, J. (1956). The grassland of North America: Its occupance and the challenge of continuous reappraisals. In W. L. Thomas (Ed.), Man's role in changing the face of the earth (pp. 350-366.). Chicago: University of Chicago Press.

Manning, R. (1991). *Last stand*, New York: Penguin Books.

Manthy, R. (1978). *Natural resource commodities*. Baltimore: Johns Hopkins Press for Resources for the Future.

McNeill, W.H. (1976). *Plagues and peoples*. Garden City: Anchor Books.

Olmstead, A.L., & Rhode, P.W. (2002). The Red Queen and the hard Reds: Productivity growth in American wheat, 1800-1940. *The Journal of Economic History*, 62(4), 929-966.

Pomerantz, K. (2000). *The great divergence: China, Europe and the making of the modern world economy*. Princeton: Princeton University Press.

Primuspilus, C. (2003).Rome's Population. Accessed on July 15, 2003 from http://www.tulane.edu/august/H303/handout.../Population/html.

Principe, L.M. (2002). *History of science: Antiquity to 1700*. Virginia: The Teaching Company.

Rapp, R.T. (1976). *Industry and economic decline in seventeenth century Venice*. Cambrige, Massachusetts: Harvard University Press.

Robson, A. (2003). *A bioeconomic view of human characteristics and demographic transitions*. Accessed on July 15, 2003 from http://www.sfu.ca/robson/_.

Roosevelt, A.C. (1984). *Population, health and the evolution of subsistence*. In N.N. Cohen, G.J. Armelagos, (Eds.), Paleopathology at the origins of agriculture (pp. 559-583), New York: Academic Press.

Ruddiman, W. (2003). *Climate and early agriculture*. Accessed on July 18, 2003 from http://cires.colorado.edu/dslecture/ruddiman/ruddimanabs.html.

Runnels, C.N. (1995). Environmental degradation in ancient Greece. Scientific American, 272(3), 96-99.

Scott, S., & Duncan, C.J. (2001). *Biology of plagues: Evidence from historical populations*. Cambridge: Cambridge University Press.

Sedjo, R., & Lyon, K.S. (1996). Forests and markets. Accessed on July 15, 2003 from ftp://ftp.rff.org/pub/dpapers/9615.exe.

Smith, V.L. (1975). The primitive hunter culture, pleistocene extinctions, and the rise of agriculture. *Journal of Political Economy*, 83, 727-56.

Steckel, R.H., & Rose, J.C. (2002). *The backbone of history: Health and nutrition in the western Hemisphere*. Cambridge: Cambridge University Press.

Tian, H., & Cairns, R.D. (2002). Evolution of population and a resource with a maximin objective. *Presented at the Canadian Resoure and Environmental Economics Study Group*, Universite de Montreal, October, 2002.

United Nations Development Programme (2002). *A guide to world resources*, 2002-2004. Washington, D.C.

Usher, D. (1989). The dynastic cycle and the stationary state. *American Economic Review*, 79(5), 1031-1044.

Williams, M. (2003). *Deforesting the earth*. Chicago: University of Chicago Press.

Wrigley, E.A. (1986). Urban growth and agricultural change: England and the continent in the early modern period. In R. Rotberg, T. Rabb, (Eds.), *Population and economy: From traditional to the modern world (pp. 123-168)*. New York: Cambridge University Press.

# CHAPTER 9

# LIMITATIONS OF SUSTAINABLE FOREST MANAGEMENT: AN ECONOMICS PERSPECTIVE

## WILLIAM F. HYDE

*Senior Associate, Centre for International Forestry Research,*
*Bogor, Indonesia.*
*Email: wfhyde@aol.com*

**Abstract.** This chapter reviews the pattern of economic development in forestry and uses that pattern as a basis for commenting on sustainability. It concludes that sustainability in its narrowest sense, a "permanent forest estate" with unchanging boundaries, is a futile objective. It is more reasonable that we determine what to sustain—critical habitat, characteristics of global climate, perpetual options on the use of forest resources, or whatever—and then consider the feasible means for achieving each objective. Each objective requires its own measure of forest resources, and few of those measures are consistent with the standard measures of national forest inventories that are readily available at present. The chapter concludes by returning to the pattern of economic development in forestry as a means for instructing policy to achieve some of these objectives.

## 1. INTRODUCTION

Sustainable forest management is a common topic in the history of professional forestry, a topic that has recently assumed broader profile as an important issue of global environmental policy. Today, we can probably all agree that sustainable forestry has merit for its contribution to long term human welfare and that, at a minimum, it implies guaranteeing the full range of options for the future use of all forest resources. However, agreement on a more specific description of sustainable forestry is problematic. I intend to address that problem from the perspective of the general course of forest development, its meaning for the sustainability of different forest resources, and the data we rely on to examine this sustainability.

For some foresters, sustainable forestry has meant the assurance of a "permanent forest estate". That is, a forest with inviolable physical boundaries, boundaries that are not subject to change over time. It may also imply a perpetual "even flow" of timber harvests from this estate. For others with modern environmental concerns, sustainable forestry has a meaning similar to the foresters' permanent forest estate, but it also suggests special attention to tropical and developing country forests where the greatest threats to a permanent estate seem to exist.

*Kant and Berry (Eds.), Institutions, Sustainability, and Natural Resources: Institutions for Sustainable Forest Management, 193-210.*
© 2005 *Springer. Printed in Netherlands.*

This inflexible definition of sustainable forestry has its problems. Surely it is difficult to prevent the poor rural inhabitants of many countries from harvesting timber and converting forestland to agricultural production. And surely their harvests of timber and fuel wood and their conversions of forestland to agricultural use decrease the size of the natural forest and decrease the flow of those environmental services that originate from the forest: outdoor recreation and biodiversity, the provision of carbon sequestration, and the control of erosion. Furthermore, if we take a broader and more aggregate view of social welfare then, in many cases, we may not even desire to deter some timber harvests and land conversions—whether for subsistence agriculture, or for roads, water impoundments and other activities associated with development. These other uses of the land often create social benefits greater than the costs associated with the losses of forest-based environmental services—as the timber harvested from the land becomes a source of capital for economic growth and the land itself becomes a base for more productive alternative activities.

A second problem, at least in the traditional perception of many foresters, is that "even flows" of periodic harvests (constant harvest volumes over time) are virtually impossible to maintain. That is, the harvest volume from biologically mature trees on a hectare of natural forest is almost always greater than the harvest volume from a hectare of younger, economically mature, replanted and sustainably managed forest. Large old-growth trees and forest stands simply contain more volume per hectare than the younger trees that grow in sustainably managed stands and that are harvested before they attain the ages and volumes of old growth stands.

Is there a solution to these problems? Is there an alternative way of looking at the world's forests and trees and an alternative definition of sustainable forestry that can accommodate the objective of maximizing long-run human welfare from the broadest current perspective, while still insuring the full range of future options for the use of all forest resources? I contend that there is. I contend that by matching our objectives with the full range of available trees and forests we can come to a better understanding of the requirements of forest sustainability. However, this requires that we reconsider the common understanding of the course of forest development—and that is where I will begin.

## 2. THE GENERAL PATTERN OF FOREST DEVELOPMENT

A common pattern of forest development emerges from observations taken almost anywhere around the world and for almost any period of time. This is a pattern of rural development, deforestation and increasing scarcity of forest products, subsequent rising prices, and the eventual substitution of silvicultural investment for harvests from depleting natural forests. This pattern of forest development follows the general pattern of economic geography first proposed by von Thunen in the 19th century. Figures 9.1-9.3 capture its basic elements. These figures also provide the key reference points necessary for further reflections on investment timing, institutional constraints, and the markets and policies affecting forestry.[1]

Figure 9.1 describes a simple landscape of agriculture and forests. Consider agricultural land first. The value of agricultural land is a function of the net farmgate price of agricultural products—which is greatest when the farmgate is near the local market at point A. Land value in agricultural use declines with decreasing access (which is closely related to increasing distance from the market) as described by the function $V_a$. That is, the periodic crop value per hectare minus the cost of growing that crop creates a net value function that declines as we move to less and less accessible land farther and farther from the value center at A.[2]

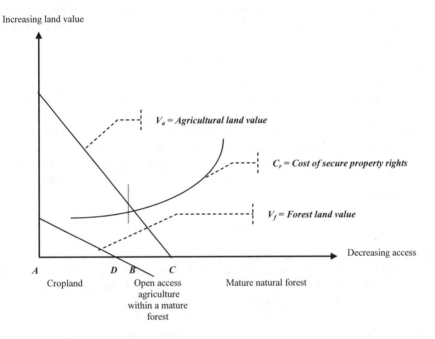

**Figure 9.1.** *A New Forest Frontier*

The function $C_r$ describes the cost of establishing and maintaining secure rights to this land. This is a transaction cost that is not a part of the net value calculation represented by $V_a$. It includes private costs such as the costs of registering a deed, fencing, and patrolling the property perimeter. The transaction cost increases as the level of public infrastructure and effective control declines and the cost of excluding trespassers expands as we move farther and farther from the value center at A (Anderson & Hill, 1975; Alston, Libecap & Mueller, 1999).

The functions explaining agricultural land value and the cost of secure property rights intersect at point B. Farmers manage land between points A and B for permanent and sustainable agricultural activities. They use land between points B and C (where agricultural land value declines to zero) as an open access resource to be exploited for short-term advantage. Local households and communities may protect some lands beyond point B to a limited degree—as by sending children out to manage their grazing livestock. Nevertheless, the function $C_r$ continues to increase after point B until eventually no reasonable number of forest guards can fully exclude illegal loggers, trespassing livestock, and other open access users of remote forests.

Local consumers harvest the products that grow naturally in this region between B and C, crops like fodder for their grazing animals, native fruits and nuts, and fugitive resources like wildlife. They do not invest even in modest land improvements in the region between B and C because the costs of protecting their investments would be greater than the return on these investments. Their use of this open access region is unsustainable except for periodic removals from pulses of regrown natural vegetation.

At the time of initial settlement, the mature natural forest at the frontier of agricultural development at point B has a negative value because the forest gets in the way of agricultural production and its removal is costly. The first settlers remove trees whenever the agricultural value of converted forest plus the value of the trees in consumption exceeds the cost of removal. In fact, farmers in some frontier settlements farm in and around the trees they have not yet removed in preference to absorbing this cost. Therefore, the function $V_f$ describing forest value must begin below the agricultural value gradient and, in the initial stage of development, it does not extend as far as the intersection of the agricultural value gradient with the horizontal axis.

## 3. DEVELOPING FRONTIERS

Both market demand and subsistence household demand justify the removal of some forest products, and they continue to justify additional removal at each new moment in time. Therefore, the forest frontier must gradually shift outward, away from the market center. The most accessible forest resources are always the first to be removed. This is true whether those resources are timber, fuel wood, bamboo, fruits, nuts, latex, or whatever. The forest value gradient continues to shift upward and outward over time (as shown by dashed function and the arrows in Figure 9.2) until it intersects the horizontal axis at some point like D.

The price of the delivered forest product in the market at A is just equal to the sum of its costs of removal from point D and delivery to the market. The *in situ* price of the forest product at point D is zero, and this means that the value of forestland at D is also zero. The region of unsustainable open access activities now extends from point B out to points C or D, whichever is farther. The costs of obtaining and protecting the property rights insure that this region remains an open access resource. Once this region extends beyond point C, marketed forest resources, not agricultural conversion, become the source of forest degradation and deforestation. This is an important distinction. It describes cases in which commercial logging, for example, is the source of forest destruction.

Some governments protect some lands past point B but they must absorb the increasing cost of protection—and even then trespass occurs. For example, some amount of illegal logging occurs almost everywhere in the world and an almost unlimited number of well-trained and well-motivated forest guards cannot prevent it entirely. It is an issue of serious policy concern, for example, in Indonesia but it occurs in the US and Canada as well. British Columbia alone suffers annual losses of US$200-300 million annually from timber theft and fraud (Smith, 2002) and as much as ten percent of all harvests from National Forest land may be illegal (Mendoza, 2003). Some local citizens illegally harvest Christmas trees from the well-managed national forests in the eastern US. The US Forest Service does not extend great effort to prohibit this theft because the costs of enforcement would be greater than the potential gain—which is another way of saying that, where illegal logging occurs, the cost function for property rights $C_r$ is above the net value function for forest resources $V_f$.

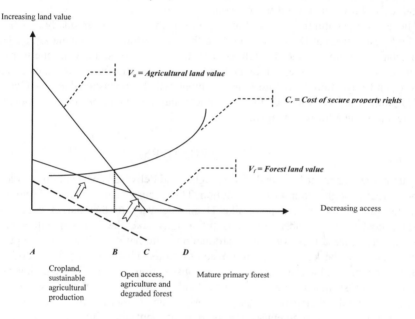

**Figure 9.2**. *A Developing Frontier*

The construct of Figure 9.1 conforms to the common description of any initial settlement. Trees actually impede agricultural development and the forest rent gradient is very low. Net forest resource values are sufficiently low that point D, where the forest rent gradient intersects the horizontal axis, is to the left of point C. This describes new settlement in the Ohio Valley in the US in the early 19th century (Richter, 1966). Apparently, it describes the forest frontier in Cote d'Ivoire (Lopez, 1998) and the Bolivian Amazon today (Bowles, Rice, Mittermier, & Fonseca, 1997). It probably describes upland settlement in the Philippines (Amacher, Cruz, Grebner, & Hyde, 1998) and many transmigrant settlements in Indonesia.

In other cases, described by Figure 9.2, the region between points B and D may be large and seriously degraded (*e.g.,* Nepal's hills or India's Rajastan). The positive net value of the original resource, together with the open access character of the region, has assured removal of the best resources. Some degraded vegetation remains in the region and with time it re-grows naturally. The lowest wage households continue to exploit this resource when the scattered vegetation grows to a minimum exploitable size or as its fruits begin to ripen (*e.g.*, Amacher, Hyde, & Kanel, 1999; Foster, Rosenzweig, & Behrman, 1997). These open access regions exist in both developed and developing countries. They are more degraded in some developing countries (typically sub-Saharan Africa and South Asia, but not Latin America) only because the relative opportunity costs of labor are lower in these countries, substitute opportunities for low wage laborers are less attractive, and pursuing and extracting the lower-valued remaining resources is the best opportunity available for those who collect these resources.

The use of the natural forest continues over time, the forest is gradually depleted, and the forest margin at D slowly extends farther and farther from the market center. Deforestation continues, and the delivered costs of forest products continue to rise. Nevertheless, the incentives of higher prices remain insufficient to induce tree planting and any attempt at forest management will be unsustainable. As Godoy (1992) points out, the prices of forest products may be rising, but they are not yet sufficient to induce forest management.

## 4. MATURE FRONTIERS

Eventually the margin at D extends far enough—and delivered costs and local prices become great enough—to induce substitution. This occurs when the costs of removal from the natural forest at a point like D and delivery to the market equal the backstop cost of some substitute. Substitution may take the forms of either new consumption alternatives or new production alternatives. The consumption alternatives could be kerosene or improved stoves as substitutes for fuel wood, or brick and concrete block as substitutes for construction timber. Tree planting and sustainable forest management on some land closer to the market would be production-related alternatives. Very clearly, the evidence of planting and sustainable forest management is not ubiquitous—but the physical presence of sustainable management is not trivial either. Indonesia, for one example, has 2.6 million ha of designated forest plantation and 13.4 million ha, or 9.4 percent of its total land area, in perennial forest plantations growing products like pulpwood, palm oil, cloves, coconut, and rubber (GOI/MFEC, 1998) and China has even more. The latest global forest resource assessment reports about 187 million ha of plantations worldwide, about four percent of total global forest cover (FAO, 2001).

The forest value gradient rises with the increase in delivered costs (from the dashed line to the new solid forest rent gradient in Figure 9.3) until, at some moment in time, it intersects the agriculture rent gradient to the left of the agricultural value intersection with the cost function for secure property rights. New, sustainably managed, forests occur in the vicinity of B'B" of Figure 9.3. They may take the form of industrial timber plantations or they may take the forms of agroforestry or even of just a few trees whose growth is permitted, or even managed, around individual households.

We might call Figure 9.3 a description of the "mature" frontier of primary forest. Regions described by Figure 9.3 are "mature" in the sense that sustainable forest activities (at B'B") compete with some resource removal activities (at D'). They are still "frontiers" because some removal of natural forest stocks at the frontier remains competitive with sustainable activities. For communities in regions of mature forest frontiers, forest product prices will be sufficient to justify the substitution of managed forests for the resources of the open access natural forest. The new managed forests may take the form of tree plantations, or they may just be a few managed trees in fields, along roadsides, or around homes. The latter (household trees) are excluded from most measures of the forest stock but their economic importance can be large. In Bangladesh, for example, they account for 3/4 of all market timber and fuel wood consumption (Douglas, 1982).

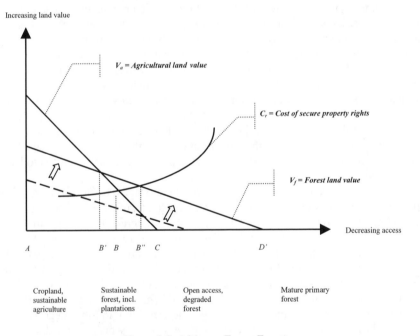

*Figure 9.3. A Mature Forest Frontier*

They are major sources of fuel wood consumption in Malawi (Hyde & Seve, 1993), timber production in Kenya (Scherr, 1995), and of positive environmental externalities in northern China (Yin & Hyde 2000) and the Murang'a region of Kenya (Patel, Pinkney, & Jaeger 1995). One Indonesian estimate suggests that they may account for tree cover on 47 percent of Java's land area--47 percent that is not part of the Ministry's official estimate of forest land (D. Garrity, pers. comm.). In North America and Western Europe, they include trees in city parks and residential backyards all over both continents. In fact, the Southeast of England, including London, is the most populated region of the country. It is also the most wooded, and the wooded area in southeast England is growing more rapidly than in any other region of the country. The city of London alone has 65,000 stands of trees and woodlands covering almost 7,000 ha. (U.K. Forestry Commission, 2001).

In all cases, removals from the mature natural stock are concentrated in the neighborhood of point D in figures 9.1 and 9.2—or D' in regions characterized by the higher forest rent gradient in Figure 9.3. Mature natural stocks in the region before D (or D') were removed in earlier times because they were open access resources. In most cases a mature natural forest of no market or subsistence value exists beyond D (or D'). Sometimes this region beyond point D (or D') is negligible (*e.g.*, in Ireland or Cape Verde). Sometimes it continues well beyond the frontier of economic activity at D (Siberia, Alaska, northern Canada, much of the Amazon, a large component of the Rocky Mountains in the US, and much of Kalimantan in Indonesia) until it becomes the largest share of reported physical stocks in some cases.

Our three figures trace an inter-temporal progression but, at any moment in time, they also describe static snapshots of local conditions. And, at any moment in time, all three local conditions may exist simultaneously in different parts of some large countries. All three can exist simultaneously because most primary forest products are either bulky or perishable and do not transport well before they reach the location of their next level of processing. Therefore, their markets are geographically contained. As a result, standing natural forest reserves remain in some regions (*e.g.*, Siberia, Alaska, northern Canada, the eastern Amazon, and Kalimantan) while, in other parts of the same country, the forests are depleted and some landowners may have begun to plant trees on their own lands (*e.g.*, the Caucasus, the US South, southern Ontario, the developed part of Brazil's Paragominas, and central Java, respectively).

In sum, this characterization identifies three stages of forest development (described by our three figures) and three categories of forest: managed forests (including industrial forest plantations, more scattered household trees, and agroforestry plantings) in the vicinity of B'B", depleted forests from point B (or B") out to point D (or D'), and an unmarketable mature natural forest beyond point D (or D'). For many uses of the forest, we might identify a fourth category: the focused region of current harvests from the mature natural forest in the neighborhood of D (or D'). Commercial timber and fuel wood, the most common wood products, generally originate from the first (managed forests) and fourth categories. Most forest policy and management is concerned with effects on the first three forest categories. Environmental and aesthetic concerns feature either the last one (*e.g.*, biodiversity and natural preserves) or select locations within any of the first three (parks, erosion control, sustainability).

## 5. QUANTITATIVE ECONOMIC EVIDENCE

This development pattern is general for all market-valued forest resources although variations may appear to exist where the natural forest provides multiple market valued products; where mountains, swamps and roads modify the locus of access;

and where land managers and governments modify the relevant bundles of property rights.

A number of empirical studies demonstrate its generality for commercial timber. For example, Berck (1979) and Johnson and Libecap (1980) demonstrate it for examples from US history, and Stone (1998, 1997) describes the pattern and its impact on the wood processing industry in Brazil's Paragominas. Recently, several studies demonstrate its relevance for fuel wood and for agricultural expansion. In particular, Hofstad (1997) confirms a pattern of expanding forest extraction for charcoal in the vicinity of Dar es Salaam, and Chomitz and Griffiths (1997) describe a similar pattern for charcoal extraction from multiple population centers in Chad. Chomitz and Griffiths also observe the substitution opportunities that constrain expanding supply regions and rising charcoal prices once the price of a backstop energy source or an alternative technology is attained. Their observation that substitution eventually constrains deforestation is consistent with empirical observations from the household economics literature that opportunities for substitution constrain the consumption of high cost fuel wood. (See Hyde, Kohlin, & Amacher, 2000.)

## 6. SUSTAINABLE FORESTRY AND THE CONTROL OF DEFORESTATION

The concern for sustainable forestry grew out of a much older concern for resource depletion. It originally focused on market-valued resources in general and, in the case of forests, on timber in particular. The Viceroy of Mexico City wrote home to the king of Spain in 1546 alerting him that North America was running out of timber. In 1876, F.B. Hough, in an address to the American Association for the Advancement of Science, described the environmental damage suffered following deforestation and that association formed a committee to encourage the US Congress to address the apparent forthcoming timber shortfall in the US. The US National Forest System was eventually created in 1891 to address this problem (and also to address watershed management issues in eastern US) (Clepper, 1977). Western Europeans have been concerned about a timber shortfall at least since Jevons (1865) wrote about the limited sources of mine props in England in mid-19th century. In addition, Europe's periodic wars over the last several hundred years have regularly depleted the existing mature timber—and, thereby, demonstrated the need for a supply of timber as a strategic material.

Of course, North America has not run out of timber. In fact, US timber stocks are greater today than they were 100 years ago. Fortunately, Western Europe seems to have entered a period with fewer large wars and the usefulness of wood as a strategic resource has declined. Furthermore, the stocks of most market-valued goods are not declining in economic terms. That is, their costs of production have not increased over time (Barnett & Morse, 1963; revised and renewed by various others).

However, residual doubts exist in the minds of many regarding the potential for timber shortages. Others, while not as concerned about depleting these market-based forest products, are concerned that we may be depleting our stock of global means to provide the non-market environmental services of forests. Taken together, these arguments are the basis for modern policy discussions of forest sustainability and controlling deforestation.

The modern discussions take a number of perspectives (Toman & Ashton, 1996). Perhaps a useful perspective for our purposes would be "sustainable options". That is, sustainability restated as a useful objective could be "maintain, in perpetuity, options for all different uses of forest resources, market and non-market, consumptive and non-consumptive, known and unknown." This would mean controlling environmental destruction. It would mean maintaining for the future the potential for all the different uses of the land and other forest resources. It would also mean using the forest to help maintain other future options.[3] It would include using the forest to help control erosion, protect critical habitat and important aesthetic resources, and limit global change. Such a statement of sustainability would allow some shifts of forestland to agriculture, others from agriculture back to forestry, and still others from natural forest to managed forest so long as both the land's productive base and also the genetic base of the forest remain undamaged. Relative values will change with time and preferred patterns of land use will change with them, but we can insure that changes in land use do not destroy opportunities for new and different land or resource uses in the future.

In the context of our model of forest development, this perspective of sustainability is consistent with minimizing the area of degraded open access forest, while locally regulating specific eroding watersheds, critical habitats, and important aesthetic resources, both within and outside the degraded area. Minimizing the degraded area is the objective because its elimination is impossible as long as secure property rights impose a cost and so long as the public agencies responsible for managing the degraded open access area have limited budgets. It is a general objective for both developed and developing countries regardless of the magnitude of their remaining natural forests because all countries contain forests with a degree of open access and, therefore, some degree of degraded habitat, some level of erosion, some loss of aesthetic resources, etc.

The fundamental means for minimizing the degraded area involve a) reducing the cost of the property rights and b) attracting human activity away from the forest. The first requires finding the most appropriate bundle of property rights and the institutions that can provide this bundle at least cost. This insures the lowest cost function $C_r'$ in figures 9.1-9.3. Of course, the appropriate bundle of rights and the most effective institution will vary with local values. Various arrangements of private rights, local community rights, or state ownership will be appropriate in different local situations but none will be a universal solution. (See, for example, Dangi & Hyde, 2001 and Hyde, Xu, Belcher, Yin, & Liu, 2003 for discussions of effective local variations in forest property rights in Nepal and China, respectively.)

We indicated the effect of the opportunity cost of labor on the natural forest. Forest users with lower opportunity costs can afford the time to travel farther into the natural forest to extract its products. Because their costs are low, they can also justify removing material in the degraded area down to a low level. Providing low wage or low opportunity cost forest users with improved employment opportunity outside the forest causes some of them to change from extractive activities in the forest to the higher wage employment. In terms of the mature economy depicted in Figure 9.4, the net forest value may remain unchanged at the market, but the forest value function shifts inward along the horizontal axis and becomes steeper as some users leave the forest for higher wage opportunities, frontier labor becomes scarcer and frontier wages increase, and the labor opportunity costs of removing additional resources from the frontier forest also increase.

The combined effects of these two fundamental improvements are a decline in the degraded open access area from $B_1"D_1$ to $B_2"D_2$ and an increase in the forest density in the remaining open access area.

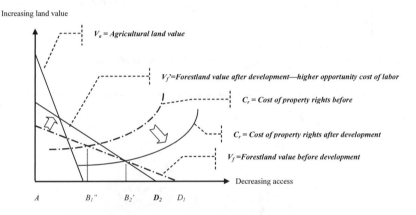

**Figure 9.4.** *Sustainable Forestry and the Control of Deforestation*

Stating the argument a different way: Poverty causes forest degradation and forest depletion. Economic development induces improvements in the forest environment as it shifts land into sustainable activities. In fact, economic development is likely to have a second round of beneficial effects as well. Improved wages and better labor opportunities create the first round. Then, along with improvements in overall welfare, the local institutions also tend to become more effective. They improve in their ability to insure property rights and in their ability to manage economic transitions and provide economic stability. Both improved institutions and a more stable economy lower the transactions cost function and cause a second round of reductions in the degraded area.

We can illustrate this argument further with figures 9.5a and 9.5b. Figures 9.5a and 9.5b contrast two regions, one more developed than the other in terms of overall economic welfare. In this illustration, both are in the third stage of forest development, although we could make similar comparisons for regions in the first and second stages as well. Agricultural land values are comparable in both regions. The cost function for property rights is lower and alternative wage opportunities for forest users are greater in the more developed region. Therefore, the degraded area between points B" and D is smaller in this region.

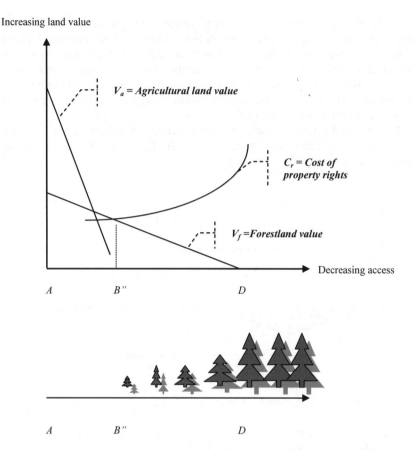

**Figure 9.5a.** *Sustainability: The Effects of Development*

The depictions of the open access degraded area and the mature natural forest (beyond point D) under the figure show the contrasts in degradation between regions of less and greater overall development. In the more developed region, Figure 9.5b, only a small area of open access forest is degraded and even this forest in not heavily degraded. It is smaller in area and also better stocked because the rewards of open access trespass onto lands with formal title are small compared with the risks incurred for local populations whose incomes, while modest, are well above those of many forest users in the poorer and more degraded region, Figure 9.5a.

In fact, we know that an area of open access forest exists, even in economically developed regions of the most developed countries. We previously indicated the large value of illegal timber harvest, even in western US and Canada. Open access removals of non-timber products can be important too. The value of open access harvests of ginseng and Christmas greenery in southwest Virginia in the US exceed one million dollars annually (Hammett & Chamberlain, 1998). These activities are often difficult to detect and their effects may even be unnoticeable to casual observers. This is one reason they can continue, even in forests with identifiable (but incompletely enforceable) formal property rights.

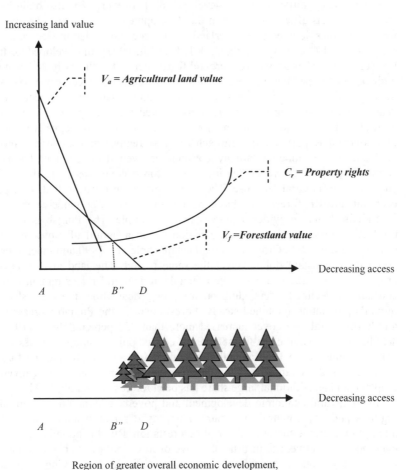

*Figure 9.5b. Sustainability: The Effects of Development*

The volume of illegal activity is greater in less developed regions and countries. Countries experiencing sharp and substantial economic instability and decline provide confirming evidence. In the late 1990s, as general economic well being in the countries of the former Soviet Union suffered serious decline, many of the formal institutions in these countries also suffered decline. Their budgets declined and the respect for their authority declined as well. Illegal logging increased dramatically and simultaneously. In Estonia, for example, up to three-fourths of the timber harvested between 1998 and 2002 may have been in violation of legal regulations (Hein & Ahas, 2004).

The broader historical evidence is also consistent with these arguments. Countries draw down their stocks of natural resources like trees and forests as they enter periods of initial economic development. However, they also build back their stocks of forests after some point in the development process. For example, natural forest cover has doubled in Switzerland and France and tripled in Denmark since the early or mid-19th century (Mather, 2001; Kuechli, 1997). In developed northeastern US, forest cover has grown from around fifteen percent in the early 20th century to as high as ninety percent in some states in the early 21st century (USDA Forest Service, 2000). Forest cover has increased by more than thirty percent over the 25 years since the first agricultural reforms marked the beginning of China's double digit annual economic growth and six-fold increase in rural wages (China Forestry Development Report 2000).[4] In each of these regions and countries, agricultural land use has remained relatively constant or even declined over the period of economic growth. However, the increase in forest cover has more than offset any change in agricultural land use. The only explanation must be that, in a period of economic growth, forest cover has expanded into areas of previously degraded land.

India's Punjab provides a more specific example. The Punjab is India's most productive agricultural region. The region began a period of rapid and sustained development in 1960. Crop yields per hectare tripled by 1990 and income per capita doubled (in constant dollars) over the same period. The land area in agricultural crops more than doubled while the principle agricultural prices remained relatively constant or declined—depending on the crop. Meanwhile, the rural share of the region's population remained steady. Forest cover in the Punjab increased six-fold and horticultural tree cover increased more than 250 percent. Before 1960 a large area had been cleared of its forest cover. It existed only as an open access wasteland. It has declined since 1960 as both the land area in agriculture and the forest stock have increased. A large share of the open access lands has been converted into cropland and an additional large share has been reforested (Singh, 1994).[5]

In sum, rural economic development and poverty alleviation are central to any region's and any country's program of improved forest sustainability and to any attempt to decrease the rate of global deforestation. Rural economic development is everyone's objective, not just the objective of forest policy. Accomplishing it is not an easy task, but it is certainly no more difficult than trying to accomplish sustainable forestry and decrease the rate of deforestation through the imposition of government regulations on the uses of relatively dispersed forest resources by a scattered and poor rural population.

## 7. CONCLUSION

We must recognize that natural forests are unlikely to be sustainable and regulations on their use are unlikely to be effective until we attain a higher state of general development than is common to substantial parts of the world today. Economic development and stability are the means for limiting deforestation and for improving general forest sustainability. Improved employment opportunities and stronger institutions are products of such stable development. They provide alternatives to greater reliance on forest resources and they also provide a better means of insuring the desired behavior in the forest. The result would be an improved forest.

This is a fundamental point and it has a crucial corollary. Those who choose to encourage forest sustainability have a difficult task. They must encourage development in the poorer forested regions of their own and other countries. Until these regions do develop, those from wealthier regions must provide the support necessary to assist the institutions of less developed regions and countries with the responsibility for protecting their forest resources, and the developed regions must provide this assistance on a long-term basis. Encouraging the global trade that will improve local employment opportunity will also be helpful.[6]

Nevertheless, we can be sure that development assistance will not result in universally sustainable forestry. Therefore, the assistance must be targeted, and targeting requires a clear understanding of objectives. Sustainable forestry itself sounds like a noble objective, but the underlying reasons for sustainable forestry provide better focus. These include:

i. carbon sequestration to protect against global climate change,

ii. protection of biodiversity and its critical habitat,

iii. protection of the resources that provide for outdoor recreation and environmental tourism, and

iv. erosion control and general watershed protection.

General forest sustainability, as sustainability of all measured forests, has some positive effect on all four of these objectives. However, the available general measures of existing forests stocks do not adequately represent any of the four. For example, all trees sequester carbon, but trees outside of forested plots are excluded from all official measures of any country's forests. Furthermore, there are more trees outside the forest than within the official estimates of some country's forests. Therefore, the available estimates of forest volume and forest cover provide inadequate estimates of the amount of carbon sequestered. For a second example, the critical habitat for many endangered species occurs outside the forest and many hectares, even many hectares of old growth natural forest, are not unique habitat. Therefore, the official measures of forest stocks are also inadequate as indicators of critical habitat. Similarly, some forest is more important than other as a resource for outdoor recreation and environmental tourism, and some forest is more important than other for its watershed value.

Finally, regardless of these inadequacies, the official standards for what to include in forest stocks vary from country to country by as much as a factor of 10,000 (Lund, 2000)! Surely this alone is reason to use international estimates and comparisons of forest degradation, deforestation, and forest sustainability only with the greatest caution.

These policy instruments that could most successfully address each of these underlying policy objectives, and the inadequacy of the common sources of forestry data for addressing any of them, are major topics in their own right. We will only identify them here, and save their greater discussion for other forums.[7] I would conclude, however, that a better understanding of the pattern of development of property rights for all forest resources would improve both the discourse about forest sustainability and any policy intending to accomplish it. I hope this chapter can be a step in that direction.

## NOTES

[1] The forthcoming book, The Global Economics of Forestry, contains greater detail on this pattern of development, as well as discussion on its variants.

[2] We can assess land value in terms of a period of time, perhaps a year or a series of years. Both the revenues from the use of the land and the costs subtracted from them to create the net value function $V_a$ refer to the same period. In the case of a series of years, the net value is the sum of discounted net values for those years. In this paper further reference to value within our figures refers to the same period— unless explicitly indicated otherwise.

[3] This perspective is consistent with the idea of "sustainable livelihoods," and with the definitions selected by the Ministerial Conference on the Protection of European Forests and the "Helsinki Process" (1993) and also by the CSCE Seminar and the "Montreal Process" (also 1993). The definitions of sustainable forest management emerging from these meetings focused on maintaining the diversity and productivity of forests while ensuring future opportunities from the forests (FAO, 2002). This perspective is not consistent with those other statements of sustainability that seek a "permanent forest estate". In fact, a forest estate with permanent boundaries will be a futile objective, as we have seen that forest boundaries will change as forest development proceeds through its three stages, and as local relative prices adjust with development.

[4] See Hyde et al. (2003) for a review of China's data and this experience.

[5] The examples of the last two paragraphs suggest a turning point for forest recovery above some level of regional welfare. (Is there an environmental Kuznets curve for forests?) The data are not satisfactory for easy assessment of this point. The evidence of similar turning points for other natural resources and for various environmental pollutants suggests a turning point at income levels in the neighborhood of $5,000-$8,000 per capita (Dasgupta, Laplante, Wang, & Wheeler, 2002). See the discussions in chapters 2 and 6 in Hyde (2004).

[6] Hyde (2004) outlines the conditions under which trade had deterministic impacts on the global forest environment.

[7] See, for example, Hyde, 2003 and Hyde, 2004.

## REFERENCES

Alston, L., Libecap, G., & Mueller, B. (1999). Titles, conflict and land use: The development of property rights and land reform on the Brazilian Amazon frontier. Ann Arbor: University of Michigan Press.

Amacher, G., Cruz, W., Grebner, D., & Hyde, W. (1998). Environmental motivations for migration. Land Economics, 74(1), 92-101.

Amacher, G., Hyde, W., & Kanel, K. (1999). Nepali fuel wood production and consumption: Regional and household distinctions, substitution, and successful intervention. Journal of Development Studies, 35(4), 138-64.

Anderson, T., & Hill, P. (1975). The evolution of property rights: A study of the American West. Journal of Law and Economics, 18, 163-179.

Barnett, H.J., & Morse, C. (1963). Scarcity and growth: The economics of natural resource availability. Baltimore: The Johns Hopkins University Press.

Berck, P. (1979). The economics of timber: a renewable resource in the long run. *Bell Journal of Economics,* 10(2), 447-62.

Bowles, A., Rice, R., Mittermier, R., & Fonseca, G. (1998). Logging and tropical forest conservation. *Science,* 280, 1899-1900.

China Forestry Development Report. (2000). Beijing: China Forestry Press. As quoted in W. Hyde, J. Xu, B. Belcher (Eds.), *Introduction to China's forests: Global lessons from market reforms.* Washington: Resources for the Future.

Chomitz, K. & Griffiths, C. (1997). An economic analysis of woodfuel management in the Sahel: The case of Chad. *World Bank policy research working paper,* 1788. Washington, D. C. The World Bank.

Clepper, H. (1977). *Professional forestry in the United States.* Baltimore: Johns Hopkins University Press for Resources for the Future.

Dangi, R., & Hyde, W. (2001). When does community forestry improve forest management? *Nepal Journal of Forestry,* 12(1), 1-19.

Dasgupta, S., Laplante, B., Wang, H., & Wheeler, D. (2002). Confronting the environmental Kuznets curve. *Journal of Economic Perspectives,* 16(1), 147-68.

Douglas, J. (1982). Consumption and supply of wood and bamboo in Bangladesh. *Field document no. 2,* UNDP/FAO project BGD/78/010, Bangladesh Planning Commission, Dhaka, Bangladesh.

Food and Agriculture Organization of the United Nations. (2001). Global forest resources assessment 2000. Rome: FAO forestry paper 140.

Food and Agriculture Organization of the United Nations. (2002). *Proceedings: Second expert meeting on harmonizing forest-related definitions for use by various stakeholders.* Rome: FAO.

Foster, A., Rosenzweig, M., & Behrman, J. (1997). *Population and deforestation: Management of village common land in India.* Draft Manuscript., Department of Economics, University of Pennsylvania.

Godoy, R. (1992). Determinants of smallholder commercial tree cultivation. *World Development,* 20(5), 713-25.

Government of Indonesia, Ministry of Forestry and Estate Crops. (1998). *1997/1998 Forest utilization statistical yearbook.* Jakarta: D.G. of Forest Utilization.

Hammett, A., & Chamberlain, J. (1998). Sustainable use of non-traditional forest products: Alternative forest-based income opportunities. *Proceedings, Natural Resource Income Opportunities on Private Lands.* pp. 141-147.

Hein, H., & Ahas, R. (2004). *The structure and estimated extent of illegal forestry in Estonia.* Unpublished Manuscript. Institute of Geography, University of Tartu.

Hofstad, O. (1997). Deforestation by charcoal supply to Dar es Salaam. *Journal of Environmental Economics and Management,* 33(1), 17-32.

Hyde, W. (2003). Economic considerations on instruments and institutions. In Y. Dube, F. Schnithusen (Eds.), *Cross-sectoral policy impacts between forestry and other sectors.* Rome: FAO Forestry Paper 142.

Hyde, W. (2004). *The global economics of forestry.* Draft Manuscript.

Hyde, W., & Seve, J. (1993). The economic role of wood products in tropical deforestation: The severe experience of Malawi. *Forest Ecology and Management,* 57(2), 283-300.

Hyde, W., Kohlin, G., & Amacher, G. S. (2000). Social forestry reconsidered. In W. Hyde, G. Amacher (Eds.), *The economics of forestry and rural development in Asia (pp.243-287).* Ann Arbor: University of Michigan Press.

Hyde, W., Xu, J., Belcher, B., Yin, R., & Liu, J. (2003). Conclusions and policy implications. In W. Hyde, B. Belcher, J. Xu (Eds.), *China's forests: Global lessons from market reforms.*(pp.195-214) Washington: Resources for the Future.

Jevons, W.S. (1865). The coal question: an inquiry concerning the progress of the nation, and the probable exhaustion of our coal-mines. London: Macmillan.

Johnson, R., & Libecap, G. (1980). Efficient markets and Great Lakes timber: A conservation issue reexamined. *Explorations of Economic History,* 17(3), 372-385.

Kuechli, D. (1997). *Forests of hope: Stories of regeneration.* London: Earthscan.

Lopez, R. (1998). The tragedy of the commons in Cote d'Ivoire agriculture: Empirical evidence and implications for evaluating trade policies. *World Bank Economic Review,* 12(1), 105-132.

Lund, H.G. (2000). *Definitions of forest, deforestation, afforestation, and reforestation. Manassas, VA: Forest Information Services.* Available at http://home.att.net/~gklund/DEFpaper.html.

Mather, A. (2001). The transition from deforestation to reforestation in Europe. In A. Angelson, D. Kaimowitz (Eds.), *Agricultural technologies and tropical deforestation* (pp. 35-52). Wallingford: CAB International.

Mendoza, M. (2003). Timber thieves in the U.S. saw forests for the trees. *Denver Post* (May 18) p. 9A.

Patel, S., Pinkney, T., & Jaeger, W. (1995). Smallholder wood production and population pressure in East Africa: Evidence of an environmental Kuznet's curve? *Land Economics*, 71(4), 516-30.

Richter, C. (1966). *The trees*. Athens: Ohio University Press.

Scherr, S. (1995). Economic factors in farmer adoption of agroforestry: patterns observed in western Kenya. *World Development,* 23(5), 787-804.

Singh, H. (1994). The green revolution in Punjab: The multiple dividend, prosperity, reforestation and the lack of rural out-migration. *Unpublished student paper*, JFK School of Public Policy, Harvard University.

Smith, W. (2002). The global problem of illegal logging. *Tropical Forest Update*, 12(1), 3-5.

Stone, S. (1997). Economic trends in the timber industry of Amazonia: Survey results from Para state, 1990-95. *Journal of Developing Areas,* 32, 97-121.

Stone, S. (1998). The timber industry along an aging frontier: The case of Paragominas (1990-1995). *World Development,* 26(3), 443-448.

Toman, M., & Ashton, M. (1996). Sustainable forest ecosystems and management: A review. *Forest Science,* 42, 366-77.

UK/Forestry Commission. (2001). *Forestry statistics 2001*. Edinburgh. Available at: http://news.independent.co.uk/uk/environment/story.jsp?story=280241.

USDA Forest Service. (2000). 1997 RPA: The United States forest resource current situation. *Forest inventory and analysis*. Retrieved from, http://fia.fs.fed.us/rpa.htm.

Yin, R., & Hyde, W. (2000). Trees as an agriculture sustaining resource. *Agroforestry Systems,* 50, 179-194.

# CHAPTER 10

# SUSTAINABLE FORESTRY IN A WORLD OF SPECIALIZATION AND TRADE

ROGER A. SEDJO

*Resources for the Future*
*Washington, DC. USA.*
*Email:sedjo@rff.org*

**Abstract:** Since Adam Smith economists have recognize that specialization provides the basis for a modern economy since it promotes increased productivity, lower costs and intra regional and international trade. Industrial forestry seems to have recognized this economic reality and in the past fifty years has been moved from obtaining almost all of its industrial wood from the logging natural forests to the production of over one-third of society's industrial wood production from a trees cropping regime of planting, growing and harvesting intensively managed forests. However, much of the modern environmental movement is opposed to specialization and stresses the concept of individual forest sustainability for a spectrum of outputs, an approach directly the opposite to that of economic specialization. This paper attempts to reconciled these conflicting approaches by recognizing the substantial differences in the outputs mix generated by different forests, referring to the commonly accepted Brundtland Commission definition of a sustainable system and applying this concept to the multiple outputs of the various forest.

## 1. INTRODUCTION

Sustainability is a difficult concept. A definitive definition of the term remains elusive. The Brundtland Commission (WCED, 1987) defined a sustainable system as one that is capable of meeting current needs without compromising the ability of the future to meet its needs. In effect, the system will have as many resources in the future as it had in the past. But, does sustainability refer to the parts, or to the whole? Is sustainability important for a community, or only for the nation state, or, even more broadly, is the relevant unit the global system? For forests there is also a question of scale. Does forest sustainability refer to an acre, a stand, the landscape or the global forest?

The literature on sustainability raises other related questions. Does sustainability depend on capital and what is the degree of substitution between natural and human-

*Kant and Berry (Eds.), Institutions, Sustainability, and Natural Resources: Institutions for Sustainable Forest Management, 211-231.*

made capital? Does future well-being depend on maintaining some critical level of natural capital? Or are natural and human capital highly fungible, allowing for the widespread substitution of human-created capital—including knowledge and capital plant and equipment—for lost natural capital? How do increases in human capital, e.g., knowledge and training, fit into the analysis of sustainability? It is clear that humans have altered the character of natural capital. Planted forests, while not identical to natural forest, replace natural forests for many purposes. Furthermore, while forests may be lost, land usually is not. There are fewer forests and grasslands today than there were 2000 years ago. However, there are substantially more croplands, and the earth is capable of providing for a greater sustainable human population than it could have 2000 years ago. Natural capital has been complemented by human capital and technical knowledge.

Another issue deals with the question of specialization. As an economist, I know that the industrial revolution was driven to a large extent by specialization. Adam Smith in his famous *Wealth of Nations* (1776) argued that specialization allows for greater productivity. Specialization generates trade and there are "gains" from interpersonal, interregional and international trade. Specialization, among other things, has allowed for the high levels of productivity that many of our societies enjoy today. On the basis of different endowments of productive factors and the resulting specialization, some sites or regions have a comparative advantage in the production of some products, while other regions have the advantage in producing other goods.

Do specialization and comparative advantage only apply in the world of manufacturing and agriculture, but not in the world of forestry? Much of the discussion of forest sustainability argues in the opposite direction from specialization. Forests are expected to produce a host of outputs. Specialization is viewed as simplifying and thus reducing the range of outputs of some of these ecosystems. Smith would argue that in reducing the range of outputs each unit (read forest) produces, the productivity on the outputs chosen for specialization increases. Might some forests have a comparative advantage in one output, e.g., timber, while other forests have the advantage in biodiversity? I find it a bit curious that society expects a variety of environmental services from forests, but not from cropped fields or pasture.

In the first section of this chapter I examine some concepts of sustainability. Next, I look at the outputs of forests and examine the role forests have played in human sustainability. I then briefly present an overview glimpse of the global forest system. I conclude that the concept of a global forest, parts of which specialize in providing different economic and environmental services, is more useful both from a practice perspective and as an ideal than the idea of a sustainable individual forest. The second section of the chapter I briefly reviews the current situation and recent changes in global land uses, with a focus on forestlands.

## 2. CONCEPTS OF SUSTAINABILITY, SPECIALIZATION, AND FOREST MANAGEMENT

### 2.1 Background

Forests have at least three distinct roles in contemporary society. One is to provide humans with an important commodity—wood. This commodity can readily be traded in markets and is quite mobile. There exist very active local, national, and global markets for this commodity. Specialization, trade, and markets are an inherent part of our production system. Sustainability of an individual forest, however, is not critical for the sustainability of the wood market system. Production may shift from one forest to another. Furthermore, due to trade, a country may enjoy the industrial wood commodity without producing a single stick of wood. A second and different role for the forests, is to provide humans with a host of useful, indeed essential, local environmental goods and services. These are highly localized, not mobile, and not easily transacted in markets, e.g., watershed protection. A third role for forests is the provision of global environmental goods, e.g., biodiversity. The sustainability of forests that provide global environmental goods is important, not only as part of a global system, but also for their individual parts (forests). Unique biodiversity is highly site specific.

Thus, I would argue, that at least three different but sustainable forest systems are needed. The first to focus on timber production, using the agricultural cropping model that relies on the specialization and intensive management, which emerged from the industrial revolution and is so much a part of the technology currently in use today. Such a productive system can shift geographically over time, and, in fact, we are seeing that shift today as wood production shifts to the planted forests of the subtropics. In this case, while the overall productive system may need to be sustainable, the individual production forest need not. The second forest system is that which provides important local environmental services, mostly nonmarket and highly localized. This system is immobile and tends to require forests that are stationary, persistent and indeed sustainable by site. The third sustainable forest system is that which provides for global public environmental goods, for example, biodiversity and carbon sequestration. Since carbon sequestration is provided by most forests through their sequestration of carbon in the biomass, such a service need not be site specific. However, if the focus is on biodiversity, site becomes important. These three roles for forests, indeed types of forests, are not always mutually exclusive, but may be. They may exist separately but in some cases all the outputs can be provided by a single forest. Society has an interest in a forest system that produces all these sets of outputs. Nevertheless, it may not have an interest in sustaining each individual forest on continuous bases.

### 2.2 Some Concepts of Sustainability

The ideas associated with the concept of forest sustainability have varied and evolved over time. Sustainability in forest management is both a biological and socioeconomic concept, and, as noted by Fedkiw (2003), it can be a goal, a process, or both. Originating in 18th century Europe with the aim of avoiding social and economic disruptions associated with timber shortages, sustained-yield forest management evolved to a highly technical process of modeling growth, mortality, and risk in order to determine a maximum biological level of timber removals that could be maintained in perpetuity. Out of this objective developed the concept of a regulated forest, including the notion of an optimal harvest rotation. The biological harvesting rule, however, was shown by Faustmann (1849) not to maximize financial returns. When the financial maximizing rule was applied to the regulated forest, it dictated a somewhat shorter optimum harvest rotation. Subsequently, there was a lengthy period of tension between the devotees of the biological rule and those of the financial rotation rule.

Early humans, of course, had no concern for sustainability or a harvest rule. Rather, they applied the simple "hunter-gatherer" mode of collection, gathering timber when needed and when and where available. Where the natural system was inadequate to regenerate the timber, humans could move on to the next forest. Centuries ago, in some parts of the world, as in Europe and parts of Asia, a "husbandry-stewardship" approach began to replace the earlier hunter-gatherer mode. From this perspective, the early concept of forestry, regulated forests, and harvest rotation lengths were developed. In much of the New World however, forestry, at least in its initial stages, could disregard notions of sustainable yield since the region was awash with forests. Indeed, attempts to apply European concepts to American forestry generally failed, due largely to the huge overhang of forest stocks. In the United States, serious concerns about sustainability did not arise until well into the 19th century. These concerns were expressed as fear of a "timber famine." Beginning in the late 19th century and continuing into the early 1970s, a timber famine was regularly forecast although never realized (Clawson, 1979).

Concerns about a timber famine in the United States were addressed in two very different ways. One approach was to establish the National Forest System "to provide the American people with a continuous supply of timber into the future." The Forest Service was then created to manage this system. In essence, a major rationale of the NFS was to insure against myopia on the part of the private sector, which was expected to excessively draw down the U.S. timber stock without consideration of future needs.

The second response to concerns about inadequate timber supply was to launch major efforts at forest regeneration, largely through tree planting. This effort began as make-work jobs during the depression of the 1930s. However, the planting thrust gained real momentum only after the middle of the 20th century, when it was predominantly driven by private sector recognition of the country's future needs for timber. By far, the major portion of the tree planting of the 20th century in America occurred after 1960 and was undertaken by the private sector (Sedjo, 1991).

If sustainability is both a process and a stock phenomenon, one could argue that the United States is responding well with respect to the industrial wood commodity, both by instituting a process whereby the stock is being replenished and by ensuring that the replenishment is sufficient so as to maintain (in fact, increase) the nation's timber stock.

However, the current view is that sustained-yield harvests do not provide sustainability of the full range of forest outputs and services. Ambiguities regarding forest management are not new. In the late 1960s and early 1970s, a debate ensued over whether the U.S. Forest Service ought to follow a management path of "dominant use" or "multiple use." The notion of multiple use prevailed, and was codified in law. Or did it prevail? Must multiple use be practiced in a manner where every acre produces every output? Or can the various components of the forest be managed in such a manner that each specializes in what it does best? In fact, the Forest Service subsequently used a type of zoning system that recognized differences in the land (e.g., its productivity, uniqueness, terrain, and so forth), and management was adjusted to recognize those conditions. In fact, the private sector behaves in a similar manner in that, it specializes intensive timber growing on the most suitable lands with remaining lands more lightly managed or used primarily for other purposes.

It is only in the past decade or so that the notion was developed that timber producing forests should be certified as being "sustainably managed." However, it soon became obvious that even the experts, who were often selected in part for their similar overall philosophies, did not agree on just what constitutes sustainable management. And, indeed, the term sustainably managed has generally been dropped and replaced with the concept that the forest is certified as being "well managed" or "managed in accordance with the specifications of the, e.g., Forest Stewardship Council (FSC), the Sustainable Forestry Initiative (SFI) or the Pan European Forest Certification (PEFC) certifying group" (Sedjo, Goetzl, & Moffat, 1998; Sedjo, 2004).

Although we may not be sure of what characteristics are necessary for sustainability, we do have a pretty clear agreement as to what the final outcome should be. The Brundtland Commission seems to capture the essence of the objective when it stated that future generations should have access to the same amount of resources as the current generation. If applied to forestry, however, does that mean every forest needs to continue producing its mix of services forever? If so, should this same standard be applied to agriculture so that productivity of every agricultural field continue forever? I would argue that, for forestry, the only reasonable interpretation is not that a particular forest produces the entire range of outputs forever, but rather that the "system," broadly defined, is capable of producing the desired sets of outputs.

By this definition, the scope of the effort, or what constitutes the "system" becomes important. Is it the forest stand, the forest landscape, the regional forest, or the global forest that should be sustained? I submit that the answer to that question is "it depends." There is, in essence, a global timber market and, for industrial wood purposes, the global forest is probably the unit that needs to be entrusted with meeting sustainable wood requirements. However, some desired forest outputs may have a spatial component. Erosion mitigation, wildlife habitat, and many other environmental outputs are local and have strong site-specific aspects. Thus, the relevant sustainable units will not generally be global, but will have regional, landscape, or very localized dimensions. In short, the sustainability desired of the forest will vary by output. This is not too much different from the notion that the amount of habitat required to sustain fauna depends on the creature, and will be very different for mountain lions than for field mice. Finally, biodiversity in a forest is often localized but it is viewed as providing global benefits. However, generally there is no need to transport these benefits as their continued existence in their natural location is generally sufficient to fulfill their social function.

## 2.3 Specialization and Ecosystem Models

Even as plantation forestry is rapidly displacing natural forests as the major source of industrial wood,[1] there is a heightened concern over the notion of forest sustainability. This concern seems to stem from our changing scientific understanding of the ecological functioning of forest ecosystems, as well as evolving attitudes toward human-made and natural ecosystems. This broader view has correctly challenged the notion that a sustained yield of timber is equivalent to sustainability of all the components and natural processes of the natural system.[2]

Many modern concepts of sustained forests move away from specialization. Bowes and Krutilla (1989), when discussing publicly owned and managed forests, suggested an optimal forest model to be used to achieve legally mandated multiple-use management. Developed in a time before the term "sustainability" had achieved such prominence, their model is driven by the notion of economic efficiency, broadly interpreted. The concept is one of the formalizations of multiple use. Their "forest factory" produces an array of joint products. Some, like timber, are marketed; others, like biodiversity protection, are not. The social planner would manage such a forest factory in a way that maximizes the discounted present value of the stream of value generated through time by the array of outputs and services provided by the forest. One explicit element the Bowes and Krutilla model lacked was a value for the condition of the forest factory itself. If the forest factory condition is valued sufficiently highly, such an approach could explicitly provide for sustainability. The high valuation placed on forest condition would prevent the process of maximization of the discounted present value from drawing down the forest. In any event, a drawdown result is very unlikely where the intertemporal nonmarket values provided by the forest are substantial and continue to be valuable into the indefinite future and where the opportunity costs are low.

In the 1990s, the concept of multiple-use forestry evolved into one of ecosystem management. Ecosystem management is a concept where the value of the "forest factory" is paramount, even as the forest factory produces other outputs, which in concept can be either marketed or not. Again, in economists' jargon, the high-value output of this system is the forest condition, which dominates concerns regarding individual outputs. For the production of some outputs, the forest condition and the volume of outputs are positively related, but not for all outputs all the time. Moreover, although forest condition may be viewed by some as the dominant value, this value is often in the eye of the beholder, and a social consensus of what constitutes optimal condition may be absent. What is the optimal condition? Are we looking for a forest that looks like the pre-colonial American forest, is the desired forest one that is never harvested, or is it a forest managed in a "tidy" manner such as was fashionable in much of Europe until quite recently?

Variants of the ecosystem theme have been picked up by Gordon (1994) and Thomas (1994), where both the multiple outputs and long-term nature of the sustainable management problem were recognized. Finally, it should be noted that most of these optimum forest proposals were suggested for the U.S. National Forest System, a public entity with an express legislative mandate to practice multiple-use forestry, specifically the Multiple Use Sustainable Yield Act (1960) and the National Forest Management Act (1976).

*2.4 Achieving the Outputs of the Forest*

The earlier section suggests that a careful look at the outputs of the forest reveal that they can be put into a number of different categories. These would include: market goods and services, local environmental outputs and global environmental outputs. For example, timber, like most agricultural commodities, is a marketed commodity that can be widely traded. By contrast, erosion control is typically a nonmarket service that is highly localized. Biodiversity, while local in its individual occurrences, can be viewed as a global nonmarket resource, viewed as benefiting humankind as a whole. Other outputs will fall somewhere in between. The question arises again as to whether there may not be a place in sustainable forestry for specialization both in the production of industrial wood and in some other outputs of multiple output forestry. Plantation forestry has a comparative advantage in providing industrial wood. Why not let plantations specialize in wood production, while realizing simultaneously some of there environmental external benefits, while regulating against serious external harms? Concurrently, other forests can "specialize" in the production of the various local and global environmental and ecological outputs.

The concept here is that all forests need not be managed for multiple outputs. Also, just as all corn fields or pastures need not be indefinitely sustainable, neither do all forests need to be indefinitely sustainable. The important thing is that the system as a whole be sustainable through time. Some forests could produce an array of environmental and nonmarket outputs, while others specialize on industrial wood. Some may produce some mix of both environmental and commodity outputs.

There is a small problem with this model, however. Assuming that the social value of all the outputs, market and nonmarket, justify the maintenance of the forest, there is still the question (or perhaps problem) of paying for the continuance of forests that provide nonmarket outputs. Obviously, such a forest would require a subsidy for the maintenance of its valued but nonmarket outputs.

*2.5 Forest Management and Specialization*

The industrial revolution was fueled to a large extent by specialization. Specialization and interregional and international trade has allowed the advantages of this process to be further manifest. As argued above, the benefits of specialization and comparative advantage apply not only in the world of manufacturing, but also in the world of agriculture and forestry. Regions differ in the fertility and topography of their lands, their climates, and the alternative uses (opportunity costs) to which lands can be placed. Thus we observe specialization in world granaries, as in the U.S. Midwest; great pasturelands, as in Argentina; and areas particularly suited to rice production, as in parts of Asia.

In the process of specialization, modern agriculture has, in fact, replaced complex natural ecosystems with much simpler biological systems. Modern agriculture involves management for a simplified ecosystem that attempts to focus all of the productivity of the land into a single crop (e.g., wheat, corn, or apples). Specialized systems, by definition, do not produce an array of outputs.

The advantage of specialization is that it allows productivity to be dramatically increased for the item in question; in some case substantial scale economies are an important source of that productivity increase. This approach allows high levels of output to be realized from relatively modest areas of land. Despite the additional needs for other (modern) inputs, it also tends to achieve the lowest overall resource cost per unit of output. Land, which can be viewed as the scarce factor, is managed to increase its productivity in the desired output.

Specialization in forestry has reached its zenith in the intensively managed plantation forest. This approach mimics cropping agriculture in the intensive attention to all phases of forest growth, from site preparation to planting, tending, and harvesting (Sedjo, 1983; Binkley, 1997). In the U.S. the private sector has responded to concerns about the sustainability of the timber commodity in large part by investing in intensively managed planted forests. Simultaneously, the role of the National Forest System has declined to the point where it contributes less than 3% of the industrial wood production of the United States.   And the phenomenon of looking to planted forests to meet society's need for the sustainable production of industrial wood has not been limited to the United States. Over the past 50 years, forest plantations have been established in much of the world. In a world of commodity trade and intensively managed agriculture, why should society continue to rely on natural forests to provide the bulk of its industrial wood? In fact, it is not. As recently as 50 years ago, almost 100% of the world's industrial wood was produced by natural forests. By the year 2000, it is estimated that about 34% of the world's industrial wood harvests came from planted forests. Furthermore, it is estimated that, by 2050, up to 75% of the world's industrial wood harvests will be from planted forests (See Figure 10.1).

However, despite the success of forest plantations, or perhaps because of it, there is growing concern for the continued existence of the natural forest. Changing scientific understanding of the ecological functioning of forest ecosystems, together with evolving attitudes toward natural ecosystems, has challenged the notion that a sustained yield of timber is equivalent to sustainability of all the components and natural processes of the natural system.

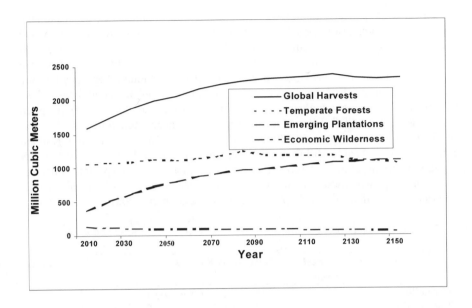

**Figure 10.1.** *Projection of Annual Timber Production*

*Source:* Sohngen, Mendelsohn, & Sedjo 1999

## *2.6 Toward a Sustainable Forest "Model"*

Today in forestry there are apparently conflicting forces moving in opposite directions. On the one hand, there is intensively managed plantation forestry, which is essentially an intensive agricultural cropping mode. On the other hand, there is concern for the host of nontimber outputs of the forest, with a focus on biodiversity preservation. This second set of concerns moves in the direction of establishing forest set-asides, where commercial harvests would occur rarely, if ever, and management would be for forest condition, not outputs in the usual sense.

Let me suggest that sustainable forestry requires not a single model, but rather at least three complementary models, if not a spectrum of models. The first model is drawn from the industrial revolution and modern agriculture and focuses predominantly on timber production. This is the intensely managed cropping system, where the other outputs of the forest are of minimal interest and the forest can be located in many places, provided the locations lend themselves to production and subsequent processing and marketing. The second model focuses on nontimber and nonmarket outputs, with the focus of providing ecosystem services, largely to a particular location. The third model relates to maintaining habitat that is conducive to the provision and continuity of biological diversity, largely native biodiversity.

At one level, these models appear to be largely independent, but in a broad global context, they are importantly interrelated. There is a growing recognition that the global system needs to move toward maintaining some stock of sustainable forests. These forests are needed not only for purposes of providing industrial wood, but also to ensure that the requisite flow of environmental services continue to be forthcoming. In fact, it can be argued that in the Northern Hemisphere temperate countries, this objective has largely been accomplished, although there may be some issues surrounding actual management of some of the various included forests. Also, importantly, intensively managed planted forests from the Southern Hemisphere might assist in could fulfilling the industrial wood needs and function. A recent article by Victor and Ausubel (2000) laid out a vision of a global forest and croplands system (Table 10.1). As management move towards intensively managed plantations, the total area of forest managed for timber can be expected to decline. This general view comports nicely with a system suggested on a regional level for ecosystem protection in Canada (Messier, Bigue, & Bernier, 2003).

**Table 10.1.** Global Forests Today and Tomorrow: One View (Area in billion hectares)

|                      | Forest area* | Industrial Forests                                  | Croplands |
|----------------------|--------------|-----------------------------------------------------|-----------|
| **Circa 2000**       | 3.2          | 0.9 *(Managed Forests)*                             | 1.5       |
| **Potential for 2050** | 3.4        | 0.4 *(Intensively Managed Forest Plantations)*      | 1.1       |

* Note: The forest area figures used by Victor and Ausubel are somewhat lower than those found and used elsewhere in this paper. This probably reflects a slightly different definition of forest that disregards lightly forested areas.

*Source:* Victor & Ausubel, 2000

It should be noted too that in this system plantation forests have two major functions: they provide much of the world's industrial wood, and, in the process of providing industrial wood, they take harvesting pressure off the natural forests (Sedjo, 1990; Sedjo & Botkin, 1997; Binkley, 2000). Over the last 50 years, the shift from natural forests to plantations has been driven by two major factors. First, the economics of plantation forestry improved dramatically as a) marginal agricultural sites were abandoned by agriculture and so were available at low cost for forestry, b) exotic species with high yields were introduced in some regions, c) yields increased dramatically through tree improvement accomplished through tree breeding, and d) technology reduced harvesting costs on accessible sites. Additionally, the economics of harvesting natural forests deteriorated as a) the better sites were already harvested, b) environmental pressures raised forest practices and harvesting standards, increasing the costs of management and harvests, and c) more forests were set aside from harvest.

Any comprehensive global forest model must incorporate both natural and plantation forests. For example, the Timber Supply Model (Sedjo & Lyon, 1990), as refined and expanded (Sohngen, Mendelsohn, & Sedjo, 1999), provides for the spectrum of industrial wood production running from plantation forests to harvests from natural forest to the retention of large areas of lightly harvested and unharvested forest. Indeed, this approach can readily be adapted to address nontimber environmental issues such as forest carbon sequestration values (Sohngen & Mendelsohn, 2003) and, potentially, biodiversity issues also.

In a globally sustainable forest system planted forests would have the role of producing the bulk of the world's industrial wood, while maintaining practices within acceptable environmental boundaries. The world's natural forests, which today cover about 3.8 billion hectares of the earth's surface, would have the responsibility of providing some industrial wood while maintaining environmental outputs. Such an approach does not threaten environmental outputs; rather, it provides the opportunity for natural forests to specialize in the production of just those values.

In summary, then, I believe that it is fruitless to try to encapsulate sustainable forestry into a single forest management template. What is needed is the recognition that the relevant sustainable system is a global one. A sustainable society requires many things, and forests can provide many of them. There are 3.8 billion hectares of forest worldwide and much more that could be returned or converted to forest. We need to examine sustainability in the context of that global system, and not simply a piece of it.

## 3. GLOBAL LAND USES

### 3.1 Global Land Resources: An Overview

With some limitations, land is fungible across uses. Table 10.2 presents Food and Agriculture Organization (FAO) estimates of some land uses at recent selected dates. The data show a gradual rise in recent years in lands under permanent crops and an associated increase in arable lands. Also, permanent meadows and pastures are increasing, Forestlands have remained roughly constant under their definition, while irrigated land also has increased substantially.

*Table 10.2. Global Land Use: Selected Years (Area in million hectares)*

| Year | Arable Land and Land under (Permanent Crops) | Permanent Meadows and Pasture Land | Forest and Woodlands* | Irrigated Lands |
|---|---|---|---|---|
| 1971 | 1,457 (89.328) | 2,987 | 4,041 | 167.399 |
| 1989 | 1,477 (103.398) | 3,304 | 4,087 | 232.828 |
| 1998 | 1,512 (131.116) | n.a. | n.a | 271.432 |

*There are some differences between the area of forest and woodlands reported in the production yearbook and that reported elsewhere by the FAO due to definitional differences.

*Source:* FAO Annual Production Yearbook, selected issues.

## 3.2 Croplands and Pasture

Table 10.2 (above) presents FAO data on land use covering a 27-year period. While the data is aggregated and some data has not been collected in recent years, perhaps what is most remarkable is the relative stability of the figures. There is, however, a gradual increase in the arable land and land under permanent crops category, due to a large extent to increases in the area of permanent crops. Also, there are significant increases in the area of irrigated lands.

## 3.3 Forestlands

The FAO Forest Resources Assessment (2000) indicates that the world's forests covered 3.86 billion hectares in 2000, or 29% of the world's land area. Table 10.3 shows forest as distributed among tropical forests (47%); subtropical forests (9%), temperate forests (11%), and boreal forests (33%). The area of temperate forest worldwide covers a land area roughly the size of North America, while the area of tropical forest covers an area roughly the size of South America, which is roughly 10% smaller than North America.

*Table 10.3.* Forestlands, by Category

| Total | Tropical | Subtropical | Temperate | Boreal |
|-------|----------|-------------|-----------|--------|
| 3.86 billion ha | 47% | 9% | 11% | 33% |

*Source:* FAO Forest Resources Assessment (2000)

There has been much concern over the extent of deforestation in recent years, and numerous studies have tried to develop accurate estimates of changes. The Forest Resources Assessment estimated that the net annual change in forest area worldwide in the 1990s was –9.4 million ha. Most of this decrease occurred in the tropics (humid and dry). This was 5.2 million ha less annual deforestation than had earlier been experienced in the 1980s (FAO, 1993). For the humid tropics, the 1993 data found a net annual loss of 6.4 million ha. It should be noted, however, that the measurement of rates of forest decline (and expansion) appear to have substantial error terms and are still subject to discussion and revision. One may argue that the tropical forest appear to be the least sustainable in the system. Furthermore, many of these forests are not expendable in that the biodiversity specialization that occurs within them is unique.

From the data, two conclusions emerge: First, that deforestation is occurring in the tropics and this should be of some concern due to the unique biodiversity function of much of this habitat. Second, the temperate and boreal regions have experienced net reforestation, often due to the abandonment of agricultural activities in those areas. The total net result is that global forests are experiencing some degree of net deforestation, although the precise amount is still the subject of debate and the impacts do not appear to be captured in many of the overall statistics. Additionally, some portion of the net reforestation is in the form of planted forests, although this is still a small fraction of total forested areas, perhaps 5%.

## 3.4 Pressure on the Resources

Much of human activity in the past several hundred years has been oriented to increasing the area of tillable land and increasing human ability to till large areas (Hayami & Ruttan, 1985). This has been particularly true in land-abundant regions. However, Richards and Tucker (1986) report that in North America and Europe, essentially all of the land expansion ended before 1920. For East Asia, most of the expansion ended by 1920. In other regions however, including the Soviet Union, South Asia, South America, and Africa, substantial expansion of areas in regular cropping continued well into the 20th century. This finding is consistent with the data reported above by the FAO.

The effect of the continuation of substantial expansion of regular cropping into undeveloped areas has been to reduce the area dedicated to other uses, including wild forest and other natural vegetation, as these lands were converted. Land conversion to agriculture is still regarded as the major force driving tropical deforestation.

Pressure on forestlands comes primarily from land use conversions away from forestry and to other uses. Forests have traditionally been viewed as the lowest use value land, at least in the European context. They were viewed as being available for conversion to pasture and cropping as well as providing lands for development of various types. In addition to land use conversion, concerns have been raised in recent decades about the effects of intensive and excessive logging. Traditionally, the concern has been that the reduction in the area of forest will impact negatively on the supply of wood. However, despite recent predictions of rising prices and growing scarcity (e.g., Barney, 1982), in recent decades overall production of industrial wood has stagnated (Figure 10.2)[3] while real prices generally have remained flat.

Finally, the advent of intensively managed planted forests, with growth capacities ten and more times that of natural forests, give promise of more wood producing capacity even as the harvesting of natural forests is declining (Sedjo, 1991; Sedjo & Botkin, 1997). In fact, one may argue that intensive management is a response to decreased natural forest availability (Hyde, Newman, & Seldon, 1992). Furthermore, the recent FAO report has indicated that the rate of forest plantation establishment has been very rapid, with close to 50 million ha of plantation, or nearly 30 percent, being 10 years old or less. Many projections of future timber supply (e.g., Figure 10.1, Table 10.1) forecast a growing role for plantation wood from nontraditional producers of timber.

## 3.5 Economic Growth

Economic growth is normally associated with increased demand for products of all types. Interestingly, this has not been the case for industrial wood over the past couple of decades. While the world's economies generally experienced significant economic growth over the past two decades, the demand for industrial wood has remained essentially constant (Figure 10.2). This stagnation of demand has occurred in a world during a period that has seen not only continuing though modest economic growth in the industrialized nations, but also the emergence of China and more recently of India as rapidly growing economies. Nevertheless, the total demand for industrial wood has barely changed over the two decade period. Furthermore, during this period wood prices have been somewhat volatile, but the trend has remained flat.

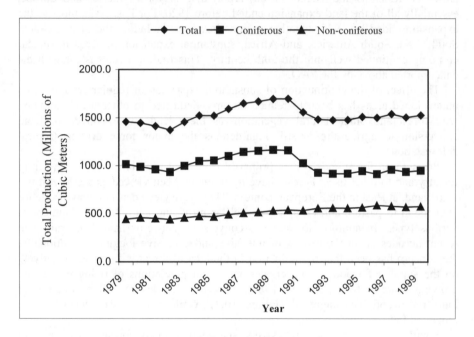

*Figure 10.2.* World Industrial Roundwood Production

*Source:* FAO, Rome

How might this stability of wood demand be explained? Casual observation suggests that in many areas non wood materials are increasingly substituting for wood. These include construction, where steel and other materials have replaced wood in many uses, and chapter, where newsprint production in much of the world has been flat. Furthermore, in much of the developing world construction utilizes cement and concrete for purposes which in the western world would have been met by wood products. One might argue the data suggest that over the past two decades wood has been an inferior good, that is, as income has grown, demand for wood was declined. This is not to argue that demand will not rise in the future, for undoubtedly it will to some extent. Rather, the point here is that even during the recent period of relatively buoyant global economic growth, including the emergence of some of the world's largest populated countries, demand for industrial wood has remand flat and an extension of that trend suggests only modest increases in demand at best. This view is increasingly becoming the conventional wisdom among those projecting future wood supply and demand (e.g., Figure 10.1).

*3.6 Population*

Finally, there is the question of the future of world population. The past several decades have seen a rapid growth of global population. That growth is now decerlerating and some projections (Figure 10.3) anticipate an absolute decline in world population by or before the middle of the 21st century. Based on existing trends, it could be argued that the world's forests appear more than adequate to the task of providing for the world's wood needs into the indefinite future. It is also likely that existing agricultural lands, with the application of appropriate technology, are adequate to meet growing world population needs for food product through the middle of the 21st century and beyond. For example, in the United States, agricultural productivity has grown at close to 2% per year over last several decades.

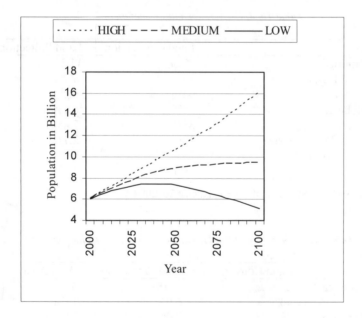

**Figure 10.3.** *UN World Population Projections (High, Medium, Low)*

*Source:* World Population Prospects 1998 Revision, *United Nations.*

However, while fertility rates are declining in many regions, some areas of the world (e.g., south Asia, Latin America, and much of Africa) are still experiencing fertility rates well above those associated with stable populations (McNeil, 2004).

Thus, even if the future world experiences stable populations, the composition of this population will certainly undergo dramatic change.

A world with stable or declining populations with the mix of populations changing, in an environment of continuing technological improvement, is likely to look quite different from a world with growing populations, with which we have grown so accustomed over the past two centuries. Populations will age, requiring different goods and services. Regions with aging and declining populations are unlikely to require large volumes of wood products. However, growing economies, as in China, will undoubtable increase their demands for wood products. These changes, no doubt, will have substantial implications for land use.

*3.7 Protected Areas*

The WRI recently prepared data on land conversion and land protection (Table 10.4). Not surprisingly, land conversion has been substantial in much of the globe. However, in every region, the majority, often the vast majority, of the land area has not been converted.

*Table 10. 4. Conversion and Protection of Land*

| Region | Land Conversion | Land Protection |
|---|---|---|
| North America | 27% | 11.1% |
| Europe and Russia | 35% | 4.7% |
| Asia | 44% | 6.0% |
| CA Caribbean | 28% | 6.1% |
| South America | 33% | 7.4% |
| Middle East/ North Africa | 12% | 2.1% |
| Sub-Saharan Africa | 25% | 6.0% |
| Oceania | 9% | 7.1% |

*Source:* WRI, as reported in the NYT, August 20, 2002, p. D4

*Table 10.5. The Extent of Wild Areas in the World by Major Regions*

| Region | Land Conversion | Land Protection | Wild Area |
|---|---|---|---|
| North America | 27% | 11.1% | 37% |
| Asia | 44% | 6.0% | 14% |
| South America | 33% | 7.4% | 21% |
| Oceania | 9% | 7.1% | 28% |
| Europe | | | 3% |
| Soviet Union | | | 34%* |

* Note that Tables 10.4 and 10.5 are not entirely consistent since the former Soviet Union includes countries not now part of Russia.

*Source:* See Table 10.4 and McCloskey and Spalding (1989).

Table 10.5 takes the material of Table 10.4 and adds the McCloskey and Spalding estimates of wild areas. This perspective suggests that there are large areas that have not been designated as protected areas, but that nevertheless retain the features of wild areas.

It is notable that the combination of Europe and Russia into a single category, as in Table 10.4, gives the impression of minimal protection areas across the two entities. However, reporting the wild areas separately, as done by McCloskey and Spalding, reveals the limited wild area in Europe (3%) but the large amounts of wild areas in the former Soviet Union (34%).

The above estimates on wild areas are consistent with the notion that habitat protection is provided through inaccessibility. The FAO estimates that about one-half of the world's forest area is economically inaccessible and therefore unavailable for timber harvests under normal circumstances (Table 10.6). This includes huge forest areas in all of the regions of the world, including the Organization for Economic Cooperation and Development (OECD) countries. Europe, however, has the smallest inaccessible area. Such a situation affords a large measure of protection from commercial harvests or development in these forests unless the areas are opened. The extent to which these areas become more accessible via road-building will, no doubt, affect future exploitation.

More generally, the number of protected areas worldwide is growing rapidly. Additional data indicates that the area in protected status has increased dramatically in recent years. Reid and Miller (1989) report that the cumulative world area under protected status has risen from a negligible area in 1900 to about 5 million square kilometers by the late 1980s, almost 4 million square kilometers of which was added after 1970. This area has surely experienced additional growth since then.[4]

*Table 10.6. Inaccessible Forest*

| Region | Area of Forest Unavailable for Timber Supply (million ha) |
|---|---|
| Africa | 233 |
| Asia | 177 |
| Oceania | 61 |
| Europe | 20 |
| Russia | 166 |
| North America | 238 |
| Central America | 49 |
| South America | 709 |
| Total | 1,653 |

*Source:* FAO Global Fibre Supply Model, 1996. Rome.

Traditionally, comparative advantage in raw forest materials was based on their natural availability, their accessibility, and the location of markets. Wood, being a generally low-cost high-volume good, tended to be logged relatively near the markets in which it would ultimately be sold. Nevertheless, important international markets developed centuries ago, including wood trade from North America to Japan and Asia as well as wood flows from the Nordic countries to Central Europe

Notions of renewability and management systems to promote regeneration and growth were developed in Europe as early as the 1800s. Until the mid part of the 20$^{th}$ century, most regeneration was largely the result of natural processes. However, it is well known that the Germans, who viewed conifers as valuable for commercial uses, promoted and planted large areas in conifer species, replacing the deciduous forests that preceded them. However, it was only in the last half of the 20$^{th}$ century that forestry began to take on the characteristics of modern agriculture.

## 3.8 Sustainable Forestry

Today we observe a world with a variety of land uses, including agriculture, forests, converted areas, wild areas, protected areas, and inaccessible areas. Within the broad array of lands and land uses, what is the meaning and significance of sustainable forestry?

Let us look at the actual experience of forestry a little closer. In recent decades, forestry has followed a path very similar to that of modern agriculture. A particular output is selected, based both on market characteristics and the capacity of the land. Seed stock is improved through selection and breeding. Planting of the improved seed stock is undertaken on lands appropriate for that species and provenance. Fertilizers and herbicides are applied, and other activities are undertaken to ensure that the productivity will be captured by the desired outputs, the planted trees.

Furthermore, specialization imparts a comparative advantage to certain regions. Brazil, for example, has transformed itself from a major wood product importer to a major exporter, not through commercial logging of its native forests, but through the creation of highly productive planted forests, usually of exotic species, and the processing of these trees, when harvested, into wood pulp and other products. Additionally, this transformation has occurred with relatively modest impacts on its native forests. The lands being established in plantation forests are, to a large degree, agricultural lands that are being bid away from agriculture by forestry, which is expected to provide a higher financial return. Granted, many of these lands were once native forests that have been cleared to provide lands for agriculture. But the most recent transition has been from agriculture back to forestry, this time modern planted forests.

Two polar approaches to agricultural and forestry production can be identified. One is to follow the path of specialization, as has most of agriculture today. Alternatively, one can continue to jointly produce a host of outputs, including timber, from various parts of the globe's 3.8 billion ha of forest. Logging could be done in a manner that meets some "certification" standards. The timber would be collected and the other joint products could be collected separately and/or be allowed to provide services to the surrounding areas. There is still a role in part of the agricultural world for multiple outputs. Agroforestry comes to mind, particularly in some parts of the tropics. In this process, a number of food and fiber items are cropped in a common area. Also, some crops, such as forest mushrooms and other forest foods, are still produced via a gathering mode, including hunting. The best ginseng, for example, is collected out of the forest, although most ginseng is cultivated today.

Can any of these approaches be clearly demonstrated to be superior from a sustainability perspective? I submit that the answer is that "it depends." From a narrow forest stand-by-stand perspective, the management of the natural forest provides a broader array of outputs and may be expected to continue to produce indefinitely, in the absence of improper management. However, much larger areas of forest must be impacted to meet human industrial wood needs. Natural forests typically yield about 2 cubic meters of wood per ha per year. Thus, for a world consumption level of 2 billion cubic meters per year, roughly 1 billion ha of forest, or about 30% of the world's forests, would need to be involved in producing timber for periodic industrial wood harvests. By contrast, intensively managed plantations produce about 10 times the natural rate, or 20 cubic meters per ha per year. Thus, to meet the world's industrial wood needs of 2 billion cubic meters annually, only 100 million ha need be involved.

The advantages here are several. First, most of the earth's 3.8 billion ha of natural forest need not be disturbed for harvesting purposes. Second, large harvesting costs associated with access and road-building to these somewhat inaccessible areas need not be undertaken. Third, the environmental damages associated with accessing and harvesting in these sites would not be incurred. Thus, the advantages are twofold. Wood financial costs are generally lower due to the smaller and more accessible areas involved in planted forests, and the nonmarket environmental damages are almost surely less.

## 4. CONCLUSION

This chapter examines the concept of a sustainable forest system and argues that, to truly address the range of needs of humankind, a multiple-faceted global approach is required. Such a system would allow for specialization among the various parts in order to generate a sustainable whole global forest system. Within this perspective, any particular forest could be destroyed. Globally, however, the overall system would be sustainable, although some individual parts may not be. Parts of that global system would be devoted to the production of industrial wood, local environmental services and global environmental goods.

Today, the global concern appears to be focused largely on the provision of ecosystem and global biodiversity, with industrial wood production as a given. The second section of the paper suggests, I believe, the global forests are today meeting a reasonable sustainability objective. This objective is consistent with the Brundtland Commission definition of sustainability. As the data suggest, large areas of forest are experiencing little or no disturbance due to either inaccessibility or legal protection. And protected areas are being expanded at a fairly rapid pace. Thus, it is difficult to argue or demonstrate that damage to natural forests is large in a global context, although admittedly, tropical forests continue to be negatively impacted. Overall, one can conclude that the natural capital in forest is largely intact, and that in recent decades general public behavior has been in the direction of protection and conservation of large forest areas, with some notable exceptions.

## NOTES

[1] In 2001 plantation forests accounted for 34 % of the world's industrial wood harvest (FAO, 2001).
[2] It is noteworthy that this same concern does not seem to lead to the suggestion of a return en mass to heterogeneous agriculture.
[3] Data on fuel wood production is much less reliable than that of industrial wood. However, recent FAO data indicated that fuelwood production/consumption may have peaked and is beginning to decline.
[4] Obviously new protected areas are being created regularly. For example, Russian protected areas are reported to increase recently by 25% (WWF, Russian Program Office, 2002). Also, on August 22, 2002, Brazil announced the creation of the largest rainforest park, covering 3.8 million hectares.

## REFERENCES

Barney, G.O. (1982). *The global 2000 report to the President*. New York: Penguin Books.
Binkley, C.S. (1997). Preserving nature through intensive plantation management. *Forestry Chronicle,* 73, 553-558.
Binkley, C.S. (2000). Forestry in the new millennium: Creating a vision that fits. In R.A. Sedjo (Ed.), *A vision for the U.S. Forest Service (pp. 83-96)*. Washington, D.C.: Resources for the Future.
Bowes, M., & Krutilla, J. (1989). *The economics of public forestlands*. Washington, D.C.: Resources for the Future.
Clawson, M. (1979). Forest in the long sweep of American history. *Science,* 204(15), 1168-1174.
Faustmann, M. (1849). Calculation of the value which forest land and immature stands possess for forestry. Allgeneine Forst—und Jagd-Zeitung, Vol. 15. Reprinted in the *Journal of Forest Economics*, 1(1), 1995.
Fedkiw, J. (2003). The forest management pathway to sustainability. Presented to the XII World Forestry Congress, September 25-29, 2004, Quebec, Canada.
Food and Agriculture Organization of the United Nations (FAO). (1993). *The state of the world's forests*. Rome: FAO.
Food and Agricultural Organization of the United Nations (FAO). (2000). *Forest Resources Assessment*. Rome: FAO.
Food and Agricultural Organization of the United Nations (FAO). (2001). *The state of the world's forest resources*. Rome: FAO.
Gordon, J. (1994). The new face of forestry: Exploring a discontinuity and the need for a new vision. Pinchot Distinguished Lecture. Milford, PA: Gray Towers Press.
Hayami, Y., & Ruttan, V.W. (1985). *Agricultural development: An international perspective*. Baltimore: Johns Hopkins Press.
Hyde, W.F., Newman, D.H., & Seldon, B.J. (1992). *The economic benefits of forestry research*. Ames, Iowa: Iowa State University Press.
McCloskey, J. M. & Spalding, H. (1989) A Reconnaissance-level inventory of the amount of wilderness remaining in the world. *Ambio,* 18(4), 222-226.
McNeil, D.G. Jr. (2004, August 29). Demographic 'Bomb' May Only Go "Pop!" *The New York Times*.
Messier, C., Bigue, B., & Bernier, L. (2003). Using fast-growing plantation to promote forest ecosystem protection in Canada. *Unasylva*, 54(214/215), 59-63.
Reid, W.V., & Miller, K.R. (1989). *Keeping options alive: The scientific basis for conserving biodiversity*. Washington, DC: World Resources Institute.
Richards, J.F., & Tucker, R.P. (Eds.) (1988). *World deforestation in the twentieth century*. Durham: Duke University Press.
Sedjo, R.A. (1983). *The comparative economics of plantation forestry: A global assessment*. Baltimore, Maryland: Johns Hopkins Press for Resources for the Future.

Sedjo, R.A. (1990). Economic wood supply: Choices for Canada's forest industry. *The Forestry Chronicle*, 66(1), 32-34.

Sedjo, R.A. (1991). Toward a worldwide system of tradable forest protection and management obligations. RFF Discussion Paper ENR91-16, August 28, 1991.

Sedjo, R.A. (2003). Introduction. In R.A. Sedjo (Ed.), *Economics of Forestry*. Part of the T. Tietenber, W. Morrison (Eds.) series, *International library of environmental economics and policy*. Trowbridge, Wiltshire, UK: Ashgate Publishing Limited, Cromwell Press,.

Sedjo, R.A. (2004). "Challenges to Sustainable Forestry: Management and Economics." 2004, in *Forest Futures*, Joe Bowersox III and Karen Abeese editors, pp. 68-83, Rowman & Littefield, New York.

Sedjo, R.A., & Botkin, D. (1997). Using forest plantations to spare natural forests. *Environment*, 30(10), 14-30.

Sedjo, R.A., & Lyon, K.S. (1990). *The long-term adequacy of the world timber supply*. Washington, D.C.: Resources for the Future.

Sedjo, R.A., Goetzl, A., & Moffat, S.O. (1998). *Sustainability in temperate forests*. Washington, D.C.: Resources for the Future.

Sohngen, B., & Mendelsohn, R. (2003). An optimal control model of forest carbon sequestration. *American Journal of Agricultural Economics*, 85(2), 448-457.

Sohngen, B., Mendelsohn R., & Sedjo, R. (1999). Forest management, conservation, and global timber markets. *American Journal of Agricultural Economics*, 81, 1-13.

Thomas, J.W. (1994). *Ecosystem management: A national framework*. Washington, D.C.: USDA Forest Service.

United Nations. (1998). *World Population Prospects 1998 Revision*. New York: United Nations.

Victor, D., & Ausubel, J. (2000). Restoring the forests. *Foreign Affairs*, 79(6), 127-144.

World Commission on Environment and Development. (1987). *Our common future*. Oxford, UK: Oxford University Press.

WWF, Russian Program Office (2004.) http://www.forest.ru/eng/old-growth/ (accessed September 13, 2004.)

# CHAPTER 11

# FOREST CARBON SINKS: A TEMPORARY AND COSTLY ALTERNATIVE TO REDUCING EMISSIONS FOR CLIMATE CHANGE MITIGATION

## G. CORNELIS VAN KOOTEN AND ALISON J. EAGLE[1]

*Department of Economics, University of Victoria*
*Victoria, British Columbia, Canada*
*Email:kooten@uvic.ca*

**Abstract.** The Kyoto Protocol (KP) requires signatories to reduce $CO_2$-equivalent emissions by an average of 5.2% from 1990 levels by the commitment period 2008-2012. This constitutes only a small proportion of global greenhouse gas emissions. Importantly, countries can attain a significant portion of their targets by sequestering carbon in terrestrial ecosystems in lieu of emission reductions. Since carbon sink activities lead to ephemeral carbon storage, forest management and other activities that enhance carbon sinks enable countries to buy time as they develop emission reduction technologies. Although many countries are interested in sink activities because of their presumed low cost, the analysis in this paper suggests otherwise. While potentially a significant proportion of required $CO_2$ emission reductions can be addressed using carbon sinks, it turns out that, once the opportunity cost of land and the ephemeral nature of sinks are taken into account, costs of carbon uptake could be substantial. Carbon uptake via forest activities varies substantially depending on location (tropical, Great Plains, etc.), activity (forest conservation, tree planting, management, etc.), and the assumptions and methods upon which the cost estimates are based. Once one eliminates forestry projects that should be pursued because of their biodiversity and other non-market benefits, or because of their commercial profitability, there remain few projects that can be justified purely on the grounds that they provide carbon uptake benefits.

## 1. INTRODUCTION

Global climate change constitutes a long-term threat to the earth's ecosystems and to the way people lead their lives. Some of the most serious threats include damages to agriculture, particularly subsistence farming in developing countries, and to coastal dwellers, who could lose their homes and livelihoods as a result of flooding caused by sea level rise. Climate change also poses a threat to forest ecosystems, resulting in changes to species composition and potentially threatening preservation of plants and biodiversity more generally. It will have impacts on sustainable forest management, creating challenges for foresters and decision makers.

*Kant and Berry (Eds.), Institutions, Sustainability, and Natural Resources: Institutions for Sustainable Forest Management, 233-255.*
© 2005 *Springer. Printed in Netherlands.*

Most scientists are convinced that the discernible rise of 0.3 to 0.6 °C in the earth's average surface temperature over the past century (Wallace et al., 2000) is related to the significant increases in carbon dioxide ($CO_2$) and other greenhouse gas (GHG) concentrations in the atmosphere. While the full extent of the potential damages from climate change remains unknown, scientists have argued that action should be taken to mitigate its potentially adverse consequences.

Does that mean that global society should immediately undertake activities to mitigate climate change? Economic principles dictate that mitigation activities should be implemented as long as the marginal benefits of so doing (i.e., the damages avoided by mitigation) exceed the marginal costs of actions to reduce atmospheric $CO_2$. However, while the (marginal) costs of mitigation measures tend to be unclear, estimation of the (marginal) benefits is even more problematic and controversial. Damages from climate change are expected in the more distant future and remain speculative, partly because they affect future generations and may be largely nonmarket in nature (e.g., affecting recreational activities, scenic amenities and biodiversity). Uncertainty about these damages (and thus the benefits of mitigation) exists in both the economic and scientific spheres.

Through the Kyoto Protocol (KP), the international community has prepared a policy response to global climate change as it relates to the emissions of GHGs. Although it is seriously flawed, the KP attempts to aid the international community in slowing or even preventing global anthropogenic emissions of greenhouse gases from rising in the future. For some countries, forest ecosystem sinks play an important role in KP compliance. Carbon uptake in forest ecosystems could be a potentially cheaper means of achieving compliance than decreasing $CO_2$ emissions (Obersteiner, Rametsteiner, & Nilsson, 2001; Sohngen & Alig, 2000). Our purpose in this chapter is to investigate in greater detail the potential role that forestry might play in helping countries achieve their KP targets. Our results indicate that, while forest carbon sinks can indeed reduce atmospheric $CO_2$, their role in enabling countries to meet their emission reduction targets is extremely limited, mainly because the creation of carbon sinks that are 'additional' is much more costly than initially recognized and, further, that such sinks are ephemeral.

Before examining carbon sinks as they relate to forestry activities in more detail, we begin by outlining the Kyoto Protocol in section 2, and in section 3, we explore how carbon sinks have been considered in lieu of $CO_2$ emission reductions. Potential carbon sinks allowed in forestry are discussed in section 4, while the question of discounting physical carbon and its impacts on estimates of the costs of carbon sequestration are the topic of section 5. This is followed, in section 6, by a more-detailed investigation into the costs of creating carbon credits in forest ecosystems through land use, land use change, and forestry (LULUCF) activities, and their limitations. In section 7, we discuss some additional difficulties related to the creation and trading of carbon offset credits. Policymakers have generally ignored landowners in their rush to create KP implementation plans, but owners may be reluctant to plant trees. This issue is discussed in section 8, because if landowners are not receptive to tree planting programs, their reticence will increase carbon uptake costs. The conclusions follow in section 9.

## 2. CLIMATE CHANGE AND THE KYOTO PROTOCOL

As a result of international concerns over anthropogenic emissions of GHGs, the Intergovernmental Panel on Climate Change (IPCC) was formed in 1988. The IPCC's first published report in 1990 led to the signing of the United Nations' Framework Convention on Climate Change (FCCC) in Rio de Janeiro in June 1992 by 174 countries. This agreement committed industrial countries to control greenhouse gas emissions to the 1990 level by year 2000, but subsequent Conference of the Parties (COP) meetings modified this target and further clarified how emissions were to be controlled.

In order to stabilize atmospheric concentrations of $CO_2$ and other GHGs, $CO_2$-equivalent emissions will need to be reduced by 50% or more from 1990 levels (Coward & Weaver, 2003).[2] Though falling far short of this target, industrial countries crafted the Kyoto Protocol at COP3 in December 1997, agreeing to reduce $CO_2$ emissions by an average of 5.2% from the 1990 level by 2008-2012. This implied a total reduction of 250 megatons (106 metric tons) of carbon[3], denoted Mt C, per year from 1990 levels. The KP will come into effect 90 days after it has been ratified by 55 states, as long as the industrialized countries that ratify account for 55% of the $CO_2$ emitted by industrialized countries in 1990. As of 26 November 2003, 120 countries had ratified, with ratifying industrial countries accounting for 44.2% of the 1990 emissions.[4] The United States, with 36.1% of industrial countries' emissions withdrew support for the KP during COP6 at The Hague in late 2000, citing high costs. Therefore, without the United States' participation, it is essential that Russia, accounting for 17.4% of 1990 industrial countries' $CO_2$ emissions, ratify the KP in order for the Protocol to come into effect.

Environmental externalities play a large role in the KP, necessitating government action to address the associated market failure. Three economic coordination methods that attend to this market failure are outlined by economists: i) Command and control (C&C), ii) common values and norms, and iii) market incentives. C&C consists mainly of standards (e.g., specifying fuel efficiency requirements of automobiles or the quality of insulation in new construction), bans and regulations (e.g., spelling out the amount of $CO_2$ a source may emit). Common values and norms constitute those elements of civil society that facilitate voluntary action, and are most often found in countries with a highly homogenous population (e.g., The Netherlands, Singapore). As to market incentives, it is well known that market instruments, such as carbon taxes or tradable emission permits (quotas), result in lower costs than C&C, because prices in the form of taxes or permits cause firms to seek the lowest cost means of reducing emissions (see Field & Olewiler, 2002). International trading of $CO_2$ emission and offset permits, and substitution of the most economical means of reducing emissions, would allow the most economic gain, while putting a value on the environmental externalities caused by $CO_2$ entering the atmosphere.

With this in mind, the KP outlines the following ways for a country to meet its commitments:

i.  Countries can simply reduce their own emissions of GHGs to the target level, say R in Figure 11.1.

ii. Rather than reducing domestic CO2 emissions to R (Figure 11.1), a country can achieve R by sequestering an equivalent amount of carbon in domestic terrestrial ecosystems. These activities are discussed in more detail on the following pages.

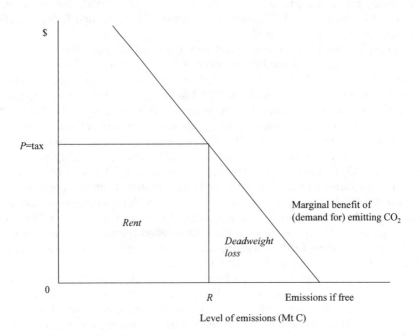

*Figure 11.1. Controlling $CO_2$ Emissions using Economics Incentives*

iii. Joint implementation (JI) is encouraged under KP Article 6. JI allows an industrial (Annex B) country to participate in emissions-reduction or carbon sequestration activities in another Annex B country (essentially in Central and Eastern Europe), thereby earning "emission reduction units" (ERUs) that are credited toward the country's own commitment.

iv. Under the "clean development mechanism" (CDM) of KP Article 12, an Annex B country can earn "certified emission reductions" (CERs) by funding emissions-reduction or carbon sequestration projects in a non-Annex B (developing) country. However, only afforestation and reforestation activities can be used to generate carbon uptake CERs, and their use is limited (in each year of the commitment period) to 1% of the Annex B country's 1990 (base-year) emissions.

v. Finally, an Annex B country can simply purchase excess emission permits from another Annex B country (Article 17). Emission permits in excess of what a country needs to achieve its commitment are referred to as "assigned amount units" (AAUs) that can be purchased by other countries. These are particularly important to economies in transition that easily attain their KP targets because of economic contraction and the concomitant closure of inefficient power plants and manufacturing facilities, thereby creating "hot air" (AAUs) to be sold at whatever price is available.

While the availability of a variety of emissions-reduction and carbon sequestration options should reduce compliance costs relative to the situation where restrictions are placed only on emissions, the addition of these options in the KP complicated matters significantly. Compared to a more simplified scheme, monitoring and enforcement authorities will need more information, such as forecasts or projections of the potential supply of carbon offsets in future years, in order to set a quota on emissions. Transaction costs of operating the trading scheme will also increase significantly.

## 3. CARBON SINKS IN LIEU OF EMISSIONS REDUCTION

Negotiations since COP3 in Kyoto have focused primarily on the so-called flexibility mechanisms, most importantly Joint Implementation, the Clean Development Mechanism and International Emissions Trading. A number of parties argued that the role for terrestrial carbon sinks as replacements for emissions reductions was inadequate, so, at COP6$_{bis}$ at Bonn in July 2001, the European Union (EU) relented to a broader role for carbon sinks, mainly to appease Japan, Australia and Canada, and the United States in absentia. This permitted countries to substitute carbon uptake from LULUCF activities in lieu of greenhouse gas emissions. The IPCC (2000a) estimates that biological sink options have the potential to mitigate some 100,000 Mt C between now and 2050, amounting to 10% to 20% of fossil fuel emissions of $CO_2$ over the same period[5]. When using the Marrakech Accords (agreed to at COP7 at Marrakech, Morocco, October/November 2001) as the basis for calculating the carbon offset potential of biological sinks, it is clear that terrestrial sinks have become an important means by which some countries can achieve their KP targets (see Table 11.1). Nearly 200 Mt of carbon credits could potentially be achieved by LULUCF activities, amounting to 80% of the 250 Mt C annual reductions that would have been required of industrial countries in 1990 but will be much higher for 2008-20012.

Under the KP, permitted terrestrial sink activities include reductions in carbon release from net land-use change and forestry in Annex B countries that had net LULUCF emissions in 1990 (Article 3.7); net removals by sinks as a result of human-induced afforestation, reforestation and deforestation (Article 3.3);[6] and net removals through changes in agronomic practices (cropland and grazing land management and revegetation actions) and from enhanced forest management (Article 3.4). The problems with terrestrial sinks are fourfold: (i) their inclusion and use under the KP are examples of political maneuvering to avoid emissions reduction; (ii) they tend to be highly ephemeral and thus not equivalent to emissions reduction (see below); (iii) the 'value' of sinks to a country is tied to the land use existing in 1990 as the base year; and (iv) carbon flux is notoriously difficult to measure.

*Table 11.1. Potential Role of Terrestrial Carbon Sinks in Meeting KP First Commitment Period Targets, Based on Marrakech Accords (Mt C per year)*

| Item | Total Annex B | Central and Eastern Europe (in Annex B) | Rest of Annex B |
|---|---|---|---|
| KP Article 3.3 (ARD) net increase in sinks | 12.28 | 0.00 | 12.28 |
| Maximum sinks due to forest management[a] | 97.87 | 38.59 | 59.28 |
| Increase in sinks due to agricultural activities | 33.56 | 3.61 | 29.95 |
| Maximum use of sinks under KP Article 12 (CDM) | 49.83 | 14.87 | 34.96 |
| **Total estimated potential of sinks to meet KP target** | **193.54** | **57.07** | **136.47** |

[a] At COP7, Russia increased its maximum sink level from 17.63 Mt C to 33.00 Mt C, thereby increasing the total here from 23.22 to 38.59. Not included is the annual 0.8 Mt C increase in permitted credits attributable to forest management as an offset against ARD debits during the commitment period, when comparing Bonn (COP6bis) with Marrakech (COP7).

*Source:* Authors' own calculations

The sequestration of carbon in terrestrial sinks will also in time encounter an equilibrium, beyond which point additional net sequestration will not be possible. Most likely, before reaching this point, the economics of continuing with sequestration as a substitute for emission reductions in other areas will no longer be feasible. Therefore, for long-term reductions in total net emissions, terrestrial carbon sinks will become less important and total emissions from fossil fuels will have to be addressed. At best, in the long-term, terrestrial carbon sinks are a stop-gap measure. The problem is that terrestrial sinks have become a distraction that prevents countries from making serious inroads with respect to emissions reductions, because it enables some countries to avoid implementing politically difficult actions.

## 4. CARBON SINKS IN FORESTRY

According to the Kyoto Protocol, while not initially included in the determination of baseline carbon emissions, afforestation, reforestation and deforestation (ARD) activities need to be considered in determining 2008-2012 emissions if forest carbon sink credits are to be claimed. Afforestation refers to human activities that encourage growing trees on land that has not been forested in the past 50 years, while reforestation refers to human activities that encourage growing trees on other land that was forested but had been converted to non-forest use prior to 1990 (IPCC 2000b). Afforestation and reforestation result in a credit, while deforestation (human-induced conversion of forestland to non-forest use) results in a debit. Since most countries have not embarked on large-scale afforestation and/or reforestation projects in the past decade, harvesting trees during the five-year commitment period (2008-2012) will likely result in a debit on the ARD account. Therefore, the Marrakech Accords permit countries, in the first commitment period only, to offset up to 9.0 Mt C each year for the five years of the commitment period through (verified) forest management activities that enhance carbon uptake, despite that fact that many of the activities can be business-as-usual (e.g., replanting, fire suppression). If there is no ARD debit, then a country cannot claim this credit, which amounts to the difference between mean annual increment (growth) and harvest on a (self-declared) managed forest. In Canada's case, the ARD debit for 2008-12 is estimated to be about 4 Mt C.

Some countries can also claim carbon credits from business-as-usual forest management that need not be offset against ARD debits. As a result of Marrakech, Canada can claim 12 Mt C per year, the Russian Federation 33 Mt C, Japan 13 Mt C, and other countries much lesser amounts – Germany 1.24 Mt C, Ukraine 1.11 Mt C, and remaining countries less than 1.0 Mt C. Japan expects to use forestry activities to meet a significant proportion of its KP obligation, while Canada can use forest management alone to achieve one-third of its emissions reduction target.[7]

In principle, a country should get credit only for sequestration above and beyond what occurs in the absence of C-uptake incentives, a condition known as 'additionality' (Chomitz, 2000). Thus, for example, if it can be demonstrated that a forest would be harvested and converted to another use in the absence of specific policy to prevent this from happening, the additionality condition is met. Carbon sequestered as a result of incremental forest management activities (e.g., juvenile spacing, commercial thinning, fire control, fertilization) would be eligible for carbon credits, but only if the activities would not otherwise have been undertaken (say, to provide higher returns or maintain market share). Similarly, afforestation projects are additional if they provide environmental benefits (e.g., regulation of water flow and quality, wildlife habitat) not captured by the landowner and would not be undertaken in the absence of economic incentives, such as subsidy payments or an ability to sell carbon credits (Chomitz, 2000).

The reason that the Kyoto negotiations have not addressed additionality explicitly is that this would disadvantage countries that have already undertaken forestry activities that generate carbon uptake benefits. For example, during the 1980s Canada invested heavily in the reforestation of not-sufficiently restocked forestland that had been harvested in previous decades but had failed to generate adequate cover on its own. The business-as-usual forest management provisions of Marrakech enabled Canada to salvage some credits for these investments, rather than penalize Canada relative to countries that had not attempted to implement sustainable forestry practices at such an early date, as would be the case under a strict additionality requirement.

## 5. DISCOUNTING PHYSICAL CARBON

Discounting implies that a unit of carbon emitted into (or removed from) the atmosphere at a future date is worth less than if that same unit were emitted (removed) today. By discounting carbon, you acknowledge that carbon sequestered in the present period has greater potential benefits than sequestration delayed until some future time. The idea of discounting physical carbon is anathema to many who would consider discounting only monetary values. However, the idea of weighting physical units accruing at different times is entrenched in the natural resource economics literature, going back to economists' definitions of conservation and depletion (van Kooten & Bulte, 2000, pp.245-47). Three approaches to discounting of carbon can be identified in the literature (Richards & Stokes, 2004; Watson, Zinyowera, Moss, & Dokken, 1996):

i. The 'flow summation method' sums carbon sequestered regardless of when capture occurs. Total (discounted or undiscounted) cost of the project is divided by the total sum of undiscounted carbon to provide a cost per ton estimate.

ii. Under the 'average storage method' the annualized present value of costs is divided by the mean annual carbon stored through the project.

iii. The 'levelization/discounting method' discounts both costs and physical carbon sequestered depending on when they occur, although costs and carbon can be discounted at different rates.

One cannot obtain a consistent estimate of the costs of carbon uptake, however, unless both project costs and physical carbon are discounted, even if different rates of discount are employed for costs and carbon. To illustrate why, consider the following example.

Suppose a tree-planting project results in the reduction of $CO_2$-equivalent emissions of 2 tC per year in perpetuity (e.g., biomass burning to produce energy previously produced using fossil fuels). In addition, the project has a permanent sink component that results in the storage of 5 tC per year for ten years, after which time the sink component of the project reaches an equilibrium. How much carbon is stored? If an annualized method (method 2) is employed, what is the annual amount of carbon that is sequestered? Is it 2 tC or 7 tC per year? Clearly, 7 tC are sequestered for the first ten years, but only 2 tC are sequestered annually after that time. Carbon sequestration, as stated on an annual basis, would either be that experienced in the first ten years (7 tC per year) or in the infinite number of years to follow (2 tC per year). Suppose the discounted project costs amount to $1,000,[8] or annualized costs of $40 if a 4% rate of discount is used. The costs of carbon uptake are then estimated to be $5.71 per tC if the higher amount of C sequestered is used, or \$20/tC if the lower amount is used. Most often the former figure is used to make the project appear more desirable.

Under the flow-summation method, the cost would essentially be zero because \$1,000 would need to be divided by the total amount of carbon absorbed, which equals infinity. To avoid an infinite sum of carbon uptake, an arbitrary planning horizon needs to be chosen. If the planning horizon is 30 years, 110 tC are sequestered and the average cost is calculated to be \$9.09 per tC; if a 40-year planning horizon is chosen, 130 tC are removed from the atmosphere and the cost is \$7.69/tC. Thus, cost estimates are sensitive to the length of the planning horizon, which is not usually made explicit in most studies (see section 6).

Cost estimates that take into account all carbon sequestered plus the timing of uptake can only be achieved under the third method. Suppose physical carbon is discounted at a lower rate (say, 2%) than that used to discount costs. Then, over an infinite time horizon, the total discounted carbon saved via our hypothetical project amounts to 147.81 tC and the correct estimate of costs is \$6.77 per tC. Reliance on annualized values is misleading in this case because costs and carbon are discounted at different rates. If carbon is annualized using a 2% rate, costs amount to \$13.53 per tC (=\$40 ÷ 2.96 tC). If the same discount rate of 4% is employed for costs and carbon, the \$10.62/tC cost is the same regardless of whether costs and carbon are annualized.

As Richards (1997) demonstrates, the rate at which physical carbon should be discounted depends on what one assumes about the rate at which the damages caused by $CO_2$ emissions increase over time. If the damage function is linear so that marginal damages are constant – damages per unit of emissions remain the same as the concentration of atmospheric $CO_2$ increases – then the present value of reductions in the stock of atmospheric $CO_2$ declines at the social rate of discount. Hence, it is appropriate to discount future carbon uptake at the social rate of discount. "The more rapidly marginal damages increase, the less future carbon emissions reductions should be discounted" (p.291). Thus, use of a zero discount rate for physical carbon is tantamount to assuming that, as the concentration of atmospheric $CO_2$ increases, the damage per unit of $CO_2$ emissions increases at the same rate as the social rate of discount – an exponential damage function with damages growing at the same rate as the social rate of discount. A zero discount rate on physical carbon implies that there is no difference between removing a unit of carbon from the atmosphere today, tomorrow or at some future time; logically, then, it does not matter if the carbon is ever removed from the atmosphere. The point is that use of any rate of discount depends on what one assumes about the marginal damages from further $CO_2$ emissions or carbon removals.

The effect of discounting physical carbon is to increase the costs of creating carbon offset credits because discounting effectively results in 'less carbon' attributable to a project. Discounting financial outlays, on the other hand, reduces the cost of creating carbon offsets. However, since most outlays occur early on in the life of a forest project, costs of creating carbon offsets are not as sensitive to the discount rate used for costs as to the discount rate used for carbon.

*Table 11.2. Carbon Content of Biomass, Various Tropic Forests and Regions*

| Region | Wet Tropical Forest | Dry Tropical Forest |
|---|---|---|
| Africa | 187 t C ha$^{-1}$ | 63 t C ha$^{-1}$ |
| Asia | 160 t C ha$^{-1}$ | 27 t C ha$^{-1}$ |
| Latin America | 155 t C ha$^{-1}$ | 27 t C ha$^{-1}$ |

*Source:* Papadopol (2000)

*Table 11.3. Depletion of Soil Carbon following Tropical Forest Conversion to Agriculture*

| Soil Carbon in Forest | New Land Use | Loss of Soil Carbon with New Land Use |
|---|---|---|
| *Semi-arid region* | | |
| 15-25 t C ha$^{-1}$ | Shifting cultivation (arable agriculture) | 30-50% loss within 6 years |
| *Sub-humid region* | | |
| 40-65 t C ha$^{-1}$ | Continuous cropping | 19-33% loss in 5-10 years |
| *Humid region* | | |
| 60-165 t C ha$^{-1}$ | Shifting cultivation | 40% loss within 5 years |
| | Pasture | 60-140% of initial soil carbon |

*Source:* adapted from Paustian et al.(1997)

*Table 11.4. Total Carbon in Tropical Ecosystems by Sink, Percent*

| Land Use | Tree | Under story | Litter | Root | Soil |
|---|---|---|---|---|---|
| **Original Forest** | **72** | **1** | **1** | **6** | **21** |
| Managed & logged over-forest | 72 | 2 | 1 | 4 | 21 |
| Slash & burn croplands | 3 | 7 | 16 | 3 | 71 |
| Bush fallow | 11 | 9 | 4 | 9 | 67 |
| Tree fallow | 42 | 1 | 2 | 10 | 44 |
| Secondary forest | 57 | 1 | 2 | 8 | 32 |
| Pasture | <1 | 9 | 2 | 7 | 82 |
| Agroforestry & tree plantations | 49 | 6 | 2 | 7 | 36 |

[a] Average of Brazil, Indonesia and Peru

*Source:* Woomer et al. (1999)

## 6. FORESTRY ACTIVITIES AND CARBON OFFSET CREDITS

In recent decades probably all of the net carbon releases from forests have come from tropical deforestation (since temperate and boreal forests are in approximate C balance[9]), thereby contributing to the build-up of atmospheric $CO_2$. Houghton

(1993) estimates that tropical deforestation was the cause of 22-26% of all GHG emissions in the 1980s. This is roughly consistent with findings of Brown et al. (1993), who report that total annual anthropogenic emissions are nearly 6.0 gigatons ($10^9$ metric tons, Gt) of carbon, with tropical deforestation contributing from 1.2 to 2.2 Gt per year. Tropical forests generally contain anywhere from 100 to 300 m$^3$ of timber per ha in the bole, although much of it may not be commercially useful. This implies that they store from 20-60 tons of carbon per ha in wood biomass, although this ignores other biomass and soil organic carbon (SOC).

An indication of total carbon stored in biomass for various tropical forest types and regions is provided in Table 11.2. The carbon sink function of soils in tropical regions is even more variable across tropical ecosystems (Table 11.3). This makes it difficult to make broad statements about carbon loss resulting from tropical deforestation. Certainly, there is a loss in carbon stored in biomass (which varies from 27 to 187 tC ha$^{-1}$), but there may not be a significant loss in soil organic carbon. While conversion of forests to arable agriculture will lead to a loss of 20-50% of SOC within 10 years, conversion to pasture may in fact increase soil carbon, at least in the humid tropics (see Table 11.3). In some (likely rare) cases, the gain in SOC could entirely offset the loss of carbon stored in biomass when forestland is converted to pasture. The conversion of forestland to agriculture tends to lead to less carbon storage, and a greater proportion of the ecosystem's carbon is found in soils as opposed to biomass (Table 11.4). To address this market failure (release of carbon through deforestation), policies need to focus on protection of tropical forests (see van Kooten, Sedjo, & Bulte, 1999).

Reforestation of deforested areas needs to take into account the carbon debit from harvesting trees, but it also needs to take into account carbon stored in wood product sinks (and exported carbon) and additional carbon sequestered as a result of forest management activities (e.g., juvenile spacing, commercial thinning and fire control). Even when all of the carbon fluxes are appropriately taken into account (and product sinks are not yet permitted under the KP), it is unlikely that 'additional' forest management will be a cost-effective and competitive means for sequestering carbon (Caspersen et al., 2000). However, as noted above, many countries can claim carbon offset credits for forest management activities that are not additional. Global data on the potential for carbon uptake via forest management are provided in Table 11.5.

Evidence from Canada, for example, indicates that reforestation does not pay even when carbon uptake benefits are taken into account (when financial returns to silvicultural investments include a payment for carbon uptake), mainly because northern forests tend to be marginal (van Kooten, Thompson, & Vertinsky, 1993).[10] The reason is that such forests tend to regenerate naturally, and returns to artificial regeneration accrue in the distant future. Only if short-rotation, hybrid poplar plantations replace logged or otherwise denuded forests might forest management be a competitive alternative to other methods of removing $CO_2$ from the atmosphere. Hybrid poplar plantations may also be the only cost-effective, competitive alternative when marginal agricultural land is afforested (van Kooten, Kremar-

Nozic, Stennes, & van Gorkom, 1999; van Kooten, Stennes, Kremar-Nozic, & van Gorkom, 2000).

**Table 11.5.** *Global Estimates of the Costs and Potential Carbon that can be Removed from the Atmosphere and Stored by Enhanced Forest Management from 1995 to 2050*

| Region | Practice | Carbon Removed & Stored (Gt) | Estimated Costs ($US x109) |
|--------|----------|------------------------------|----------------------------|
| Boreal | Forestation[a] | 2.4 | 17 |
| Temperate | Forestation[a] | 11.8 | 60 |
| | Agroforestry | 0.7 | 3 |
| Tropical | Forestation[a] | 16.4 | 97 |
| | Agroforestry | 6.3 | 27 |
| | Regeneration[b] | 11.5 – 28.7 | 44 - 99 |
| | Slowing-deforestation[b] | 10.8 – 20.8 | |
| **TOTAL** | | **60 – 87** | |

[a] Refers primarily to reforestation, but this term is avoided for political reasons.

[b] Includes an additional 25% of above-ground C to account for C in roots, litter, and soil (range based on uncertainty in estimates of biomass density)

*Source:* Adapted from Watson, Zinyowera, Moss, & Dokken (1996, pp.785, 791)

Surprisingly, despite the size of their forests and large areas of marginal agricultural land, there remains only limited room for forest sector policies to sequester carbon in the major wood producing countries (Canada, Finland, Sweden, Russia). We illustrate this using The Economic, Carbon And Biodiversity (TECAB) model for northeastern British Columbia (Krcmar, Stennes, van Kooten, & Vertinsky, 2001; Krcmar & van Kooten, 2003). The model consists of tree-growth, agricultural activities and land-allocation components, and is used to examine the costs of carbon uptake in the grain belt-boreal forest transition zone. Estimates for the study region, extended to other regions, provide a good indication of the costs of an afforestation-reforestation strategy for carbon uptake for Canada as a whole, and perhaps for other boreal regions as well. The study region consists of 1.2 million ha, of which nearly 10.5% constitute marginal agricultural land, with the remainder boreal forest. The boreal forest is composed of spruce, pine and aspen. For environmental reasons and to comply with BC's Forest Practices Code, the area planted to hybrid poplar in the model is limited only to logged stands of aspen and marginal agricultural land. Other harvested stands are replanted to native species or left to regenerate on their own, depending on what is economically optimal. Carbon fluxes associated with forest management, wood product sinks and so on are all taken into account. An infinite time horizon is employed, land conversion is not instantaneous (as assumed in some models), carbon fluxes associated with many forest management activities (but not control of fire, pests and disease) are included, and account is taken of what happens to the wood after harvest, including decay.

Results indicate that upwards of 1.5 million tons of discounted carbon (discounted at 4%) can be sequestered in the region at a cost of about $100 per tC ($27 per t $CO_2$) or less. This amounts to an average of about 1.3 t ha$^{-1}$, or about 52 kg ha$^{-1}$ yr$^{-1}$ over and above normal carbon uptake. If this result is applied to all of Canada's productive boreal forestland and surrounding marginal farmland, then Canada could potentially sequester 10-15 Mt C annually via this option in perpetuity. The total C sequestered in this manner would be about 20% of Canada's annual KP-targeted reduction of 65.5 Mt C per year. If prices for carbon offsets (or carbon subsidies) are higher, more carbon credits will be created, but marginal costs of creating additional carbon offsets rise rapidly.[11] This rapid increase in costs is partly due to the slow rates of growth in boreal ecosystems – boreal forests are globally marginal at best and silvicultural investments simply do not pay for the most part, even when carbon uptake is included as a benefit of forest management. Afforestation with rapid growing species of hybrid poplar provides some low-cost carbon, but thereafter marginal costs also rise rapidly (van Kooten, 2000, also see below).

Globally, carbon sequestration in forest ecosystem sinks is expected to play a significant role in achieving KP targets, as indicated in Table 11.1, but at what cost? Manley, van Kooten, and Smolak (2004) address this issue by employing 694 estimates from 49 studies for a meta-regression analysis of the average and marginal costs of creating carbon offsets using forestry. Estimates of the uptake costs are derived from three meta-regression analysis models: (i) a linear regression model where reported costs per tC are regressed on a variety of explanatory variables; (ii) a model where costs are converted to a per ha basis and then regressed on the explanatory variables using a quadratic functional form; and (iii) a model where per ha uptake costs are regressed on the explanatory variables using a cubic functional form. Using the estimated regression models, average costs of carbon sequestration for various uptake scenarios and regions can be calculated. These are provided in Table 11.6.

Baseline estimates of the average costs of sequestering carbon (of creating carbon offset credits) through forest conservation in the tropics are US$11-$40 per tC. Sequestering carbon in terrestrial forest ecosystems is (generally) somewhat lower in the Great Plains than elsewhere, including the tropics. Surprisingly, costs are higher in the Corn Belt than in the tropics or Great Plains. Compared to simple conservation of existing forests, tree planting increases costs by nearly double, and agroforestry activities increase costs even more while forest management is the least costly option. Needless to say, if the opportunity cost of land is appropriately taken into account, costs are 3.5 times higher than the baseline where such costs are assumed negligible or ignored.

When post-harvest storage of carbon in wood products, or substitution of biomass for fossil fuels in energy production, are taken into account, costs are at their lowest – from US$3.57/tC for a project that includes product sink carbon (Table 11.6) to US$31.18/tC for a project that takes into account fuel substitution in other regions. Accounting for carbon entering the soil also lowers costs. The reason is that the inclusion of soil and wood-product carbon sinks, or fossil fuel substitution, results in more carbon being counted for the same costs.

*Table 11.6.* Projected Average Costs from Three Models of Creating Carbon Offsets through Forestry Activities, 2002 ($US per tC)

| Scenario | Model | | |
|---|---|---|---|
| | Linear | Quadratic | Cubic |
| Baseline (Tropics/Conservation) | 11.06 | 30.22 | 40.44 |
| Tropics | | | |
| Planting | 17.98 | 55.79 | 77.46 |
| Agroforestry | 25.39 | 63.81 | 87.79 |
| Forest Management | 10.57 | 25.38 | 33.33 |
| Soil Sink | 8.02 | 14.64 | 16.29 |
| Fuel Substitution | 5.51 | 18.96 | 24.45 |
| Product Sink | 3.57 | 10.92 | 13.35 |
| Opportunity Cost of Land | 40.42 | 109.81 | 140.58 |
| Great Plains | | | |
| Conservation | 13.91 | 23.99 | 30.91 |
| Planting | 22.61 | 44.29 | 59.20 |
| Agroforestry | 31.93 | 50.66 | 67.09 |
| Forest Management | 13.30 | 20.15 | 25.47 |
| Soil Sink | 10.09 | 11.62 | 12.45 |
| Fuel Substitution | 6.94 | 15.05 | 18.68 |
| Product Sink | 4.49 | 8.67 | 10.20 |
| Opportunity Cost of Land | 50.83 | 87.18 | 107.44 |
| Corn Belt | | | |
| Conservation | 17.37 | 33.92 | 43.30 |
| Planting | 28.24 | 62.63 | 82.93 |
| Agroforestry | 39.88 | 71.64 | 93.99 |
| Forest Management | 16.61 | 28.50 | 35.68 |
| Soil Sink | 12.60 | 16.43 | 17.44 |
| Fuel Substitution | 8.66 | 21.29 | 26.17 |
| Product Sink | 5.61 | 12.26 | 14.29 |
| Opportunity Cost of Land | 63.50 | 123.27 | 150.51 |
| Other Regions | | | |
| Conservation | 18.41 | 39.92 | 51.58 |
| Planting | 29.94 | 73.70 | 98.79 |
| Agroforestry | 42.28 | 84.30 | 111.96 |
| Forest Management | 17.61 | 33.53 | 42.50 |
| Soil Sink | 13.36 | 19.34 | 20.77 |
| Fuel Substitution | 9.18 | 25.05 | 31.18 |
| Product Sink | 5.95 | 14.42 | 17.03 |
| Opportunity Cost of Land | 67.31 | 145.07 | 179.29 |

*Source:* Manley et al. (2004)

However, some of the activities (wood product sinks) are not currently admitted under KP accounting rules, are difficult to measure and monitor (soil carbon), or are not easily implemented (biomass burning).

Finally, while the average costs reported in Table 11.6 are useful to decision makers, they are not truly indicative of the potential costs of creating carbon offsets because they are average estimates only. As already noted, they ignore transaction costs but they also fail to recognize that costs rise as additional carbon is sequestered in terrestrial ecosystems. This is true not only as tree planting activities gobble up agricultural land of increasing productivity and value, but also as an attempt is made to create more carbon offset credits on the same site. Manley et al. (2004) report that, for almost all regions, marginal costs are relatively flat, but rise very steeply once the lower cost opportunities are exhausted. For example, in the Great Plains region, they rise slowly from nearly US$2/tC to US$10/tC by 6-7 tC per ha, but then increase very quickly thereafter.

## 7. TRADING TERRESTRIAL CARBON CREDITS

Some trading of carbon credits has now been initiated through trading networks such as the Chicago Climate Exchange (CCX) and the UK market for carbon emissions allowances (CO2e.com), but they involve only large industrial emitters (LIEs) in a limited geographic area. While others, such as the Winnipeg Commodity Exchange, have proposed the establishment of carbon trading, continuing uncertainty about whether the KP will indeed be ratified hampers efforts to stabilize these markets. Trading so far has been focused on industrial emissions and has not included agricultural or forestry offsets, although the potential for trading offsets exists with the CCX and the Winnipeg Commodity Exchange. However, before a market-based approach to carbon sinks can be applied in practice, certain market conditions will need to be met. For example, carbon offsets need to be certified, a method for seamless trading between $CO_2$ emissions and carbon offsets needs to be found, and an overseeing body with well-defined rules and regulations has to be established (Sandor & Skees, 1999).

Carbon rights were first created in legislation in New South Wales, Australia, but they are rudimentary at best, as indicated by a judgment by Australian solicitors McKean & Park on the potential for carbon offset trading. They indicated that trading in carbon credits is unlikely to occur before 2005 because it would take that long to establish the required rules.[12] In order to buy and sell carbon offset credits, it is necessary to have legislation that delineates the rights of landowners, owners of trees and owners of carbon, because what any one of these parties does affects the amount of carbon that is sequestered and stored. Without clear legislation, buyers of carbon offsets are not assured that they will get proper credit – their claims to have met their emission reduction targets with carbon credits is open to dispute.

Landowners need clear guidelines as to how their activities would qualify for carbon offsets and how credits are to be certified so that they have a well-defined 'commodity' to sell in the carbon market. In the case of afforestation of private land as a carbon sink, even if all conditions for trade are present, there remain concerns about the extent of landowners' willingness to plant trees for carbon uptake on large tracts of (marginal) agricultural land. Tree-planting subsidies, for example, may be inadequate because of uncertainty about future farm payments and subsidies, implications for trade, or transactions costs associated with the creation of carbon sinks on agricultural land (van Kooten, Shaikh, & Suchánek, 2002).

The other problem of mixed $CO_2$ emissions-carbon offset trading concerns the factor for converting temporary into permanent removal of $CO_2$ from the atmosphere. Compared to not emitting $CO_2$ from a fossil fuel source, terrestrial sequestration of carbon is unlikely to be permanent, particularly for carbon stored in fast-growing tree plantations on agricultural land. Yet, temporary removal of carbon is important because it (i) postpones climate change, (ii) allows time for technological progress and learning, (iii) may be a lower cost option than simply reducing $CO_2$ emissions, and (iv) some temporary sequestration may become permanent (Marland, Fruit, & Sedjo, 2001, p.262).

The ephemeral nature of terrestrial carbon uptake can be addressed in a variety of different ways. First, instead of full credits, partial credits for stored carbon can be provided according to the perceived risk that carbon will be released from the sink at some future date. The buyer or the seller may be required to take out an insurance policy, where the insurer will substitute credits from another carbon sink at the time of default. Alternatively, the buyer or seller can provide some assurance that the temporary activity will be followed by one that results in a permanent emissions reduction. For example, arrangements can be put in place prior to the exchange that, upon default or after some period of time, the carbon offsets are replaced by purchased emission reduction permits. Again, insurance contracts can be used. Insurance can also be used if there is a chance that the carbon contained in a sink is released prematurely, but it is also possible to discount the number of credits provided by the risk of loss (so that a provider may need to convert more land into forest, say, than needed to sequester the agreed upon amount of carbon). However, the risk that default will occur remains. This is especially true in the case of the KP as there is currently no requirement that countries that count terrestrial carbon uptake credits during the commitment period 2008-12 are penalized for their release after 2012.

Another method that has been proposed is to employ a conversion factor that translates years of temporary carbon storage into a permanent equivalent that can be specified. The IPCC (2000a) uses the notion of ton-years to make the conversion from temporary to permanent storage.

Suppose that one-ton of carbon-equivalent GHG emissions are to be compensated for by a ton of permanent carbon uptake. If the conversion rate between ton-years of (temporary) carbon sequestration and permanent tons of carbon emissions reductions is $k$, a LULUCF project that yields one ton of carbon uptake in the current year generates only $1/k$ tons of emission reduction – to cover the one-ton reduction in emissions requires $k$ tons of carbon to be sequestered for one year. The conversion rate ranges from 40 to 150 ton-years of temporary storage to cover one permanent ton, with median estimates around 50:1. The choice of conversion rate really amounts to a choice of a rate for discounting physical carbon. For example, if

1 tC is stored in a forest sink in perpetuity and physical carbon is discounted at 2%, then the discounted amount of this perpetual storage equals 50 ton-years. With a 2.5% discount rate on physical carbon, the exchange rate between $CO_2$ emissions and carbon offsets is 40 ton-years, while it is 100 ton-years if the discount rate is 1%. Thus, the idea of ton-years is directly linked to the rate used to discount physical carbon.

As Marland et al. (2001) note, the ton-year accounting system is flawed: ton-year credits (convertible to permanent tons) can be accumulated while trees grow, for example, with an additional credit earned if the biomass is subsequently burned in place of an energy-equivalent amount of fossil fuel (p.266). To avoid such double counting and the need to establish a conversion factor, the authors propose a rental system for sequestered carbon. A one-ton emission offset credit is earned when the sequestered carbon is rented from a landowner, but, upon release, a debit occurs. "Credit is leased for a finite term, during which someone else accepts responsibility for emissions, and at the end of that term the renter will incur a debit unless the carbon remains sequestered and the lease is renewed" (p.265, emphasis in original). In addition to avoiding the potential for double counting, the landowner (or host country) would not be responsible for the liability after the (short-term) lease expires. Further, rather than the authority establishing a conversion factor, the market for emission permits and carbon credits can be relied upon to determine the exchange rate between permanent and temporary removals of $CO_2$ from the atmosphere.

The carbon sink potential in CDM reforestation and afforestation projects exceeds that within industrial countries, making impermanence of terrestrial sinks a more pressing issue for the CDM. The issue of the impermanence of carbon sinks in CDM projects was considered by COP8 in New Delhi in October 2002. Workshops early in 2003 discussed (1) insurance coverage against the destruction or degradation of forest sinks (referred to as iCERs), and (2) the creation of 'temporary' CERs (certified emission reductions) and RMUs (removal units), denoted rCER or tRMU, whereby the certified units would expire at the end of the commitment period or after a different specified period of time. When expired, these credits would have to be covered by substitute credits at that time or reissued credits if the original project were continued. Negotiations regarding definitions and modalities continued at COP9 in Italy, December 2003, but no final resolution has yet been announced. The reason is that countries with large sink potential generally oppose solutions, such as the idea of ton-years and rental rates, that reduce the value of carbon offsets relative to emissions reduction, thereby requiring such countries to make greater efforts to reduce $CO_2$ emissions.

This method for dealing with the question of permanence does not resolve the issue of higher (transaction) costs related to contracting. It is our view that the least cost option would be to tax emissions when they occur, whether these are emissions from LULUCF activities or fossil fuel burning, and to provide a subsidy of the same amount as the tax when carbon is sequestered through some LULUCF activity. The tax revenue should be more than adequate to cover the needed subsidies.

## 8. ARE LANDOWNERS WILLING TO CREATE CARBON SINKS?

A land-rich country such as Canada expects to rely on afforestation of agricultural land to meet a significant component of its KP commitment. As indicated in previous sections, there is a limit to the amount of carbon offset credits that can be claimed from forest management activities on existing forestlands. Thus, the focus will shift to afforestation of agricultural land, where the role of private landowners is more important as most forestland in Canada is publicly owned. Griss (2002) estimates that roughly 1.1-1.4 million ha of agricultural land in Canada could plausibly be converted to tree plantations for carbon uptake purposes, while the Sinks Table of Canada's National Climate Change Process suggested that 843,000 ha of agricultural land could be afforested. The problem of tree planting is not related to biophysical possibilities, however, but to the willingness of landowners to create carbon credits.

It is imperative to identify methods by which landowners are willing to create carbon credits and their capacity to create and market carbon offsets. Landowner preferences for different carbon sequestration methods are likely influenced by the available information and methods, institutional support and structure, and relative risk and uncertainty with regards to maintaining a profitable enterprise and remaining eligible for government programs.

Of course, farmers are generally interested in receiving carbon credits – that is, subsidies – for activities that result in soil conservation, such as a change in agronomic practices from conventional to conservation tillage or a reduction in the proportion of tillage summer fallow, both of which increase SOC by retaining organic matter. In addition, agricultural landowners may be willing to change land use by afforestation of previously cultivated land. If sinks are to be used as a flexible mechanism for meeting $CO_2$ emissions goals, it is important to understand landowners' incentives, motivations and preferences, as well as the transaction costs of implementing tree-planting programs. These issues have been studied using a survey of landowners in western Canada conducted in 2000 (Shaikh, Suchánek, Sun, & van Kooten, 2003; Suchánek, 2001; van Kooten, Shaikh, & Suchánek, 2002).

When asked about tree planning, landowners in west Canada generally express a preference for shelterbelts rather than large-scale afforestation (Suchánek, 2001). The survey also shed light on landowners' willingness to engage in carbon offset trading (see Table 7). Respondents stated that they preferred contracts with governments and large industrial emitters to change land use (or take on certain activities) over the sale of carbon credits per se (Suchánek, 2001). Contracts with government and LIEs shift responsibility for the carbon offsets away from the landowner to the government or LIE. Specifically, the landowner as agent does not have an incentive to produce carbon offsets beyond switching land use (and might even cut trees for firewood), thereby adding to transaction costs as the principal needs to monitor the contract (see van Kooten et al., 2002). Interestingly, survey respondents indicated that they preferred contracts with government and LIEs, and carbon trading, to contracts with environmental NGOs (Table 11.7). Perhaps this is because environmental NGOs are perceived to be more likely to enforce contracts and penalize agents for acting with guile than will government or LIEs.

It is also worth noting that van Kooten et al. (2002) found that past land use may affect the willingness of landowners to plant trees on a large scale. In particular, in regions that had previously been treed and where landowners or their forbears had incurred substantial sacrifice to carve out farms, there is a reluctance and even refusal to take part in tree planting programs.[13]

*Table 11.7. Western Canadian farmers' ranking of means for establishing carbon sinks*

| Governance structure | Normalized Rank |
|---|---|
| Tree-planting contracts with government/state agency | 1.00 |
| Tree-planting contracts with private firms (large $CO_2$ emitters) | 0.87 |
| Sell carbon credits in markets established to allow trade | 0.71 |
| Tree-planting contracts with ENGOs | 0.44 |

*Source:* van Kooten et al. (2002)

Finally, on a positive note, landowners who did indicate a willingness to participate in tree planting programs (and 25.3% would not consider planting trees under any circumstances) were willing to accept a payment below the opportunity cost of the next best alternative land use. Using survey data, willingness to accept compensation for block tree planting was estimated to be between $14.32 and $22.27 per hectare, while the opportunity costs of land were calculated to be $17.00/ha for pasture land, $19.12/ha for land in hay and $29.08/ha for land in grain production (Shaikh et al. 2003). It is likely that forested land provides benefits to some landowners that are not captured in the market. These include benefits from greater scenic diversity, increased wildlife habitat, water conservation and soil conservation.

## 9. CONCLUSIONS

While terrestrial carbon sinks do have potential to sequester carbon from the atmosphere, they are not the 'silver policy bullet' that many people are expecting, and they are more likely to be a distraction from the real goal of reducing fossil fuel $CO_2$ emissions. Because of their temporary nature, transaction costs to maintain the sinks are ignored. The use of sinks as a replacement for reducing $CO_2$ emissions during the earlier KP commitment periods may make it more difficult to reduce emissions in the future, when sinks are nearing their economic maximum level, because of the lack of investment in technology. The uncertainties with respect to carbon trading, additionality and leakage of projects, and the actual costs of sequestration are also of concern.

Although carbon sinks have some value, especially in the short term as countries seek to implement appropriate emission reduction policies, our view is that their value is highly overrated. It is true that carbon uptake considerations are likely an important impetus for sustainable forest management, but, if sustainable forest management has merit (which we believe it does), its value cannot be justified on the basis of the carbon sink function of forest ecosystems. Likewise, soil conservation (reduced tillage) and tree planting on agricultural lands cannot be justified solely on the basis of their carbon uptake benefits. If the argument to pursue conservation (reduced/zero) tillage and afforestation cannot be justified on the basis of their on- and off-farm (and nonmarket) benefits, it is highly unlikely (with some exceptions) that the addition of carbon offset benefits will prove a good enough reason to pursue them in any event.

## ABBREVIATIONS

**AAUs** – Assigned amount units – emission permits in excess of what a country needs to achieve KP commitment. Can be purchased by other countries.

**ARD** – Afforestation, reforestation, and deforestation

**CDM** – Clean Development Mechanism, where an Annex B country earns "certified emission reductions" by funding emission reduction or carbon sequestration projects in non-Annex B (developing) countries

**CER** – Certified emission reductions

**iCER** – CERs for which insurance coverage shall be maintained for a specified period

**rCER** – Removal CER, which is related to a tRMU

**COP** – Conference of Parties, followed by a number to indicate which meeting is referenced (e.g. COP6)

**COP6$_{bis}$** – The continuance of COP6 in Bonn in Spring 2001 after the breakdown of COP6 in The Hague the previous Fall ("bis" meaning "Part II").

**ERU** – Emission reduction unit – earned as credit for a country that participates in JI activities in another country

**EU** – European Union

**FCCC** – The United Nation's Framework Convention on Climate Change, signed in Rio de Janeiro in 1992

**GHG** – Greenhouse gas

**IPCC** – Intergovernmental Panel on Climate Change

**JI** – Joint Implementation, where an Annex B country participates in emissions reduction or carbon sequestration in another Annex B country

**KP** – Kyoto Protocol

**LIE** – Large industrial emitter of greenhouse gases

**LULUCF** – Land use, land use change and forestry

**NGO** – Non-governmental organization

**RMU** – Removal unit for carbon sinks

**tRMU** – Temporary RMU

**TECAB** – The Economic, Carbon and Biodiversity forest management model of the Forest Economics and Policy Analysis (FEPA) Research Unit at UBC

**SOC** – Soil organic carbon

## OTHER DEFINITIONS

**Annex I** – Countries listed in Annex I of the United Nations' Framework Convention on Climate Change of 1992: Australia, Austria, Belarus, Belgium, Bulgaria, Canada, Croatia, Czech Republic, Denmark, Estonia, Finland, France, Germany, Greece, Hungary, Iceland, Ireland, Italy, Japan, Latvia, Liechtenstein, Lithuania, Luxembourg, Monaco, Netherlands, New Zealand, Norway, Poland, Portugal, Romania, Russian Federation, Slovakia, Slovenia, Spain, Sweden, Switzerland, Turkey, Ukraine, United Kingdom and the United States. These agreed to limit GHG emissions to the 1990 level by 2000.

**Annex B** – Countries listed in Annex B of the Kyoto Protocol of December 1997 include those of Annex I minus Belarus and Turkey. These countries agreed to achieve self-imposed limits on GHG emissions by 2008-12 relative to 1990.

**Carbon offsets** – Carbon credits created via an approved terrestrial sink activity, and referred to as RMUs.

**Commitment period** – The KP commits countries to attain self-declared emission control targets by 2008-12. This period is referred to as the first commitment period in anticipation of successful future negotiations to limit $CO_2$ emissions even further by targeted dates.

**Economic efficiency** – Maximizing aggregate economic benefits which consist of consumer plus producer surpluses

## NOTES

[1] Research funding from the BIOCAP/SSHRC joint initiative and the Canada Research Chairs Program is gratefully acknowledged. Subject to the usual proviso, the authors wish to thank David Boulter, Albert Berry and Shashi Kant for their insightful comments and suggestions.

[2] In this chapter we consider only $CO_2$, because $CO_2$ is the most important anthropogenic greenhouse gas from the perspective of climate change. This is reported throughout the chapter in units of carbon (C), where $1 \text{ tC} = 3.67 \text{ t } CO_2$.

[3] The word "ton" is used to refer to "metric ton", as opposed to imperial ton.

[4] From the following website (accessed 18 February 2004): http://unfccc.int/resource/kpthermo.html

[5] This is an overly optimistic estimate of the role that carbon sinks might play because it ignores the ephemeral nature of sinks and continued deforestation in tropical regions. In fact, as of 1990, land use change in the tropics (mostly deforestation) represented C emissions ranging from 20 to 37% of global emissions from fossil fuel burning (Brown et al., 1993).

[6] Not included is the COP6$_{bis}$ (COP7) provision that a country can offset in any year of the commitment period an accounting deficit under Article 3.3, say from clear cutting, with a net increase in sinks due to forest management under Article 3.4 to a maximum of 8.2 (9.0 at COP7) Mt C. This is discussed in the next section.

[7] Excluding the ARD debit, since its emissions (along with those of most other countries) have risen dramatically since 1990, Canada needs to reduce emissions in 2008-2012 by 65.5 Mt C, with forest management to account for 18.3% of the targeted amount. Additional credits will be claimed for afforestation programs (see van Kooten, 2003).

[8] All monetary values are in Canadian dollars, unless otherwise indicated.

[9] Scientists are unable to identify all of the components of the annual $CO_2$ flux – a carbon sink appears to be 'missing'. Some analysts believe that this missing carbon sink can be explained by the expanding biomass in boreal forests, which is mainly due to the aging of these forests.

[10] $CO_2$ emission reductions are expected to trade for $55-$110 per tC ($15-$30 per t $CO_2$) in international markets (van Kooten, 2003). Carbon offset credits will sell for about one-tenth of that amount because of their ephemeral nature. The research reported here finds that, even for carbon offset prices as high as $110/tC, investments in reforestation do not pay.

[11] Recall from the previous endnote that carbon offset credits, being ephemeral, are likely to trade for no more than a few dollars per tC, and not near the $100/tC reported in the study using TECAB.

[12] Their ruling could be found on April 30, 2003, but not as of February 20, 2004, at the website: http://www.mckeanpark.com.au/html/enviroprop/epcarbtrd/epcarbnav.htm#carboncredit. It might have been removed for political reasons, but that is pure speculation.

[13] Forestland continues to be cleared for agriculture. For the 2-year period 1995-1997, for example, 0.7% of Alberta's forestland (some 200,000 ha) was converted to agriculture (Alberta Environmental Protection 1998).

# REFERENCES

Brown, S., Hall, C.A.S., Knabe, W., Raich, J., Trexler, M.C., & Woomer, P.L. (1993). Tropical forests: Their past, present, and potential future role in the terrestrial carbon budget, in terrestrial biospheric carbon fluxes. In J. Wisniewski, R.N. Sampson (Eds.), *Quantification of sinks and sources of CO₂.* (pp.71-94) Dordrecht, NL: Kluwer Academic.
Caspersen, J.P., Pacala, S.W., Jenkins, J.C., Hurtt, G.C., Moorcroft, P.R., & Birdsey, R.A. (2000). Contribution of land-use history to carbon accumulation in U.S. forests. *Science*, 290, 1148-1151.
Chomitz, K.M. (2000). *Evaluating carbon offsets from forestry and energy projects: How do they compare?* World Bank, Development Research Group (2000), Washington, DC. Retrieved June, 2003, from http://econ.worldbank.org.
Coward, H., & Weaver, A. (2003). *Climate change in Canada.* Waterloo, ON: Wilfrid Laurier University Press.
Field, B., & Olewiler, N. (2002). *Environmental economics* (2nd Canadian ed.). Toronto: McGraw-Hill Ryerson.
Griss, P. (2002). *Forest carbon management in Canada.* Final Report of the Pollution Probe Forest Carbon Management Workshop Series, Pollution Probe, Canmore, AB.
Houghton, R.A. (1993). The role of the world's forests in global warming. In K. Ramakrishna, G.M. Woodwell (Eds.), *The World's Forests for the Future: Their Use and Conservation.* New Haven: Yale University Press.
IPCC (2000a). *Land Use, Land-Use Change, and Forestry.* New York: Cambridge University Press.
IPCC (2000b). *The Marrakesh Accords and the Marrakesh Declaration.* Marrakesh, Morocco: COP7, IPCC. (Available online at: http://unfccc.int/cop7/documents/accords_draft.pdf).
Krcmar, E., Stennes, B., van Kooten, G.C., & Vertinsky, I. (2001) Carbon sequestration and land management under uncertainty. *European Journal of Operational Research*, 135, 616-629.
Krcmar, E., & van Kooten, G.C. (2003). *Timber, carbon uptake and biodiversity tradeoffs in forest management: A compromise programming approach.* FEPA Working Paper. Vancouver, BC: University of British Columbia.
Manley, J., van Kooten, G.C., & Smolak, T.M. (2004). *How costly are carbon offsets? A meta-analysis of forest carbon sinks.* REPA Working Paper. Victoria, BC: University of Victoria.

Marland, G., Fruit, K., & Sedjo, R. (2001). Accounting for sequestered carbon: The question of permanence. *Environmental Science & Policy*, 4(6), 259-268.

Obersteiner, M., Rametsteiner, E., & Nilsson, S. (2001). *Cap Management for LULUCF Options*. IIASA Interim Report Ir-01-011. Laxenburg, Austria: IIASA.

Papadopol, C.S. (2000). Impacts of climate warming on forests in Ontario: Options for adaptation and mitigation. *The Forestry Chronicle*, 76(1), 139-149.

Paustian, K., Andren, O., Janzen, H.H., Lal, R., Smith, P., Tian, G., Thiessen, H., van Noordwijk, M., & Woomer, P.L. (1997). Agricultural soils as a sink to mitigate $CO_2$ emissions. *Soil Use and Management*, 13, 203-244.

Richards, K.R. (1997). The time value of carbon in bottom-up studies. *Critical Reviews in Environmental Science and Technology*, 27, S279-S292.

Richards, K.R., & Stokes, C. (2004). A review of forest carbon sequestration cost studies: A dozen years of research. *Climatic Change* (in press).

Sandor, R.L., & Skees, J.R. (1999). Creating a market for carbon emissions opportunities for U.S. farmers. *Choices*, First Quarter, 13-18.

Shaikh, S.L., Suchánek, P., Sun, L., & van Kooten, G.C. (2003). *Does inclusion of landowners' non-market values lower costs of creating carbon forest sinks?* REPA Working Paper (2003-03). Victoria, BC: University of Victoria.

Sohngen, B., & Alig, R. (2000). Mitigation, adaptation and climate change: Results from recent research on U.S. timber markets. *Environmental Science & Policy*, 3, 235-248.

Suchánek, P. (2001). *Farmers' willingness to plant trees in the Canadian grain belt to mitigate climate change*. M.Sc. Thesis. Faculty of Agricultural Sciences, University of British Columbia, Vancouver, BC.

van Kooten, G.C. (2000). Economic dynamics of tree planting for carbon uptake on marginal agricultural lands. *Canadian Journal of Agricultural Economics*, 48, 51-65.

van Kooten, G.C. (2003). Smoke and mirrors: The Kyoto protocol and beyond. *Canadian Public Policy/Analyse de Politique*, 29(4), 397-415.

van Kooten, G.C., & Bulte, E.H. (2000). *The economics of nature: Managing biological assets*. Oxford, UK: Blackwell Publishers.

van Kooten, G.C., Thompson, W.A., & Vertinsky, I. (1993). Economics of reforestation in British Columbia when benefits of $CO_2$ reduction are taken into account. In W.L. Adamowicz, W. White, W.E. Phillips (Eds.), *Forestry and the Environment: Economic Perspectives (pp.227-247)*. Wallingford, UK: CAB International.

van Kooten, G.C., Sedjo, R.A., & Bulte, E.H. (1999). Tropical deforestation: Issues and policies. In H. Folmer, T. Tietenberg (Eds.), *The International Yearbook of Environmental and Resource Economics (pp.198-249)*. Cheltenham, UK: Edward Elgar.

van Kooten, G.C., Krcmar-Nozic, E., Stennes, B., & van Gorkom, R. (1999). Economics of fossil fuel substitution and wood product sinks when trees are planted to sequester carbon on agricultural lands in Western Canada. *Canadian Journal of Forest Research*, 29(11), 1669-1678.

van Kooten, G.C., Stennes, B. Krcmar-Nozic, E., & van Gorkom, R. (2000). Economics of afforestation for carbon sequestration in Western Canada. *The Forestry Chronicle*, 76(1), 165-172.

van Kooten, G.C., Shaikh, S.L., & Suchánek, P. (2002). Mitigating climate change by planting trees: The transaction costs trap. *Land Economics*, 78, 559-572.

Wallace, J.M., Christy, J.R., Gaffen, D., Grody, N.C., Hansen, J.E., Parker, D.E., Peterson, T.C., Santer, B.D., Spencer, R.W., Trenberth, K.E., & Wentz, F.J. (2000). *Reconciling observations of global temperature change*. Washington, DC: National Research Council.

Watson, R.T., Zinyowera, M.C., Moss, R.H., & Dokken, D.J. (1996). *Climate change 1995: Impacts, adaptations and mitigation of climate change: Scientific-technical analysis*. Cambridge, UK: Cambridge University Press.

Woomer, P.L., Palm, C.A., Alegre, J., Castilla, C., Cordeiro, D., Hairiah, K., Kotto-Same, J., Moukam, A., Risce, A., Rodrigues, V., & Van Noordwijk, M. (1999). *Carbon dynamics in slash-and-burn systems and land use alternatives: Findings of the alternative to slash-and-burn programme*. Working Paper. Nairobi, Kenya: Tropical Soil Biology and Fertility Program.

# CHAPTER 12

# THE INTERNATIONAL TRADE AND ENVIRONMENTAL REGIME AND THE SUSTAINABLE MANAGEMENT OF CANADIAN FORESTS

HARRY NELSON AND ILAN VERTINSKY

*FEPA Research Unit, University of British Columbia*
*Vancouver, BC, Canada.*
*Email: ilan.vertinsky@commerce.ubc.ca*

**Abstract:** The paper explores the structure and impact of the international regime that governs forest management. We pay special attention to three components of the regime: (1) regional and international trade agreements and multilateral environmental agreements; (2) international criteria and indicator processes; and (3) international forest certification systems. The first two components represent subsystems that are molded and enforced by governments, while the third component represents largely a private regulatory system enforced by market behaviour. We show how these three components interact with each other and with the domestic regulatory system within Canada to directly affect sustainable forest management (SFM). We examine how international agreements and processes have introduced environmental issues into domestic policy-making, and assess whether there has been any conflict between the economic and environmental objectives of these agreements. We show that market access concerns have been and continue to be the primary drivers of changes in the regime even as rules designed to facilitate trade have been strengthened through limiting the discretion of countries to enact trade barriers. We conclude with some observations on what insights the Canadian experience offers to the larger debate about the impact of environmental outcomes on trade.

## 1. INTRODUCTION

Canada is a member and a signatory of numerous international and regional agreements that have or might have an impact on forest practices and on the benefits and costs that may accrue to the various stakeholders in the forest. Various trade agreements contain some measures that directly affect trade in forest products (e.g., the General Agreement on Tariffs and Trade (GATT), the North American Free Trade Agreement (NAFTA), the past Softwood Lumber Agreement) and indirectly affect forest practices (e.g., harvesting and silvicultural methods) and the environment. Various environmental agreements can serve to restrict forest practices or create incentives to change practices. Such agreements include the 1992 Convention on

*Kant and Berry (Eds.), Institutions, Sustainability, and Natural Resources: Institutions for*
*Sustainable Forest Management, 257-295.*
© 2005 *Springer. Printed in Netherlands.*

Biological Diversity, the 1992 U.N. Framework Convention on Climate Change, and the 1997 Kyoto Protocol to the latter convention.

These agreements contain legal obligations that directly affect signatory governments. They may restrict actions governments may take or limit or curb the discretion of the government to adopt certain policies. These "hard law" obligations may also be supported by enforcement mechanisms and dispute resolution procedures, as is the case in the trade agreements to which Canada is a signatory.

Not all international institutions that impact forests have the power of "hard law". International agreements may also articulate a common set of ideals or goals that countries will pursue, as is the case for many multilateral environmental agreements (MEA's) such as the Convention on Biological Diversity and the UN Framework Convention on Climate Change (UNFCCC). These agreements may mould and energize highly influential processes through which international consensus is built and legitimized. For example, the International Dialogue on Criteria and Indicators for the Conservation and Sustainable Management of Temperate and Boreal Forests (the Montreal Process), influences the policy debates within Canada that attempt to define Sustainable Forest Management (SFM). Together these agreements and processes create a "soft law" that can influence policy development within countries. These agreements may even require the development of implementation strategies that involve specific legal obligations, as is the case for the Kyoto Protocol. They may require countries to implement polices at the national level in order to meet national commitments. They may cause countries to pay greater attention to those issues raised within the agreements when developing their own policies to address the issue. Political pressure within the regime, exerted by other countries and international NGO's, may also highlight particular issues raised in these agreements that also affect the development and implementation of national policies.

Another, no less important layer of the institutions that define the international regime affecting forest management, consists of certification initiatives backed by non-governmental institutions (NGOs). For example, the Forest Stewardship Council, backed by buyers' groups and environmental non-government groups is having an increasing impact on forest practices around the world through its certification systems. Canada, with one of the largest forest estates in the world, receives especially close scrutiny.

This "international regime" is having an increasing influence on the management and the economic and environmental outputs of Canadian forests. In this paper, we 1) describe the three principal components of the international regime that affect forest management in Canada, 2) examine how these institutions interact with domestic policy-making as well as where they support one another, where they may be in conflict, and 3) show how these institutions have modified and affected firm behaviour and forest management within Canada.

## 2. THE INTERNATIONAL REGIME

*2.1 Definition of International Regime*

There exist a number of different definitions, but in general international regime describes the set of institutions that consist both of international organizations and agreements in which countries set mutual goals that govern the interaction between countries (Porter, Brown, & Chasek, 2000). These agreements may simply enunciate general principles and exhort countries to aspire towards those goals, or more concretely, provide a consultative framework and/or process for further discussion/negotiation and, at its fullest force, develop rules to achieve those objectives. These rules may contain general legal obligations for parties and can range in stringency, most importantly in the means and intensity of enforcement (such as trade restrictions and sanctions). These agreements may also include formal dispute resolution procedures. Participating countries may develop single-purpose organizations to support these agreements, such as the World Trade Organization (WTO), or may rely on existing international organizations (such as United Nations Environmental Programme (UNEP) or United Nations Commission on International Trade Law (UNCITRAL). Membership in these agreements is voluntary, but there may be some negative consequences of being a non-party to an agreement (e.g. loss of legitimacy, reputation and exposure to sanctions from ENGOs). Generally speaking, those agreements resulting in specific legal obligations for signatories are known as "hard law". More general agreements without specific obligations but which provide general statements of principles or objectives or create frameworks to address the problem are known as "soft law". To date, most trade agreements have taken the form of hard law while most environmental agreements consist mainly of soft law. However, many of the framework agreements lead to ongoing processes that provide for discussion and articulation of the issues, and these can result in more specific obligations similar to those found in "hard law" agreements. In addition, there are efforts through voluntary approaches to labeling and certification to create private regulatory systems outside government processes.

*2.2 The establishment and functioning of international regime*

Clearly a wide range of factors can explain why countries may pursue particular sets of actions: these reasons would include political imperatives and pressures from domestic constituencies; past history; and cultural norms (Bernauer, 1995; Sprinz & Vaahtoranta, 1994). Porter, Brown, and Chasek (2000) argue that there have been four traditional approaches to answering the questions of regime development: (i) the structural approach in which strong states define the rules of the game to their benefit; (ii) a game theoretic approach in which coalitions of states form and bargain over mutual benefits (or avoidance of costs); (iii) institutional bargaining in which nation states develop international institutions to act as intermediaries through which they can interact because they are fundamentally incapable of interacting directly; and (iv) epistemic communities in which international learning from scientific research establishes common values and goals that shape the evolution of the regime. They claim that a common weakness of these approaches is the treatment of states as unitary agents and that not enough attention is paid as to why states develop coalitions to block regime formation or how these coalitions may reduce the performance of these regimes (also see Bernauer, 1995).

While economic models that focus on the benefits and costs that accrue to countries offer insight into the motives for cooperation (see, for example, Dasgupta, 1997), the political economy perspective supports the idea that relying only economic calculations is not sufficient when modeling the choice of trade policies. Treating countries as unitary actors cannot explain the choice of sub-optimal policies measured by sacrifices in the terms of trade (Zhou & Vertinsky, 2002). Instead, the presence of lobbies and other interest groups appears to be an important determinant of outcomes but cannot explain why sometime they may only achieve part of their desired goals (Levy, 2003). Part of the explanation may reside in the fact that beliefs play an important role in explaining outcomes as shown by new research in experimental economics. While the strict calculus of economic rationality suggests people will free ride if given the opportunity to voluntarily provide a public good or overexploit a resource, experimental games often show surprisingly strong results in which people voluntarily choose more cooperative outcomes than those predicted by economic theory. There is also empirical evidence to support the idea that voluntary efforts and moral suasion can be powerful organizing forces in the provision of public goods (Klein, 2002). Such outcomes, however, appear achievable only when there is a well-established system of rights and responsibilities and the effective enforcement of rights (Dasgupta, 1998). Other authors have noted that people's willingness to pay or engage in voluntary actions to improve the environment will increase if they adopt a more altruistic viewpoint encompassing a shared responsibility rather than considering only their economic self-interest (Nyborg, 2000). Therefore, norms and beliefs may explain in part the ways in which these agreements might work, especially those concerned more with "soft law" processes and objectives. In fact, Young (2002) suggests that there are two sharply divergent theories explaining why countries may cooperate: one is that they rationally choose to act collectively to avoid free-rider problems, while the other involves a view of shared norms and social responsibilities that motivate cooperation.

*2.3 The impact of the international trade and environment regime*

Figure 12.1 shows the framework used to assess the impact of the international regime upon Canadian forestry. In it, we consider the importance of economic factors (the distribution of costs and benefits) but also the role norms and values can play. We elaborate upon the interaction between national policies and international subsystems, how those linkages can affect forest product markets, and the resulting effect on firm behaviour. In the model, the international regime consists of a layer including international trade and environmental agreements, international criteria and indicator (C&I) processes, and certification initiatives. It is within this international layer that countries and the policies they develop interact. In some cases, countries may use the regime to try to directly influence the policies of other countries.

Countries may develop collective agreements or obligations that then require the development of national policies. These are then filtered through national and provincial policy-making processes. Different components of the international regime may constrain or shape these policy choices, and the policies that are adopted can affect forest product manufacturing, forest product markets, and forest practices and management. In the case of certification, international ENGO's attempt to tighten this system of public regulation by governments by creating a private regulatory system, acting directly to prescribe forest practices and management using their influence on forest product markets. These public and private channels of regulation together, then, link the international regime to the actual way in which Canadian forests are managed and utilized.

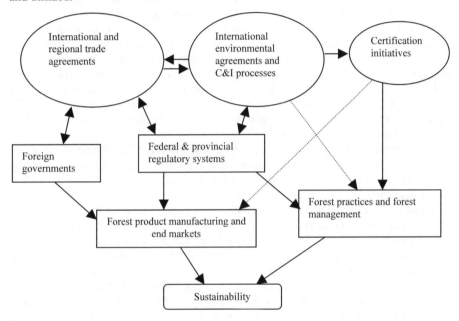

*Figure 12.1. A Conceptual Model*

The components of all these systems can reinforce or contradict one another across a number of dimensions. How conflict is resolved between these systems is equally important. The interaction of these systems may distort prices. International agreements may lead to trade liberalization that can deepen or create new markets. Trade liberalization can also place greater emphasis on differences in domestic policies between countries (Burfisher, Norman, and Schwarz, 2001. The emergence of new competitors may sharpen competitive pressures in export markets and lead to increased demand for trade protection where domestic industries were previously sheltered. Countries may adopt trade polices in response to these concerns. The effect of these policies may not be straightforward. Tariffs designed to reduce imports in one country may perversely lead to increases in harvesting rates in exporting countries as firms attempt to maintain export revenues as net prices fall. Log export restrictions designed to prevent deforestation may instead hasten it by encouraging the conversion of forestland to other higher-valued uses. Differential tariffs or import restrictions may shift harvesting pressure from one region to another less able to accommodate the increased activity (that is either more environmentally sensitive and/or lacks the institutions to effectively manage or mitigate the environmental impacts associated with increased production). Certification may open new markets and reward environmentally preferred products or could potentially become a trade barrier if it is a condition of market access and thus devalues the resources and reduces the incentives for proper stewardship of them.

Governments may develop new regulations governing forest practices and management, or firms may adopt new harvesting practices in order to meet certification requirements. All of these can affect prices and costs in existing markets, as well as change the scope and nature of those markets. This, in turn, can affect what firms cut, where they cut, and when they cut. They can determine how the forest is managed, what it is managed for, and how decision-makers determine the location and intensity of forest activities. They can influence investment decisions and alter the long-run composition of the forest products industry and consequently of the forest. All of these decisions have ecological, social and economic consequences that affect the sustainability of forests.

The international regime does not exist in isolation. It arises from the efforts of national governments to respond to scientific and public opinion and issues raised by NGO's (where we use the term broadly to include industry associations as well as Environmental NGO's (ENGO's)).[1]

Assessing the impact of the international regime upon Canadian forests has to take into account the federal structure of Canada and the fact that most decisions affecting natural resources are made at the provincial level within the Canadian system. Under the constitution, Canadian provinces enjoy a great deal of autonomy. Hoberg and Harrison (1993) note that the provinces have been quick to assert provincial jurisdiction in areas affecting natural resources, including environmental protection. Therefore, the Federal government has historically confined itself to a role in which it facilitates and coordinates policy development, while forest management policies are developed at the provincial level.[2] The Federal government, however, has primary responsibility for international trade and for negotiating MEA's but is reluctant to exercise its powers in forest-related trade issues unless provinces are in agreement.

*2.4 Analyzing the Impact of Trade*

In assessing the potential impact of trade, the analysis is often framed by the adoption of endpoints that either involve no trade whatsoever (autarky) or complete free trade. Where a country may have abundant resources, such as the forest resources found in Canada, the implications of either position can lead to unrealistic outcomes. Under autarky, there is no trade with other countries and unless there is sufficient domestic demand for the resource (perhaps even for alternative uses such as firewood) the resource may be devoid of market value. Forestland may then be converted to higher valued uses, or is as often the case, converted to agricultural uses. At the other endpoint, unfettered free trade with no constraints whatsoever increases the size of markets for exporters, who can then realize higher values than if they were restricted to only meeting demand within their own country (through higher prices in world markets or taking advantage of any economies of scale that may exist). Under this scenario the full market value of the resource is realized.

Different trade rules can have different economic impacts, then, upon the nature of the markets in which the good is sold and the value of the resource. There may also be related impacts both upon national wealth and local communities that derive economic and other benefits from regional forest resources.

Governments establish policies that can influence trade for a number of different reasons. They may enact tariffs or other border restrictions to protect their domestic industry from harm caused by other countries' unfair trade practices, which may consist of subsidies or dumping. A government may establish trade measures to protect an infant industry that is unable to compete with more established firms until it climbs its learning curve or reaches its optimal size. Domestic policies can indirectly influence trade. Governments may enact standards and regulations designed to protect the health and safety of their citizens or address domestic environmental issues. These standards may affect traded goods if they are subject to such measures. Environmental considerations have traditionally not entered into trade policy other than the establishment of standards or regulations to protect the domestic health or safety of consumers (e.g. banning foods that contain toxic residues). [3]

The concerns over these policies acting as non-tariff barriers (or NTB's) arise because firms (and governments) may adopt trade and domestic polices designed to protect domestic industries. These may be in response to lobbying in which firms (or other groups) seek to enact barriers to obtain economic rents. [4] Governments may simply seek to shelter domestic firms from more efficient foreign competitors. [5] Governments may even deliberately pursue a strategic trade policy, taking advantage of the relative size of an industry and resulting market power to alter the terms of trade to the benefit of domestic producers or consumers. [6]

A number of authors have suggested that the use of non-tariff barriers has grown as average world tariffs have fallen (Barbier, 1996; Levy, 2003). Gandolfo (1998) suggests that these barriers offer more discretion and are less overt and that their use has grown for three principal reasons: (i) they permit countries to ostensibly comply with new trade rules which emphasize a reduction in tariffs while still offering protection to domestic industries; (ii) barriers implemented through such measures are easier to enact since they are less transparent and visible (than are tariffs or quantitative restrictions); and (iii) interest groups and politician find such measures more palatable again because such measures (and the costs they entail) are less apparent. [7]

## 3. THE INTERNATIONAL TRADE REGIME

The international trade regime consists of the array of international rules and agreements that govern the trade policies countries can develop. Generally the international rules are designed to facilitate trade and reduce the ability of countries to establish trade barriers. What affects trade patterns is the resulting mosaic of tariffs and non-tariff barriers as well as the degree of security offered by the international trade regime in terms of protection from arbitrary and opportunistic measures by other countries.

Tariffs faced by Canadian forest product firms in foreign markets are generally low across most product categories with several noticeable exceptions (these are in selected product categories, such as plywood, and among developing countries). The main non-tariff measures Canadian firms face, principally health and safety standards, apply to solid wood products, primarily logs and lumber, and Canadian firms are not generally singled out for especial scrutiny in this regard (Schwab, 2002). There are concerns, however, that the development of certification may become a non-tariff barrier for forest products in general (New Zealand Forest Research Institute Limited, 1999; New Zealand Institute of Economic Research, 2000).

The development of market barriers (through tariff and non-tariff barriers) and opportunistic trade actions are the biggest threat facing exporters. The only available protection is through multilateral agreements that reduce the risk from these threats through two important components: (1) agreements that free trade and offer protection from arbitrary moves by importing countries and (2) dispute resolution mechanisms that allow enforcement.

Canada is a member of two important trade agreements that directly affect trade in forest products (the General Agreement on Tariffs and Trade (GATT) and NAFTA). One agreement, GATT, is international in scope, governing trade with most developed countries. The other, NAFTA, is a regional trade agreement that covers three-way trade between Canada, Mexico, and their most important customer, the US. These agreements contain three sets of rules that have a major impact on the operation of the forest sector. One set of rules permits countries to only apply protective measures to ensure fair competition and protection of their industries from sudden surges in imports[8]. These rules have been frequently used to justify protectionist measures against the Canadian forest products industry (see Nelson & Vertinsky 2004). The second set of rules permits countries to establish environmental measures designed to protect animal, human, and plant health. These are modified by a third set of rules that restrict the application of such rules where they do affect trade, requiring that the rules must be for legitimate reasons, scientifically justified, and chosen so as to be the least trade-restrictive as possible.[9]

In additions to these sets of rules, both agreements also contain dispute resolution procedures. Both utilize quasi-legal processes in which panels are struck to hear disputes; both are similar in that only parties to the agreements (i.e. member countries) can have any formal standing in the process; and both allow for an appeal process. They differ in the eligibility of who can serve on the panels as well as in the enforcement mechanism.[10] As well, NAFTA contains a set of rules regarding the treatment of investments (Chapter 11) that are not found in GATT. Under Chapter 11, private parties from member states can sue host governments over government policies or actions that diminish or expropriate the value of their investment. Finally, the most recent GATT Agreement established a formal organization, the WTO that monitors and enforces the trade rules under the agreement as well as facilitating ongoing discussion of modifying those rules. NAFTA does not have a similar organization (aside from a secretariat to administer the dispute resolution process) nor is there an institutionalised procedure in place to modify the agreement.

## 3.1 International Trade in Forest Products

Historically much of the analysis of forestry related issues and trade focused primarily on the environmental effects of the tropical timber trade and the impact of log export bans on forest products trade in the Pacific (Tomberlin, Buongiorno, & Brooks 1998). The main concern surrounding tropical forests has been deforestation and trade has been raised as a possible explanatory factor. Research results, however, suggest that institutional factors such as patterns of ownership, the strength of those rights (including customary rights), political stability, and government policies regarding other land use activities such as agriculture are more important in explaining deforestation than trade in such products (Southgate, Salazar-Canelos, Camacho-Saa, & Stewart, 2000; Deacon, 1995). More recently, there have been a series of studies looking at the impact of tariffs (and reductions in tariffs) on trade flow patterns. [11] Sedjo and Simpson (1999) consider the effect of further tariff liberalization (post-Uruguay round) and conclude that as the most significant reductions have already taken place, there would only be a small aggregate increase in forest products traded, and changes in production and consumption would also be small. However, there would be a change in the composition of trade, as more value added products will be traded, and increased trade from countries with significant plantations such as Chile and New Zealand will be realized.[12] There has been little investigation of the impacts of institutional arrangements on trade patterns in forest products. In one of the few studies that investigate this issue, Southgate et al. (2000) consider the actual effects of trade liberalization in Ecuador on the domestic forest products industry. They show that improved market access can theoretically lead to improved prices that would be expected to lead to better timber management and stewardship; however local market imperfections (oligopsonistic markets and weak institutions) prevent the benefits of higher timber prices from flowing through back to the landowners.

Some authors have suggested that there has been a proliferation of non-tariff measures affecting forest products, citing the application of sanitary and phytosanitary standards to imported logs and solid wood, and the development of performance standards that may discriminate against certain types (and hence suppliers) of wood (Bourke & Leitch, 2000; Barbier, 1996; and the NZ Institute of Economic Research, 2000)[13]. Sampson (2000, p. 66), however, states that "[o]f these measures, tariff escalation is believed to be the main source of trade restriction and distortion in this sector".[14]

## 3.2 The Softwood Lumber Dispute

Despite the presence of a free trade agreement with US under which Canadian forest products can enter duty free, Canadian softwood lumber exports have for much of the past quarter-century faced some kind of border restrictions entering the US. Nelson and Vertinsky (2004) provide a narrative describing the various rounds within the dispute and the motivation for US trade action. In general, US timber interests have been highly successful in capturing government and Congressional support to press for trade barriers. These restrictions have taken a number of different forms; these include province-specific export tax rates in which some regions have been exempted; the Softwood Lumber Agreement, under which firms received individual quotas and faced volume constraints on the amount of lumber they could ship to the U.S duty-free; and countervailing and antidumping duties today. All of these various border restrictions had an impact upon prices and the market and upon producers' decisions. Estimates of economic impact of these various restrictions generally show the same pattern: American consumers have been harmed while Canadian and US producers have benefited. These studies show that there are significant transfers: Wear and Lee (1993) found that American producers saw a gain of $2.6 billion while American consumers saw a loss of $3.8 billion (all in $1982) during the period of the MOU, and Zhang (2001) estimated that the SLA had increased lumber prices by just under $59/mbf in the US, with producers benefiting by $7.7 billion and American consumers facing a loss of consumer surplus of $12.5 billion (all in $1997). Van Kooten (2002) estimates that the SLA benefited Canadian consumers by $109 million annually and that producers on both sides of the borders benefited at the expense of US consumers.[15]

### 3.3 The Impacts of Trade

Given the export orientation of the Canadian industry, it is access and prices firms face in those markets that can have the most significant impact on firms' decisions. In Canada, changes in market access and prices have been significantly affected by the softwood lumber dispute. Various resolutions of the dispute have had different effects on the prices of softwood lumber, influencing firms manufacturing and harvesting decisions. Over time, border restrictions have shifted harvesting activities towards provinces originally not covered by the agreement, such as the Maritimes and the Prairie provinces, and towards species used to manufacture products not covered by the dispute, such as the utilization of aspen to produce OSB. There has also been an incentive to shift towards products not covered by the dispute. Under the border restrictions, value-added product such as roof trusses and pre-fabricated housing components were exempt. Firms that could had an incentive to produce and ship higher valued goods to the US, the most prominent example being Western Red Cedar lumber producers on the BC Coast. Within Canada, firms in regions that were not constrained by the SLA responded to the higher prices created in the US by the agreement by increasing lumber production. This led to an increase in demand for logs and increased harvest rates on private forestland in Eastern Canada. The higher harvest levels over this period has raised concerns about the ability of this sector, historically a locally important component of the timber supply, to contribute to future timber needs (Jaako Poyry Consulting, 2002).

More recently, the high duties have reduced net prices to Canadian exporters to the US, and this has led to significant improvements in productivity as Canadian mills have proven adept at cost cutting (Hamilton, 2004). Indeed, the US duties have had the effect of increasing the competitiveness of Canadian mills, as they have pursued economies of scale through rationalizing production by closing down smaller, uneconomic mills (this has been facilitated by mergers that permit companies more flexibility to redirect their log supply to more efficient mills). The increased efficiency was achieved largely through rationalization (plant closures) resulting in higher unemployment rates in several forestry-dependent communities.

## 4. THE INTERNATIONAL ENVIRONMENTAL REGIME

The international environmental regime consists of the rules countries have chosen to govern their interaction in addressing shared issues of environmental concern. Historically, countries pursued bilateral or regional agreements to cover issues of concern involving shared transboundary resources (Canadian examples include agreements with the US on migratory birds, shared watersheds with the US, and North Pacific fisheries with other countries). More recently, those concerns have broadened to encompass a broader set of environmental and social issues to larger, more global issues of pollution with an international dimension such as ozone depletion and global warming as well as the environmental and social consequences of economic growth. There is no one international agreement that deals directly with temperate forests and the complex set of environmental, economic and social issues they raise. Instead, an international environmental agreement may directly or indirectly address forestry-related issues depending upon the nature of the problem; for example, deforestation may be addressed as part of efforts to combat desertification. Therefore, it is necessary to identify those MEA's whose scope or objectives can encompass forestry-related issues and then assess to what extent they may influence Canadian forest management. [16]

Canada is a signatory to a number of such agreements of which several early MEA's have the potential to affect forest management. These include: the Convention on International Trade in Endangered Species (CITES), signed in 1973, designed to protect endangered and threatened species through restrictions on trade; the Convention on Wetlands of International Importance Especially as Waterfowl Habitat (the Ramsar Convention), signed in 1971, where the goals are to protect and facilitate the wise use of wetlands; and the International Tropical Timber Agreement (ITTA), signed in 1983, that addresses the use and conservation of tropical forests through developing consultative framework between consumers and producers and identify criteria for sustainable tropical forest management (Whiting, 2001).

To date, these early agreements have not had an impact upon Canadian forest management because they are either not applicable to Canadian forests or the issues addressed by these agreements have not arisen in Canada. For example, while CITES can affect trade in certain tree species that are considered at risk or endangered (e.g. tropical mahogany), as none of the commercially harvested species in Canada are considered endangered or threatened, it currently has no impact upon Canadian forest management. Although Ramsar can affect forest management by encouraging national policies that take into account the importance of forested wetlands for waterfowl habitat, its focus to date has been on the individual selection and designation of protected sites in Canada, some of which happen to lie in forested areas. The ITTA, while it does not directly address Canadian forest management or practices because of its focus on tropical forests, does provide a potential model for other international processes (both in its attempts to bring together ENGO's and governments, along with consumers and industry groups, and the attention it pays to using C&I to measure sustainable forest management).

More recently, Canada has made commitments under several international processes and new MEA's that can potentially have a more direct impact on forest management. Many of these were only initiated in the past decade and a half and have come out of a series of international discussions that started with the Report of the Brundtland Commission on the Environment and Development (Bruntland, 1987).[17] Indeed, the concept of "sustainable development" gained its modern context from the Bruntland Commission and it was from that beginning, and the recognition that resources must be managed more carefully, that the term "sustainable forest management" emerged. The Commission also led to the United Nations Conference on Environment and Development (UNCED) in Rio in 1992, also known as the Earth Summit, where numerous countries, including Canada, adopted several convention documents of which the two most important agreements for Canada's forests were the UN Framework Convention on Climate Change (UNFCCC) and the Convention on Biological Diversity (CBD). At the same time, however, participants were unable to achieve one of the major goals of the summit, a consensus on how to manage the world's forests.

Despite the failure to achieve an international forestry agreement, however, among the outcomes of the summit were the Rio Declaration; Agenda 21; and Statement of Forest Principles. All of these consensus documents contained passages that were directly related to the future management of forests throughout the world. Agenda 21 specifically called for the formulation of scientifically sound criteria and guidelines for the management, conservation and sustainable development of all forest types.  As a result, more than 140 countries became involved in the development of eight different, yet similar, sets of international "Criteria and Indicators of Sustainable Forest Management" during the mid-1990's.  One such set was developed through the "Montreal Process" which involved the twelve member countries that are home to the temperate and boreal forests outside of Europe.[18] The Montreal Process influences the policy debates within Canada that attempt to define Sustainable Forest Management (SFM). In addition, negotiations to establish an international forest convention commenced at UNCED, and continue to this day through the UN Forum on Forests (UNFF).

Of the MEA's that address forestry-related issues, only two to date have had some direct impact on forest management and use within Canada - the Convention on Biological Diversity (CBD) and the Kyoto Protocol. At the same time, ongoing efforts through the Montreal Process and the UNFF also have an impact in that they provide legitimacy for ENGO's demands for domestic legislation regarding the need to protect biodiversity and other environmental values and can also help frame the issues as perceived by both the public and policy-makers. We can identify several areas where there have been (or will likely be) specific changes in provincial policy attributed to these agreements and associated with these processes. We first consider the CBD.

*4.1 The Impact of the CBD*

The Convention articulates several important objectives. These include the conservation in situ of biodiversity through the establishment of parks and protected areas; the promotion of the idea of sustainable use incorporating the conservation of resources while pursuing economic development; and the equitable sharing of resources with local communities (including indigenous communities). The CBD is generally aspirational in nature, urging countries to recognize the importance of biodiversity and to develop national strategies. It encourages countries to develop procedures to assess the environmental impacts of proposed development, gather information about the flora and fauna found within the country, and monitor biodiversity. It does not contain any general prescriptions or legal obligations, and it does not provide significant financial support (despite calling for significant financial expenditures for developing countries to develop the capacity to meet their commitments) (Whiting, 2001).

In response to commitments made under the CBD, Canada has formulated a national biodiversity strategy (Canadian Forest Service, 2002), as have some provinces.[19] An important part of the strategy involved a national legislation to protect endangered species along with reporting on biodiversity in different jurisdictions. After reaching a federal-provincial accord in 1996, which committed the provinces and Federal government to introduce complementary legislation, Canada recently enacted the Species at Risk Act (SARA). The Act provides a mechanism for listing endangered species and does not permit killing, harm, or the destruction of their habitat (narrowly written as dwelling places such as nests, dens). Under the Act, the Federal government has lead authority on federal land while provincial governments have the lead role on provincial lands. While provinces retain the option to list their own species they consider threatened or endangered and exercise their provincial laws, the Federal government has the authority to act if provinces do not take sufficient actions to protect federally listed species. In terms of forest management activities, this means that provinces will need to take into account the effect of forest activities on listed species at all levels and may be required to prepare species recovery plans. At the very least, provinces will have to be able to demonstrate that current forest management policies will not threaten specific species or hamper their recovery.

*4.2 The Impact of Kyoto*

Potentially the most important MEA in terms of leading to direct changes in forest management policies is the UNFCCC and the Kyoto Protocol. The 1992 UNFCCC identified greenhouse gas emissions and climate change as a source of global concern and the 1997 Kyoto Protocol established several key approaches to address the concern that are relevant to forest management. First, under the Kyoto Protocol, industrialized countries assumed binding targets (upon ratification by individual countries and entry-into-force of the Protocol) and committed themselves to establishing a series of rules over how to account for sources and sinks of carbon. The Kyoto Protocol identified forests as both a potential source of carbon emissions as well a potential sink of carbon (i.e. carbon sequestration). This had several important effects. A set of accounting rules have been developed that identify how carbon sources and sinks are to be tabulated. There are international rules spelling out how afforestation and reforestation (creation of new forest), deforestation (permanent loss of forest) and forest management are treated under the system of national greenhouse gas emissions accounting. Canada has a specific cap permitting it to use forest management as a carbon sink up to a predetermined level. Second, the Protocol allowed for the development of an international emission trading systems in which countries can elect to participate. Canada has indicated it will develop a domestic trading system that will be linked to the international system, and is exploring how sequestration and emission reductions from forest carbon projects can be traded (as offset credits) in the domestic emissions trading system.

This gives a potential value to carbon stored in the forest. The value of this carbon depends upon a number of factors. It will depend in large part upon the rules set at the international level and national level (countries have the ability to develop their own national trading systems). Proposed rules could change the relative economics of harvesting from natural forests versus plantations (i.e. afforestation/reforestation) depending upon the opportunity cost of carbon (the forgone gains from letting existing forests sequester additional carbon) versus the economic return from harvesting.

Within Canada, the Federal government has proposed that pulp and paper mills would be included within a domestic trading system that sets annual caps on emissions of large companies. Other smaller forest product manufacturing facilities (i.e. with less greenhouse gas emissions) will not be given caps. Annual allocation of emission permits to pulp and paper mills, up to the level of their cap, is expected to be free and based on targeted emissions intensity. The allocation of emission permits can clearly have an impact on firms' costs and revenues. Firms, if constrained, will either need to purchase additional permits or offset credits, or undertake internal emission reductions, or may incur financial penalties depending upon the stringency of the cap. If they reduce greenhouse emissions below their cap they will have excess permits to sell.

Canada also faces choices over determining what will count as managed forest and whether to account for forest management in its Kyoto accounting. Because there is a risk that the managed forest can be a source of emissions, due to fires and insect infestations, Canada also must decide how best to manage that risk, for example through changes in forest protection policies.

Canada's fundamental choice is whether it wants to include forest management in its Kyoto accounting. Forest management is a potential source of carbon credits, but if included Canada must also tabulate the debits arising from forest management activities and harvesting, as well as natural disturbances. If it does include forest management, then management of carbon within the managed forest would be included within the domestic trading system. Key difficulties for implementation of a forest carbon trading system are the determination of appropriate methodologies for establishing project-specific baselines, and handling permanence (e.g. who assumes the risk that credits issued for forest carbon projects are lost when the forest is burned). It is clear that buyers are likely to have little interest in purchasing forest carbon credits if they have to assume the risk, and it is likely that the risk will be shifted onto the sellers or perhaps fixed by the rules employed by the government.

In Canada, the Federal government has assumed responsibility for Kyoto and makes the policy decisions on Kyoto in its areas of jurisdiction. Yet much of the decisions, especially those over land use and forest management, rest with the provinces. Ideally, the federal government hopes to implement these rules with provincial support. Many of the decisions to be made may encourage the provinces to develop policies on aspects of forest carbon. For example, it is still not clear how ownership of carbon credits from a project on Crown land would be established (especially through a baseline approach in which the amount sequestered depends upon actions taken, most likely by a company).

The long-run consequences of these decisions are uncertain. It is still unclear as to whether or not Canadian managed forests are a source or sink for carbon. There are concerns about the permanence of carbon and how it will be treated under the trading system rules, which could reduce the value of carbon and hence interest in forest carbon. Uncertainty about future changes in forest carbon stocks may lead to a decision to exclude forest management from Canada's Kyoto accounting, and hence from a domestic trading system.

Clearly, however, those forest product firms operating in a carbon-constrained world are likely to see an impact in their costs. This, in turn, can lead to changes in their relative competitiveness and this may lead to changes in what firms produce and where they produce it. Firms might face an incentive to either change their product mix (to goods that are less GHG intensive or not covered by the trading system) or shift production facilities to a country that isn't subject to such restrictions (the US is currently not participating in Kyoto). Furthermore, regardless of whether or not carbon is included in Kyoto's accounting, it is likely to become an additional consideration in forest management planning and regulation.

### 4.3 International Efforts to Develop a Forest Convention

Despite the failure of UNCED to develop a forest convention, it sparked subsequent efforts by several countries (most notably Canada and Malaysia) to continue the dialogue that resulted in a new process through the formation of the Intergovernmental Panel on Forests (IPF) in 1995. Organized through the Commission on Sustainable Development (CSD) at the UN, the IPF had a two-year mandate to develop recommendations over a complex series of issues relating to the development of criteria and indicators for sustainable forest management and trade and environment as they relate to forest products (Humphreys, 2003). However, while a number of proposals were developed, many of these involved a high level of generality and did not specify actions or result in any commitments. Participants were unable to reach a consensus on any of the more difficult issues involving trade, the transfer of financial resources and technology, or the development of a forest convention (Porter et al., 2000). Subsequently the UN established the Intergovernmental Forum on Forests (IFF) with a three-year mandate organized within the CSD again. The IFF concluded without making significant progress in implementing any of the recommendations of the IPF and was unable to develop any further a consensus towards reaching an international convention (one of Canada's key objectives in participating in the process). Upon the expiration of the IFF, the UN Forum on Forests (UNFF) was established with a five-year mandate to conclude in 2005. One of the principal goals of the UNFF is to develop recommendations for a framework legal convention, although current expectations are low as to whether any substantial progress will be made in this regard as participants appear to be unable to resolve the issues that plagued the earlier processes (Humphreys, 2003).

*4.4 The Impact of the Montreal Process and C&I initiatives*

We noted earlier that the Rio summit highlighted the importance of identifying "sustainable forest management" and helped initiate the Montreal Process. A consensus on a precise definition of the term "sustainable forest management" has eluded most institutions concerned with forest policy, although in Canada the Canadian Council of Forest Ministers (CCFM) clearly stated that the goal of sustainable forest management was: "To maintain and enhance the long-term health of our ecosystems for the benefit of all living things both nationally and globally while providing environmental, economic, social and cultural opportunities for the benefit of present and future generations". In order to evaluate Canada's progress in reaching that goal, CCFM developed a Canadian national set of Criteria and Indicators (C&I) in 1995. In turn, the provinces are developing specific sets of C&I in order to report on the "State of the Forest" to Canadians. All of these processes are nested within each other. However there are few standards and methodologies for data collection and reporting, which has hampered the synthesis of this information into coherent reports (Montreal Process, 2003).[20]

The Federal government has played a role in both financing research and supporting efforts to coordinate the development of SFM while also representing Canada's positions in international negotiations. The Canadian Council of Forest Ministers (CCFM), a group including the provincial and federal ministers responsible for forestry, has developed the National Forest Strategy (NFS), a strategic framework intended to guide the development of national and provincial policies. In recent years, the framework has been modified to consider environmental and social values as well as involving a wider spectrum of stakeholders. The Council has developed a national Forest Accord spelling out a set of goals, commitments, beliefs, and action plans for Canada's forests (Duinker, Bull, & Shindler, 2003). One of the primary goals of the

CCFM has been the development of the criteria and indicators of sustainable forest management in a Canadian context.

Other national polices that have had an impact include strategies initiated by the federal government around sustainable development, such as the National Roundtable on the Economy and the Environment, which led to some short-term provincial initiatives resulting in large scale land use planning exercises (Dwivedi, Kyba, Stoeet & Tiessen, 2001). A number of different provinces (British Columbia, Ontario, Quebec and Saskatchewan) have adopted new forestry legislation that incorporates the principles of sustainable forest management.

### 4.5 Changes in Forest Management to Address Ecological Concerns

The CBD and various international processes such as the UNFF and Montreal Process generally call for greater emphasis paid to ecological issues and the development of policies to protect and conserve biodiversity. Provinces have responded to a number of different environmental issues within the scope of addressing biodiversity, including the establishment of protected areas. The Federal government has developed legislation for endangered species

One of the goals of the Bruntland Commission was to increase the amount of protected areas found within political jurisdictions to 12%.[21] The idea of protected areas is supported in certification systems and used as one of the major indicators for the ecological criteria. In four of the major forested provinces of Canada, British Columbia, Alberta, Ontario and Quebec, the amount of protected area has grown significantly and all provinces are developing specific policies around protected areas. In some cases, such as in Saskatchewan, provinces have developed an explicit biodiversity strategy (Natural Resources Canada, 2003). Related policies have also been developed around maintaining old growth forest (a fixed percentage of the stands to be found in older age classes), often couched in the need to maintain biodiversity.

Thus, while it has been difficult to develop operational definitions of biodiversity (that can be implemented at the ground level), the idea has been incorporated into new forest management policies in Canada that set aside protected areas and require the retention of wildlife habitat as well as through standards developed for different certification systems that highlight the need to protect and conserve ecological values.

Provinces are also experimenting with new planning processes that incorporate environmental objectives or consider the environmental impact of forest operations.[22] There is also the development of long-term plans (on the order of 100 year or more planning horizons) to simulate the effect of current policies to assess their impact on the future forest profiles (Duinker et al., 2003). There has been a shift in emphasis stressing long-run forest profiles, framed in terms of sustaining forest values (a significant shift from historical perspectives that assumed falldown as a sign of good forest planning).

*4.6 Changes in Forest Management to Address Social Concerns*

These international processes also call for greater transparency and more public involvement in decision-making. Again, there has been a significant increase in public involvement in forest planning across Canada. Public participation, especially at the local level, is also strongly endorsed by certification systems, reinforcing the need for public input to local management decisions.

Finally, we note that that greater attention is being paid to Aboriginal issues in Canadian forest management. Much of this is being driven by efforts at the international level, although at this level efforts are weaker and less organized (Bombay, 2004).[23] While local resource sharing and equity underpin some of the Forest Principles of Agenda 21 and the CBD, this idea is also being driven strongly by FSC certification.[24] There has been a rapid increase in the amount of partnerships, forest tenures, and other attempts to involve First Nations communities across Canada (NAFA, 2003; NAFA-IOG, 2000). Here too the Federal government plays a role through its responsibility for aboriginal issues, including the development of forest management plans for forested reserve lands, and it is evident that aboriginal participation in forest management and the forest industry is expected to grow.[25] One of the main areas of interest is in greater sharing in the economic benefits; this may drive changes to existing tenure systems to either accommodate new entrants (aboriginal communities as either new license holders) or new methods of collaboration between aboriginal communities and government and industry.

*4.7 Industry Responses to Addressing Environmental and Social Concerns*

At the same time as governments have made efforts to address a broader range of environmental and social values, similar efforts are underway in industry to promote more environmentally and socially friendly practices. UNCED also highlighted the idea of sustainable development, first raised in the Bruntland Report, and voluntary self-reporting by companies on social performance is on the rise (under the term Corporate Social Responsibility).[26] An important component of the reports include an assessment of the environmental impacts from their operations, including greenhouse gas emissions, and while there exists skepticism, even some critics acknowledge that by making managers pay attention to these measures, one ensures they will be paying attention to the issues (Cortese, 2002).

## 5. FORESTRY RELATED CERTIFICATION PROCESSES

There has been a growing trend in the development and use of voluntary approaches in which firms make commitments to address environmental problems (OECD, 1999). These approaches have been devised by a number of different groups; NGO's, governments, and industry associations, and have been used in pollution abatement and emissions reductions, development of best practices, and eco-labeling. Within forestry, this approach has been most strongly manifested through the development of certification and the use of market governance mechanisms. Certification is a voluntary activity that requires that an independent audit be carried out by an accredited third party on forest management systems and or forest management practices. It offers another way in which international consumers can make their voices heard and directly affects forest practices by potentially creating new markets for forest products produced in a sustainable manner. Through providing information, it offers the means by which a private regulatory system can monitor and enforce environmental performance.

In a separate set of processes, a number of "certification" schemes have been developed and are now being implemented within Canada. Among them are: ISO 14000 EMS registration; Canadian Standards Association National Certification (CSA); ForestCare in Alberta; the Sustainable Forestry Initiative (SFI) that was started in the U.S. (for which Canadian companies are eligible); and the international Forest Stewardship Council certification process (FSC).

In Canada over 147 million hectares (if ISO is included) have been certified by one of the above schemes, representing about 123.3 million m3 of the annual allowable cut in 2003 as shown in Table 12.1 (Abusow, 2003). Most operations have chosen the ISO scheme (although it certifies management systems, not areas directly as do the other three systems operating in Canada).

*Table 12.1.* Certified Forest land in Canada, by System in hectares, 2000-2003

| Year | FSC | CSA | SFI | ISO* |
|------|------|------|------|------|
| 2000 | 21,000 | 480,000 | n.m. | 15,390,000 |
| 2001 | 36,000 | 5,000,000 | 4,000,000 | 44,000,000 |
| 2002 | 973,856 | 8,820,000 | 8,350,000 | 107,785,000 |
| 2003 | 4,211,907 | 28,405,000 | 25,775,550 | 127,819,550 |

*In ISO the company is certified not the land and the area reported reflects land managed by the company. Note that an area may be certified under more than one system and areas are not additive.

*Sources:* Abusow, (2003)

Customers have yet to express a clear preference for one system over another (or even for certification in general). Yet certification has clearly become an important component of forest management practices in Canada; it has grown rapidly and indeed appears to have become a de facto requirement within the industry. All members of Canada's largest industry association are required to be certified under an independent third party system as a condition of membership by 2006 (FPAC, 2004) and Ontario and New Brunswick will also require major licensees to be certified in the next three to four years (OMNR, 2004 and New Brunswick Department of Natural Resources). As well, a number of large retailers have indicated that they will require their suppliers to have certification in place over the next few years (examples include retailers such as Home Depot, Lowes, IKEA, and others). FSC has the clear support of the ENGO's, but has also shown the slowest growth of the four major systems in work in Canada. While FSC, CSA, and SFI all require third party verification of external standards, those standards under the FSC are more focused on performance while the standards under the latter two systems are more oriented towards processes.

ISO focuses on continuous improvement and the development of management systems, rather than requiring specific practices on the ground.

The FSC has certified primarily small-scale operations and private land in Canada. Proportionally very little FSC certified operations are found in BC, which accounts for over one-third of the harvest in Canada and has also been the focus of intense environmental scrutiny over the past two decades. Most of the certification in BC has taken place under the ISO and other two domestic systems. FSC has had difficulty in developing regional standards in several regions, again most noticeably in BC. The difficulties appear to lie in developing a consensus by all stakeholders as to what standards are appropriate. This difficulty is not only at the regional level, but also developing a consensus as to what extent regionally developed standards should be accepted at the international level or further modified (McDermott & Hoberg, 2003).[27] Indeed they suggest that this difficulty in obtaining a consensus in highly politicized environments explains some of the slow pace of FSC certification in Canada. Cashore *et al.* (forthcoming) find that companies prefer local substitutes (in this case SFI and CSA) because they offer more flexibility relative to the more stringent and non-discretionary standards set by the FSC.

Certification attempts to change the economic incentives facing firms by creating markets for environmental goods in which consumers will be willing to pay a premium. The expected premium is important because there are higher costs associated with certification. These include changes in management practices that reduce timber supply from a given area (i.e. shifting to variable retention in which a portion of the timber stand is left unharvested, withdrawing areas from operations) and lead to increased timber supply costs through greater operational restrictions (greater investments in stream crossings, road construction, and in harvesting equipment that can be used on more sensitive soils). In addition, there are additional costs associated with certification (i.e. auditing, increases in planning and monitoring costs associated with meeting a broader range of environmental and social objectives). To date, however, such premium has failed to materialize. Premiums for wood products certified by the Forest Stewardship Council (FSC) are generally small or nonexistent and do not cover the added costs of certification (Baldwin, 2001; Kim & Carlton, 2001; Kiekens, 2000).

Despite the lack of such premium, however, firms have been seeking certification with the anticipation of gaining market share, or at least not losing market share (Vertinsky & Zhou, 2000; Bass, 1997a & 1997b; Forsyth, 1998). ENGO's have been organizing various buyers groups that are indicating that in the future they will be looking to purchase and sell only products that come from sustainably managed forests, and Canadian forest sector firms have sought certification in large part due to concerns over market access (Raunetsal, Juslin, Hansen, & Forsyth, 2002). If these Buyers' Groups lean towards one particular certification scheme over the others, they could be establishing a non-tariff trade barrier, beyond the control of the government. Cashore *et al.* (2004) investigate whether firms adopt a particular certification system as a way to lessen external pressure. They find it is significant in explaining the choice of ISO but not the other system chosen (either FSC or competing domestic systems within Canada, the US, and Germany) and that to date market access or the threat of market actions do not appear to be significant in explaining the adoption of a particular certification system.

## 6. INTERACTION WITHIN THE REGIME

We now consider how the three components we have identified within the international regime interact with one another. We examine the potential links between trade and the environment in the regime and consider how these have also influenced policy development within Canada.

### 6.1 Assessing the Environmental Impact of Trade

The effect of trade on the environment was originally explored in the context of trade models in which the environmental effects depended upon how the economic-environmental interaction was modeled (Pearson, 2000). Generally, these models suggest that the larger gains from trade typically outweigh the losses associated from increased pollution (see, for example, Grafton, Adamowiscz, Dupont, Nelson, Hill & Renzetti, 2003). More recently, interest in the effects of trade upon the environment has moved toward considering the implication of trade patterns and how trade policies may affect environmental resources.

Trade theorists have generally followed the same approach of viewing trade as "all or nothing" in determining the environmental impacts of changes in trade policies. Under this approach, economic growth leads to increased incomes that in turn lead to an increased demand for environmental improvements (this hypothesis is otherwise known as the Environmental Kuznets Curve). Therefore, trade policies that facilitate trade result in increased growth and national wealth and benefit environmental resources.[28] Several authors, however, have raised a number of different arguments under which trade policies designed to facilitate trade could lead to environmental degradation and harm (see, for example, Esty & Mendelsohn, 1998; Esty & Gerardin, 1998; and Sizer, Downes & Kaimowitz, 1999). This may happen for several reasons. It may be the case that increased trade results in overexploitation of the traded good (where it is assumed institutions are too weak to prevent irresponsible resource use) or countries ignore environmental damage linked to increased use of the resource. Another is the case that domestic environmental regulations may affect the competitiveness of domestic industries. Differences in environmental standards can potentially become a source of competitive advantage in trade. Therefore, countries may engage in a "race towards the bottom" in which they lower their environmental standards (or *roll back*) when trade is liberalized in an attempt to maintain the competitiveness of their domestic industries, or dirty industries will relocate to countries with laxer environmental standards (the *pollution haven* effect). A related argument is that of a *regulatory chill*-even if countries do not lower their standards, the prospect of reduced competitiveness is sufficient to preclude or limit the willingness of countries to adopt higher standards than they would in the absence of such trade. Therefore, trade policies need to be developed to take this effect into account to prevent this happening.

Neumayer (2001) suggests that the typical analyses of trade, in which economists focus on the overall welfare improvements from liberalizing trade, ignore the possibility of localized effects of environmental degradation, while NGO's ignore the beneficial effect of improved economic opportunities. In a recent survey of theoretical models and empirical work on the impact of trade on the environment, Copeland and Taylor (2003) investigate this debate and come to the conclusion that there is a strong relationship between increased incomes and better environmental quality, and find little evidence to date to suggest that there has been a strong evidence of a pollution haven effect at work. They argue that this suggests that trade policies should not be used to achieve environmental ends, since the most likely outcome of trade measures aimed at improving the environment in other countries will be to either reduce resource values or weaken property rights in those countries, potentially leading to worse environmental outcomes.

More realistic analyses of how trade and the environment interact involve understanding the multi-dimensional nature of the international regime in which a number of different systems simultaneously operate. The influence of trade upon the environment does not move in just one direction. It may also be the case that economic concerns drive the development of environmental policies, and that environmental policies influence trade.

## 6.2 Why National Environmental Policies May Affect Trade

Governments may respond to environmental concerns through national measures that affect trade. Governments can respond directly to concerns over resource use and associated environmental impacts related to trade. Countries may then adopt policies to rectify or prevent environmental harm or degradation associated with trade including restrictions on imports (such as endangered species), export bans, or other measures designed to mitigate environmental damage. Such measures may involve national regulation, or may involve measures addressed at environmental harm taking place outside the country.

DeSombre (2000) raises the possibility that instead of a "race for the bottom" dynamics, in which trade concerns lead to weaker domestic environmental regulation, the opposite may occur where concerns over the impact of increasingly stringent domestic environmental regulations on competitiveness may spur countries to establish similar international policies and standards through MEA's in other countries. DeSombre discusses the internationalization of US domestic environmental laws, arguing that when industry and environmental group's interests are aligned in that both will benefit from the adoption by other states of US polices that the US will push hardest to turn domestic policies into international policies. DeSombre argues that the US has two main tools as its disposal, multilateral diplomacy and the threat of market power through restricting access to its market, and that the ultimate success of the US effort depends upon other countries' reliance on US markets. Countries more exposed to the threat of US actions are more likely to adopt agreements that incorporate US policies. She examines endangered species, ozone and whaling, and concludes that the overall success of the internationalization effort will depend upon the nature of the coalition and to what extent the domestic industry benefits from the exclusion of the good.[29]

Countries may also decide to address these environmental issues collectively through MEA's. In this case, one of the fundamental problems in tackling international environmental issues is not only obtaining commitments but also in ensuring effective performance. Developing a consensus is a difficult task as can be seen from the difficulty in developing a forest convention. The difficulties include establishing who bears the burdens of the costs, whether or not countries may be able to credibly commit to cost-sharing mechanisms or financial transfers, and the problems in measuring the environmental benefits (which provide the motivation for action). While countries are willing to take efforts to address these issues, they are reluctant to yield any of their sovereignty over their domestic affairs, and obtaining compliance can be a difficult task.    Therefore, negotiators for environmental agreements are increasingly considering the use of trade restrictions as one of the instruments that can be used to help achieve the objectives of the agreement (Stavins & Barrett, 2002).

There are three principal reasons as to why trade measures may be employed: to deter free riding (both among members and by parties outside the agreement); to prevent leakages (for example, the agreement shifting the source of emissions to a non-member); and to directly control trade in the resource in question (Neumayer, 2001). Indeed, several MEA's (CITES covering trade in endangered species, as well as the Montreal Protocol governing ozone depletion and the Basel Convention on hazardous waste) have incorporated trade restrictions. These include clauses that restrict the ability of parties to the agreement to trade with non-members as well as require monitoring of trade in goods covered the MEA. The Montreal Protocol in particular has been regarded as the most successful MEA to date (in terms of achieving reductions in ozone-depleting chemicals), and the trade measures incorporated in the agreement (which restrict trade with non-members) are thought to have contributed significantly to its success (Victor, Rautsiala, & Skolnikoff, 1998).[30]

Jackson (2001) has identified four broad categories within which trade and environment conflicts might arise within the international regime: (1) national measures taken to protect the environment; (2) unilateral national measures taken to protect the environment outside of national jurisdiction; (3) international environmental agreements (MEA's) and the agreements under the WTO; and (4) process/product distinctions.[31] ENGO's strongly believe that the current trade regime does not take environmental values sufficiently into account and that the existing trade regime (primarily WTO rules designed to reduce countries discretion to erect trade barriers) might actually prevent countries from taking actions to improve the environment and that trade restrictions may be required in order to achieve environmental objectives (Neumayer, 2001). Indeed, some ENGO's suggest that restricting market access may be required to prevent this pressure from driving Canadian forest management practices downwards:

> Governments are increasingly reluctant to maintain or enforce effective regulatory standards to protect environmental values. This international "race to the bottom" in environmental protection has been caused, in part, by trade and investment agreements and the faster flow of investment capital across borders. As a result, Canadian

environmental and community groups have been forced to turn to alternative mechanisms
to achieve protection of environmental values.

Recent years have seen a dramatic increase in the use of these alternative mechanisms to
influence forest practices and protection of non-timber values in our forests. Market
campaigns are one of the most powerful of these alternative mechanisms, and are rapidly
becoming a focal point for the environmental movement (Global Forestwatch, 2004).

We consider, then, to what extent NAFTA and the WTO specifically address
environmental issues and to what extent conflict might arise.

## 6.3 Addressing Environmental Issues in NAFTA and WTO

We first note that both agreements recognize the importance of ecological and social
benefits in addition to economic benefits in their preambles. Sustainable development,
environmental protection and enforcement are objectives of NAFTA, while the
GATT Agreements establishing the WTO also acknowledge the objectives of
sustainable development, and that members seek to both protect and preserve the
environment. Neither agreement singles out forest products individually.

The main difference between the agreements is that a separate environmental
agreement, the North American Agreement on Environmental Cooperation
(NAAEC), was developed at the same time as NAFTA. No such agreement exists in
GATT. NAAEC serves several purposes. It provides a forum for regular meetings
between members on environmental issues. It encourages the harmonization of
standards and does not allow members to lower their standards. It also established an
independent body, the Commission for Environmental Cooperation (CEC), to monitor
member's compliance with their own domestic environmental laws. The Commission
is also meant to serve as an advisory body on both the environmental impact of
proposed trade laws as well as serve as a repository of environmental information.
The Commission can hear complaints regarding lack of compliance from private
citizens and NGO's as well as other member governments. If the complaints are
warranted, the Commission is limited to preparing a factual record of the complaint
that may or may not be made public. Member governments can bring complaints
against the other members and, if successful, fines can be assessed against the non-
compliant member through either financial assessments or trade sanctions (although
Canada is excluded from such fines) (GAO, 2001).

The WTO does have a process for ongoing negotiations of environmental issues,
the Committee on Trade and the Environment (CTE). It is through this committee that
WTO members are addressing the Doha Mandates in which members agreed to
negotiations in the next round over the relationship between trade and environmental
issues.[32] As part of the negotiations members agreed to specifically discuss the
relationship between existing WTO rules and trade obligations found in existing
MEA's.[33]

### 6.3.1 Allowing National Measures to Address Environmental Issues
We earlier noted that both agreements contain sets of rules that permit the adoption of
environmental measures. In addition, similar rules govern the treatment of

investments under NAFTA.[34] In terms of measures taken in response to international agreements, under NAFTA, obligations under a MEA can explicitly take precedence over NAFTA obligations.[35] There is no such clause in the WTO Agreement, although there have been no conflicts between WTO and MEA obligations yet. However, it is felt that if the use of trade restrictions as an instrument of implementation of MEA's grows, the potential for conflict may emerge where a country is a member of WTO but not an MEA (so that trade restrictions are challenged under GATT rules). In the Doha Round, the relationship between trade obligations in existing MEA's and the WTO is being addressed and whether GATT obligations should prevail if such a conflict does emerge.[36]

### 6.3.2 Process and Product Distinctions

We noted earlier that the rules covering the developments of standards do not permit discrimination between products based on the production method unless there are differences in the characteristics incorporated into the product. WTO members agree that if product characteristics are distinct and cause environmental harm then their imports can be restricted, but there are differences in opinions between members as to whether the agreement covers non-product related process and production-related methods (PPM's) where the production method has no impact on the product (Hirsch, 2000). This is important because many eco-labeling systems focus on the method of production to identify environmentally preferred methods for producing identical products. While voluntary private labelling systems are allowed, so long as they are not required, mandatory requirements are not, and some ENGO's feel that this has impeded the use of certification as a tool to help achieve better forest management practices (Sizer et al., 1999). Indeed some ENGO's feel that mandatory rules on eco-labeling, so long as they are not discriminatory, should apply (Hirsch, 2000). ENGO's have argued that these WTO restrictions means that countries cannot enact regulations to restrict access to their markets if they feel the good was produced in an environmentally damaging manner (if there are no physical differences between the product and similar products), thereby restricting the use of certification as an instrument.

The issue of labeling for environmental purposes is also being discussed (but not as part of the negotiations mandate) as part of the group of trade and environment issues under Doha round, but member countries remain far apart, with many viewing it as a potential barrier to market access and preferring to keep the discussion within the general context of discussions around the rules governing standards (ICSTD, 2003). Trade rules regarding labeling have not acted as an impediment to certification in Canada. Certification has grown rapidly despite its voluntary nature and trade agreements do not appear to have impeded its adoption. Indeed, at least in terms of commercial forest operations, Canada has adopted certification more quickly than the US. There are more forests certified under the American standard, SFI, in Canada than the US (Abusow, 2004).

## 6.4 Linking Trade to Environmental Issues

We noted earlier that ENGO's and others have also raised concerns that trade agreements might create a regulatory chill that could preclude member governments from adopting more stringent environmental regulation. One of the examples commonly cited is that of NAFTA's Chapter 11 governing the treatment of investments (see, for example, Cosbey, 2003). Here the argument is that member states might be precluded from adopting environmentally beneficial policies in fear of being sued by foreign investors. There is no evidence of this yet and, in fact, of the disputes brought to date under Chapter 11, the two cases involving forestry concerned policies developed in response to the softwood lumber dispute. In the first case, a US lumber producer with mills in Canada (Pope & Talbot) sued Canada over what it claimed was the discriminatory application of softwood lumber quotas during the Softwood Lumber Agreement (the claim failed on its most important points). In the second case, a Canadian company (Canfor) is currently suing the US government over the application of countervailing and anti-dumping duties in the most recent round of the trade dispute (the claim has yet to be heard but has proceeded to the stage of formal submissions).

Instead, trade has provided an avenue by which ENGO's have raised environmental issues. ENGO's have used various institutions within the trade regime to call attention to forest practices within Canada. For example, the Commission on Environmental Cooperation (CEC) under NAFTA has heard several complaints regarding Canada's failure to comply with its own laws: most of these have involved the impact of logging practices and other development on fish habitat and migratory and endangered species, all federal responsibilities. In the past, these complaints have focused on specific incidents, rather than general policies or legislation.[37] In addition, at the request of a US senator, the US Fish and Wildlife Service conducted a study to see whether Canadian logging practices were threatening several transboundary endangered species and came to the conclusion that the evidence suggested that they were not (and that other factors posed a greater threat).[38] ENGO's both within Canada and outside Canada have also used the softwood lumber trade dispute to press concerns about Canadian forest management practices. They have attempted to characterize Canadian harvest rates as unsustainable, that environmental regulations are laxer than those found on federal lands in the US, and that these differences should be recognized as subsidies (see Sizer et al., 1999; and Chase & Kennedy, 2002). More recently, aboriginal groups within Canada have sought to link their land and treaty claims to the trade dispute as well, arguing that ignoring their rights also constitutes a subsidy (since forest sector firms are avoiding payments that should be made to local indigenous communities and the forgone payments constitute a subsidy).

## 6.5 Evolving Regimes

We now turn to consideration of how the regime is evolving. Some of the major forces at work that can have a significant influence on Canadian forest management policies are policies that will be adopted in the near future in response to the outcome of the current trade dispute before the US and in the long run the increasing attention paid to environmental issues within the trade regime.

The most important determinant of Canadian forest management policies to be adopted and utilization of Canadian forests in the short-term continues to be the impact of the resolution of the current round in the softwood lumber dispute. Resolution of the dispute through either of the NAFTA or WTO processes will prove difficult. A reversal of the duties under NAFTA would mean that the U.S. had not properly applied its trade laws; the U.S. would be free to simply rewrite those laws and start over. The WTO dispute process can take several years and even at the end, after all appeals have been exhausted, the offending trade restrictions may not necessarily be eliminated and there is no mechanism by which any excess duties may be refunded. Therefore, the outcome is likely to be a negotiated settlement.

Nelson and Vertinsky (2004) show how difficult it is to reach consensus, as the economic circumstances created by the trade restraints (and proposed solutions) affect firms within provinces differently, based upon the nature of wood supply and product mix. However, there are strong political pressures within Canada to reach a negotiated settlement (such as the SLA). Since protectionist trade concerns are driving the case, however, and given the strength of the US timber lobby, the outcome will likely involve some form of restriction on Canadian lumber shipments (as has been the pattern to date). Depending upon the form this restriction takes (e.g. a border tax or volume restriction), it may reduce the value of timber within Canada, reducing incentives for more intensive management (one possible strategy contemplated under SFM) and in general leading to a reduction in the funds available for long term investment in enhancing the resource. Furthermore, if the agreement again takes the form of a short-term agreement, it will merely perpetuate the uncertainty for forest product firms, reducing further the incentives to invest in new equipment or new technologies.

There is also pressure to harmonize Canadian forest management policies with American policies. U.S. CVD duties have been assessed primarily based on allegations about provincial stumpage policies, but American timber lobbies have also complained about other Canadian forest management policies (these include long-standing features like harvesting and utilization requirements, appurtenancy which requires the operation of processing facilities in conjunction with a timber tenure, the long-term renewable nature of most Canadian timber tenures, and log export restrictions). The US and Canadian governments have discussed the possible development of a policy framework in which, under a negotiated agreement, Canadian provinces might be able to escape US duties by adopting new forest management policies (e.g. removing appurtenancy requirements and putting more timber up for sale through timber auctions). Harmonization (e.g. moving to market pricing) and market integration (e.g. removal of log export restrictions) can therefore proceed through US trade pressure that focuses on differences in selected Canadian forest management institutions, policies and practices. Harmonization may even take place through the conscious alignment of Canadian policies with US policies by Canadian policy-makers in order to reduce trade irritants. However, while these might lead to similar rules or approaches in managing both Canadian and US forests, this approach does not take into account significant environmental and institutional differences and pressures in Canada to retain sovereignty.[39]

Trade rules can also change and clearly environmental issues are playing a larger role in trade discussions (indeed, Canada along with other countries are now preparing environmental assessments as part of trade negotiations (DFAIT, 2003)). Within the WTO, the discussion of environmental issues within the Doha round negotiations, as well as the success of the US in enacting unilateral restrictions on imports in the shrimp-turtle dispute, suggest a greater receptivity to the idea of environmental considerations as a legitimate motive for trade-restrictive measures.[40] The developed countries, principally the EU and the US, have indicated their interest in incorporating environmental considerations into trade (Oxley & Osborne, 2002). However, change in the trade regime happens slowly as countries face domestic pressure and resistance when modifying trade rules. [41]

Change happens even more slowly when it comes to developing MEA's. Here there is even greater resistance to reaching agreements that require any kind of binding obligations on participants as the domestic costs typically are more obvious and therefore seem larger than the environmental benefits that are more widely diffused (and more difficult to quantify). However, the agreements that are reached do help provide the basis for further actions through legitimization of norms and clearer definition of problems to be addressed. This has been the case with biodiversity conservation and may in the future be the case with forest carbon sequestration to address climate change.

The main way in which action has proceeded has been through voluntary efforts. Certification has clearly entered the mainstream especially within Canada. Certification has placed pressure (through concerns over market access) on countries to improve environmental outcomes (in some sense a "competition" for virtue). The inability to develop an international consensus on whether or not this approach can be incorporated into trade rules does not appear to have slowed adoption or dissemination of this approach. While certification may not be sufficient (or happen quickly enough) to satisfy some ENGO's, it did help in building a consensus within countries to move towards SFM where there are institutions capable of ensuring effective enforcement. The certification processes mobilize international ENGO's offering them instruments to exert pressure on industry through market action. Indeed, US ENGO's have created and funded Canadian ENGO's to advance their positions regarding Canadian forest management practices and land use decisions (Bernstein & Cashore, 2000).

Within certification systems, the presence of a "green" certification system like the FSC helps maintain "competitive pressure" leading to an upward shift in values. Indeed, the FSC considers its approach as a way to generally lift all government standards (Cashore et al., 2003, Meidinger, 1997). The FSC process also explicitly considers harmonization of its regional standards with one another as a way to raise those standards if they are perceived as too low or lax (FSC, 2004a).

There is also another effect of certification. We noted earlier that the provision of information and public scrutiny could be a source of compliance. Simply paying attention to environmental issues and raising their prominence can help mobilize political support and create a framework in which these issues matter. International NGO's both help disseminate information and articulate norms that help advance the global debate. The Forest Stewardship Council, FSC, for example, is articulating and promoting a common understanding of what good forestry is one of three main goals of its initiatives in developing standards for the boreal forest in Canada (FSC, 2004b). By helping identify what it believes are the critical components of sustainable forest management, it helps ensure that other certification systems pay attention to these components as well.

Voluntary efforts are not only focused on certification. ENGO's are also attempting to develop a consensus on large-scale land use changes within Canada. Under the Canadian Boreal Framework, a coalition of resource companies, several First Nations groups, and some prominent ENGO's have developed a plan that calls for greatly increasing the protected area in Canada's boreal and changing forest management practices on the portion that would remain available for commercial timber operations. What is striking about such plans is that they do not involve any government representatives at the provincial or national level; rather, they are an attempt to develop a consensus among business groups, local communities, and NGO's that will then provide the basis for government implementation. A key component of the framework is the establishment of protected areas within half of the boreal region, and the use of certification (specifically the FSC) to guide harvesting practices on the remaining areas open for commercial forest activity. One of the hopes expressed by ENGO participants is that it will facilitate efforts elsewhere:

> Josh Reichert, environment director at The Pew Charitable Trusts which helped establish the Canadian Boreal Initiative and set the Framework discussions in motion added, "Not only is this the largest forest and wetlands conservation initiative ever proposed, it is also a whole new approach to balancing conservation and economic development that could provide a model for protecting other globally important ecosystems like the Amazon rainforest and the Russian Taiga." (US Newswire, 2004)

This attempt to develop a new collaborative forum involving NGO's, industry and government is one way to respond to what are perceived as more formal trade and environmental rules that are lagging behind public demand for changes.

Indeed, one approach to resolving environmental issues that might arise through trade has been through developing agreements on environmental standards (thereby eliminating the possibility of a race to a bottom dynamics contemplated by some ENGO's). Bhagwati (1996) discusses the positive role private voluntary actions may play in establishing common standards, especially where these schemes work to develop local political support within different countries. This can also happen through the development of common values or norms. If it is the case that a consensus can be developed, then, this raises the possibility that the reliance on social norms to sanction inappropriate or unacceptable behaviour may be able to play an effective role in establishing and enforcing agreement on how to address some of the more difficult environmental issues. This also permits the possibility of harmonization achieved through the voluntary adoption of common standards.

The problem with organizing markets and developing a consensus around voluntary action to support environmentally preferred goods is that they are vulnerable to manipulation and false claims. Formal private systems backed by specific market and social actions achieve the desired results faster. Indeed, the Canadian Boreal Framework explicitly notes that the majority of industrial goods produced from the boreal region, both forest products and oil and gas, are destined for US markets (US Newswire, 2004) and that market boycotts may be required to compel firms to change practices (Hamilton, 2002a). ENGO's have argued that the domestic certification systems employed by Canadian firms (e.g. CSA and SFI) may not lead to a significant difference in practices, and that therefore a particular certification system, one with ENGO support such as FSC, is necessary to achieve the appropriate environmental objectives (FERN, 2003). The problem is that even certification may not be immune to rent-seeking behaviour where domestic industries may use it as a means to advance traditional protectionist measures by arguing that foreign competitors are practicing unsustainable forest management (see Chase & Kennedy, 2002).

More generally, there are concerns about requiring a particular system that involves certain norms and values that may not be held by all parties. Indeed, coercive harmonization appears to be at odds with the development of a genuine consensus. Meidinger (2001) suggests that ethically there is a fundamental limitation to the use of certification if it cannot garner sufficient local support especially in developing countries.[42] Bhagwati (1996) suggests that the preferential route may be mutual recognition pacts, and indeed we are seeing this emerge as domestic certification systems evolve (most noticeably as the PEFC has moved towards an international framework for recognizing individual domestic systems).[43] Indeed, if voluntary efforts are going to develop new forms of governance, what emerges should be transparent and accountable.

## 7. CONCLUSIONS

The ways in which the international forest regime shapes and interacts with Canadian policy-making processes is complex, as it moves through multiple layers, filtered by national and provincial policy-making processes, interacting with a number of different factors-foreign government policies, international and domestic NGO's, and public opinion.

Canadian forest management has changed significantly over the past twenty years. Domestic legislation and policies have had to respond to a rapidly changing international regime in which trade and increasingly environmental issues play a greater role. Several observers have felt that Canada has made more significant changes in its forest management policies over the past two decades than the US in part because of its dependence on export markets (Beckley, Shindler, & Finley, 2003; Duinker et al., 2003). It is clear that within Canada the Federal role in forestry must grow, despite constitutional and political constraints, in large part due to the commitments made in the international regime. While these commitments may not directly affect provincial polices, they increasingly require provinces to adopt policies or make choices based on the implementation of international commitments, such as the Kyoto Protocol and the CBD.

The main impact on the Canadian industry to date has been through the impacts in markets created by US trade pressure and increases in regulatory costs resulting from international market pressures to protect the environment. Reduced prices and higher regulatory costs have simply provided the industry with greater incentives to rationalize production further and become even more competitive, although this has come at a high cost in terms of forest communities sustainability.

The main movement toward defining and implementing sustainable forest management strategies has been through the use of C&I developed through various international processes and certification systems. If measurement of movement toward SFM is possible then a focus on outcomes may both motivate progress and promote efficiency in achieving it. However, there is a great deal of scientific uncertainty around many of the criteria that have been developed and a significant effort is required to gather and assess the data. As Lackey (1999) shows, even if there is a consensus that we want SFM, the conflict surrounding ecosystem management depends on the underlying assertions and values that differ significantly between those proponents that envision ecosystem management as a continuation (with some modification of approaches and emphasis on different goods) of existing multiple-use management practices with those that see it as a fundamental shift in the way that we approach society and current lifestyles with their emphasis on material goods. The question of who decides and what weight should be given to public participation at different levels is a difficult one and still unresolved in Canada and is made even more difficult by changing norms over time.[44] The ambiguity and uncertainty as to what constitutes SFM and what weight should be given to different "publics" make it difficult for many countries to commit to specific obligations for many of the values embedded in SFM. Indeed, it is agreed that local level indicators are required to identify SFM (Hirsh, 2000), yet many of the criteria involve environmental values that have international dimensions. The ambiguity in the international regime on how to reconcile existing government policies with new SFM prescriptions matches the same uncertainty found within countries.

Finally, the introduction of certification has opened up the policy process to a wider range of groups within Canada than have traditionally participated in forest policy planning by incorporating to varying degrees (depending upon the system) a role for public participation and the promotion of social and environmental values. Even here we note the importance of trade, however. Because of Canadian forest products firms' reliance on access to export markets, international ENGO's have been able to press for changes in Canadian policies through raising concerns about market access. In some cases, even though these ENGO's do not have any formal standing in the trade agreements or their associated processes, these agreements have provided entry points where the ENGO's have been able to introduce their ideas and arguments into Canadian policy processes. Given the increasing attention paid to environmental issues, and the environmental scrutiny Canadian forests receive, Canadian forest policies have and will continue to incorporate a number of important ideas and values developed in international environmental agreements. These ideas and values will also be reinforced through certification systems that either incorporate the objectives of those agreements or, in the case of ENGO-supported systems, attempt to implement direct changes in policies and practices.

## NOTES

[1] While we consider how the international regime is filtered through national policy-making processes, and how NGO's influence this, we do not cover the role governments and NGO's play in developing these international agreements in similar detail. This is an area in which extensive research has been conducted and literature is available (for the interaction between international policies and national policy-making processes in Canada; see, for example, Dwivedi, Kyba, Stoeet & Tiessen (2001); for Canadian forestry in particular Howlett (2001); for the role of NGO's in international policy-making in general see, Oberthuür, Buck, Müller, Pfahl, Tarasofsky, Werksman & Palmer (2002), and Porter et al. (2000); and for the emergence of these agreements in general see Bernauer (1995); Sprinz & Vaahtoranta (1994); and Young (2002)).

[2] When the provinces and federal government can reach a consensus, however, the federal government is quite capable of implementing substantive policy changes (Feigenbaum et al. 1993: 73-74).

[3] Standards have been used to describe voluntary arrangements in which a producer can choose but is not required to meet the standard while regulations are those requirements that are mandatory. Examples of standards include building codes (where products may be required to meet specified performance levels if used for certain purposes).

[4] Gandolfo (1998) notes that there are a variety of constituencies that have to be considered in understanding when such protectionism will be successful, ranging from interest groups (including firms and consumers), politicians, and bureaucrats.

[5] There are also a host of government policies that will affect macroeconomic factors such as exchange rates, workforce skills, and capital availability that can influence the relative competitiveness of an industry, but these policies are generally applied more widely and not targeted towards a specific industry or group of firms.

[6] These may even include the formation of explicit cartels of countries, such as OPEC.

[7] Magee, Brock, and Young (1989) describe this as the "principle of optimal obfuscation."

[8] In the WTO this consists of three separate agreements found within the first annex to the WTO framework: the Agreement on Subsidies and Countervailing Measures; the Agreement on Safeguard Measures; and the Agreement on the Implementation on Article VI of the General Agreement on Tariffs and Trade 1994 (addressing anti-dumping). Chapter 19 within NAFTA permits countries to employ their own trade remedy measures while respecting their GATT obligations.

[9] These two sets are contained in the Agreement on Technical Barriers to Trade (TBT) that addresses the development of technical standards and regulations and the Agreement on the Application of Sanitary and Phytosanitary Measures (SPS). The TBT Agreement permits mandatory labelling requirements and regulations based on differences in product characteristics (i.e. structural properties of different lumber species, percentage of recycled material contained in newsprint) but does not permit discrimination based upon process or production methods. The SPS Agreement does permit discrimination against products from different countries based on risk. Again similar measures are found within NAFTA (Chapters 9 and 7 respectively). In addition, Article XX of the WTO Agreement also permits exceptions to the general GATT obligations where environmental measures are allowed if they are necessary to conserve scarce natural resources, or protect human, animal and plant life, but again these measures must be the least restrictive possible and required to obtain the results. A similar exception is also found in NAFTA (Chapter 21, Article 2101).

[10] Under NAFTA, the offending party is required to bring its measures into compliance with its obligations (so if countervailing measures are found not to be justified they must be dropped). Under the WTO, an offending party can bring its measures into compliance; alternatively, it may choose not to do so and offer equivalent financial compensation or concessions to the aggrieved party; or if it fails to take any actions, the aggrieved party has the right to establish its own measures (i.e. import duties) or withdrawals of concessions. In terms of panel members, under NAFTA panelists are drawn from countries that are party to the dispute; under the WTO, panelists cannot be drawn from parties to the dispute.

[11] These studies were motivated by efforts to liberalize tariffs for forest products through an initiative

under APEC and a topic for discussion at WTO talks in Seattle in 1999.

[12] An interagency study was also carried out by the US government and found similar results (United States Trade Representative and the Council on Environmental Quality, 1999). Both studies used partial equilibrium models.

[13] Recent examples of the use of SPS standards directed at forest products include increasing requirements for kiln-dried softwood used for palleting and packaging by a number of countries. The forest products industry in New Zealand complains of increasing frustration faced in meeting building standards that discriminate against New Zealand species, a complaint echoed by the US forest products industry (NZ Institute of Economic Research, 2000; United States Government 2003).

[14] This is supported by evidence that suggests tariffs on manufactured pulp and paper products post Uruguay round are still high in many developing countries; examples include Brazil, China, and Indonesia (Bacchetta & Bora, 2001).

[15] The SLA, by establishing a tariff rate quota (in which fixed volumes could enter duty-free but shipments above that volume were taxed at a high rate) led to a wedge between Canadian and US prices over the course of the agreement, leading to what was called the "Canadian discount" where the amount by which Canadian prices would be lower would depend upon overall demand for lumber.

[16] For example, an agreement that has a forestry-related dimension but do not affect forest management is the Agreement on the Long Range Transport of Air Pollutants (or LRTAP) to which Canada and the US are party. Although it is concerned with the effect of acid precipitation on forests in Europe and North America, it does not address the management of forests, nor does it affect forest product markets. An example of an agreement that does involve forestry is the Agreement for the conservation of the Biodiversity and Protection of priority Forest Areas in Central America (CAA), signed in 1992. However, it is not relevant to Canada since it is a regional agreement and Canada is not a signatory.

[17] Developed countries have focused on issues of the environmental dimensions of sustainable development, and developing countries, while acknowledging the importance of environmental issues, have emphasized ideas of social responsibility and the need for the transfer of expertise and technology from developed countries to help improve the economic well being of their citizens.

[18] In addition to Canada, these include Argentina, Australia, Chile, China, Japan, Korea, Mexico, New Zealand, the Russian Federation, United States, and Uruguay. The Helsinki Process, covering forests in European countries, started in 1990 and culminated in a series of criteria and indicators for SFM that were published in 1998. Those C&I subsequently became the basis for what would be required by national certification systems under the mutual recognition system under the Program for the Endorsement of Forest Certification systems (PEFC) (an umbrella system that has currently endorsed thirteen national standards, all European-based).

[19] Saskatchewan has developed a provincial biodiversity strategy (Natural Resources Canada 2003) while Quebec has prepared a draft strategy (see http://www.menv.gouv.qc.ca/biodiversite/strateg_02-07-en/). Other provinces are incorporating biodiversity objectives into their planning processes or adapting existing programs and developing monitoring systems. Examples include Alberta http://www.abmp.arc.ab.ca/Overview.htm) and Manitoba (http://www.gov.mb.ca/conservation/wildlife/managing/biodiversity.html).

[20] An update of the C&I is scheduled to be published in 2005.

[21] It is interesting to note that a prominent official at a Canadian ENGO in 1989 noted that "we have ten years or less left to protect at least 12 per cent of Canada in a wild state...[t]his is going to take considerably more political vision than currently experienced in this country." Hummel (1989:272). By 1996 British Columbia had already protected over 9%, reaching 12% by 2001 (Pedersen 1996; Scudder 2003).

[22] This includes the development of higher-level plans (as in BC) to the use of environmental assessment procedures (in Ontario, Manitoba, and Saskatchewan).

[23] The International Labor Organization (ILO),a UN agency, has developed a document, Convention 169, that addresses the recognition of indigenous rights but Canada is not a signatory.

[24] Principle 3 of the Criteria within the FSC system has evolved to require the willing participation of local aboriginal groups as a condition of meeting the criteria, and aboriginal groups have endorsed the FSC system as preferred over other systems (Collier, Parfitt, & Woollard, 2002).

[25] This will take place through treaty settlements in British Columbia and in resolving treaty and land claims elsewhere in Canada.

[26] The UN through its Global Compact is promoting the idea of CSR more broadly by developing principles and norms that companies would then internalize in their respective businesses. The nine principles address human rights, labour standards, and environmental responsibility (Pitts, 2004).

[27] There are also complaints over perceived inconsistencies between standards developed in different regions, despite similar forest types, as was the case for the standards developed for the Maritimes in eastern Canada versus those developed for the northeastern US (McDermott & Hoberg, 2003).

[28] The NAFTA Commission for Economic Cooperation suggested that this is the reason why pre-NAFTA predictions of environmental damage from the trade agreement have failed to materialize (CEC 2001).

[29] DeSombre gives an example of the U.S. endangered species law that provided the motivation for the efforts that resulted in CITES. There was no domestic industry involved in trading endangered species, and the US was unable to achieve as much as ENGO's wanted from Asian countries. More successful have been efforts at marine mammal and sea turtle protection, where there existed a domestic fishing fleet that would benefit from either restricted access to US fishing waters by foreign fleets or by restricted imports into the US and would support ENGO's efforts.

[30] This view is not universally shared. Some authors have argued that the development of substitute products in the US, the largest producer and consumer of ozone-depleting chemicals, meant that domestic producers benefited from the trade restrictions and that the costs faced by US citizens were such that the US would have acted unilaterally. Therefore, the agreement was not necessarily needed to achieve the reductions (Barrett, 1994).

[31] It should be noted that the most prominent dispute involving environmental issues and trade under the WTO has involved the application of trade measures by the U.S. under Article XX to restrict seafood imports from countries not deemed to be undertaking sufficient measures to protect endangered turtles (otherwise known as the shrimp-turtle dispute). In general, the most recent interpretation has let stand US laws restricting imports from countries whose fishers do not make efforts to reduce sea turtle bycatch (CTE, 2002).

[32] Strong support for inclusion came from the EC, Japan, Norway and Switzerland. The majority of other countries resisted inclusion of environmental issues in the Doha Round, with developing countries fearing that environmental negotiations might simply expand the range of environmental measures that could potentially be used as NTB's (ICTSD, 2003).

[33] Members also agreed to negotiations over developing better linkages with MEA secretariats, as well as information sharing and observer status, and the elimination of tariff and NTB to environmental goods and services.

[34] For example, Article 1106:6 is similar to Article XX under the WTO, while another article in chapter 11 states that nothing in the agreement should be construed as preventing countries from enacting environmental measures governing investment activities (Article 1114:1).

[35] Article 104 lists the specific MEA's to which members are party to and makes a provision for the addition of future MEA's.

[36] Current discussions are narrowly confined to examining existing agreements to see whether any clarification is required, although some countries (principally the European countries) argue that is should expand to the consideration of more general trade measures required to achieve environmental objectives (ICTSD, 2003).

[37] For example, the factual record prepared for the impact of logging practices in BC on fish habitat looked only at a cut block on southern Vancouver Island while the initial investigation of logging of migratory bird habitat in Ontario revolved around whether there was any evidence of actual nests destroyed (rather than an estimate based on expected species density and area harvested). There was a subsequent complaint initiated that is in the process of being reviewed.

[38] The report looked at marbled murrelets, grizzly bears, woodland caribou, and Bull Trout. It noted that there was no evidence that marbled murrelets migrated across the border so that they could not be assessed (GAO, 2002).

[39] It is not clear that the US approach to forest management is considered any more sustainable and, indeed, concerns have been raised over the years about whether or not an appropriate balance has been struck between environmental, social, and economic concerns and whether or not the primary mechanism used to grant access to timber-timber auctions-are the most appropriate means to achieve public objectives (on the first point see Floyd, D., Alexander, K., Burley, C., Cooper, A., DuFault, A., Gorte, R., Haines, S., Hronek, B., Oliver, C. & E. Shepard, 1999; on the second Hamilton, 2002b; Saunders, 2003; and Taxpayers

for Common Sense, 2001).

[40] See Fn. 30.

[41] Weintraub (2003) suggests that in the US foreign policy considerations are currently driving US trade negotiations, especially in choices to pursue bilateral agreements with selected countries rather than efforts to develop a consensus within the WTO.

[42] The relative unevenness in terms of certified forests by international system across different countries, and the general lack of certification in developing countries to date, raises concerns about embedding any one particular certification system such as the FSC as a requirement of market access.

[43] In fact, both the CSA and SFI are in the process of applying to the PEFC system.

[44] Indeed, the same question is still unresolved in terms of US public forestland management (see Floyd *et al.* 1999).

# REFERENCES

Abusow, K. (2004). Forest certification in Canada. *Speech at the National Conference on Aboriginal Forestry*. Thunder Bay, ON. May 11-13.

Abusow, K. (2003). *Canadian forest management certification status report*. Prepared for the Canadian Sustainable Forestry Certification Coalition.

Bacchetta, M., & B, Bora. (2001). Post-Uruguay round market access barriers for industrial products. *Policy Issues in International Trade and Commodities Study,* Series No. 12. United Nations Conference on Trade and Development.

Baldwin, S. (2001). Sustainable or certified forestry? *Timber Mart-South Market Newsletter*, 6(2), page 3.

Barbier, Edward. 1996. Impact of the Uruguay Round on International Trade in forest Products. FAO. Retrieved August 30, 2001, from http://www.fao.org/DOCREP/003/w0723e/w0723e00.htm.

Barrett, S. (1994). Self-enforcing international environmental agreements. *Oxford Economic Papers*, 46: 878-894.

Bass, S. (1997a). Introducing forest certification. A report prepared by the Forest Certification Advisory Group (FCAG) for DGVIII of the European Commission. *European Forest Institute Discussion Paper* 1, Joensuu.

Bass, S.M.J. (1997b). The principles of certification of forest management systems and labelling of forest products. *Working paper, International Institute for Environment and Development*, Oxford, UK.

Beckley, T., Shindler, B., & Finley, M. (2003). Are we there yet? Assessing our progress along two paths to sustainability. In B. Shindler, T. Beckley, M. Finley (Eds.), *Two paths towards sustainable forests: Public values in Canada and the United States*. Corvallis, Oregon: University of Oregon Press.

Bernauer, T. (1995). The effect of international environmental institutions: How we might learn more. *International Organization*, 49(2), 351-377.

Bernstein, S., & Cashore, B. (2000). Globalization, four paths of internalization and domestic policy change: The case of ecoforestry in British Columbia, Canada. *Canadian Journal of Political Science,* XXXIII, 1.

Bhagwati, J. (1996). The demands to reduce domestic diversity among trading nations. In D. Bhagwati, R. Hudec (Eds.), *Fair trade and harmonization: Prerequisites for free trade?* (Vol. 1, pp. 9-38). Cambridge, MA: MIT Press.

Bombay, H. (2004). International influences on national policies. *Speech at the National Conference on Aboriginal Forestry*, Thunder Bay, ON. May 11-13.

Bourke, I.J., and J. Leitch. 2000.*Trade Restrictions and their Impact on International Trade in forest Products*. FAO. Accessed at http://www.fao.org/DOCREP/003/X0104e/X0104e00.htm August 30, 2001.

Braga, C. (1992). Tropical forests and trade policy: The cases of Indonesia and Brazil. In P. Low (Ed.), *International trade and the environment*. Washington, DC: World Bank.

Braudo, R., & Trebilcock, M. (2002). *The softwood lumber strategy: Implications for Canada's future trade strategy*. Faculty of Law, University of Toronto. May. Mimeograph.

Bruntland, G. (ed.), (1987). *Our common future: The World Commission on Environment and Development*. Oxford, Oxford University Press.

Burfisher, M., Norman, T., and R. Schwarz. 2001. NAFTA Trade Dispute Resolutions: What Are the Mechanisms? In Loyns, R., Mielke, K., Knutson, R., and A. Yunez-Naude (eds.) *Trade Liberalization*

*Under NAFTA: Report Card on Agriculture.* Proceedings of the Sixth Agricultural and Food Policy Systems Information Workshop. University of Guelph. Canada.

Canadian Forest Service. (2002). *Canada's forest biodiversity: A decade of progress in sustainable management.* Sciences Branch, Natural Resources Canada, Ottawa.

Cashore, B.,Van Kooten, C., Vertinsky I., Auld, G., & J. Affolderbach. 2004. A comparative study of forest certification choices in Canada, the U.S. and Germany. *Forest Policy and Economics,* In press.

Chase, S., & Kennedy, P. (2002, April 18). Eco-probe of softwood trade urged. *Globe and Mail,* B2.

Collier, R., Parfitt, B., & Woollard, D. (2002). *A voice on the land: An indigenous peoples' guide to forest certification in Canada.* Ottawa: NAFA and Ecotrust, Canada.

Commission on Environmental Cooperation (CEC), 2001. *The North American Mosaic: a state of the environment report.* Accessed July 20, 2003 at http://www.cec.org.

Committee on Trade and the Environment (CTE). (2002). *GATT/WTO Dispute settlement practice relating to GATT article XX, paragraphs (B), (D) and (G). World Trade Organization (WTO).* March 8. WT/CTE/W/203.

Copeland, B., & Taylor, M. (2003). *Trade, growth, and the environment.* Mimeograph, May. Department of Economics working paper. University of British Columbia. Vancouver, BC.

Cortese, A. (2002, March 24). The new accountability: Tracking the social costs. *New York Times,* B4.

Cosbey, A. (2003). *NAFTA's Chapter 11 and the environment: discussion paper for a public workshop of the Joint Public Advisory Committee of the Commission for Environmental Cooperation of North America. March 24.* Mexico City. Retrieved from http://www.iisd.org/trade.

Dasgupta, P., Maler, K., & Vercelli, A. (1997). *The economics of transnational commons.* Oxford: Clarendon Press.

Dasgupta, P. (1998). The economics of poverty in poor countries. *Scandinavian Journal of Economics,* 100(1), 41-68.

Deacon, R.T. (1995). Assessing the relationship between government policy and deforestation. *Journal of Environmental Economics and Management,* 28, 1-18.

Department of Foreign Affairs and International Trade (DFAIT). *Initial environmental assessment: Trade negotiations in the World Trade Organization.* Retrieved from http://www.dfait-maeci.gc.ca.

DeSombre, E. (2000). *Domestic sources of international environmental policy.* Cambridge, MA: MIT Press.

Duinker, P., Bull, G., & Shindler, B. (2003). Sustainable forestry in Canada and the United States: Developments and prospects. In B. Shindler, T. Beckley, M. Finlay (Eds.), *Two paths towards sustainable forests: Public values in Canada and the United States.* Corvallis, Oregon: University of Oregon Press.

Dwivedi, O., Kyba, P., Stoeet, P., & Tiessen, R. (2001*). Sustainable development and Canada: National and international perspectives.* Peterborough, ON: Broadview Press.

Esty, D., & Geradin, D. (1998). Environmental protection and international competitiveness: A conceptual framework. *Journal of World Trade,* 5-46.

Esty, D., & Mendelsohn, R. (1998). Moving from national to international environmental policy. *Pol. Science,* 31, 225-235.

FERN. (2003). *Eco-labeling, certification and the WTO.* Retrieved May 5, 2004, from http://www.fern.org/pubs/briefs/WTOecolabel3.pdf.

Feigenbaum, H., Samuels, R., and Weaver, R.K. 1993. Innovation, coordination, and implementation in energy policy. In *Do institutions matter? government capabilities in the United States and abroad.* Edited by R.K. Weaver and B. Rockman. Brookings Institution. Washington, D.C. pp. 49-109.

Floyd, D., Alexander, K., Burley, C., Cooper, A., DuFault, A., Gorte, R., Haines, S., Hronek, B., Oliver, C. and E. Shepard. Choosing a Forest Vision. *Journal of Forestry.* May 1999. pp 44-46.

Forest Products Association of Canada (FPAC). (2004). *Certification similarities & achievements.* Prepared by Kathy Abusow, March.

Forest Stewardship Council (FSCa). *Standards harmonization.* Retrieved May 5, 2004, from http://www.fsc-bc.org/Harmonization.htm.

Forest Stewardship Council (FSCb). *Boreal initiative.* Retrieved May 5, 2004, from http://fsccanada.org/boreal/index.shtml.

Forsyth, K. (1998). *Certified wood products: The potential for price premiums.* Edinburgh: LTS international.

Gandolfo, G. 1998. International Trade Policy and Theory. Springer-Verlag. New York.

GAO. (2001). *North American Free Trade Agreement: U.S. experience with environment, labor, and investment dispute settlement cases.* Report to the Chairman, Subcommittee on Trade, Committee on Ways and Means, House of Representatives. GAO-01-993. July. 67 pp.

GAO. (2002). *Transboundary species: Potential impact to species.* Washington, DC: GAO-03-211R.

Global Forest Watch (2004). *The context of market campaigns.* Retrieved May 5, 2004, from http://www.globalforestwatch.org/english/canada/news.htm.

Grafton, Q., Adamowiscz, A., Dupont, D., Nelson, H., Hill, R., and S. Renzetti. 2003. The Economics of the Environment and Natural Resources. Blackwell. Oxford, UK.

Hamilton, G. (2004, May 1). Good times are back in B.C. forests. *Vancouver Sun*, H3.

Hamilton, G. (2002a, August 30). The north: BC's next battleground. *Vancouver Sun*.

Hamilton, G. (2002b, September 4). U.S. lumber firm accused of killing competition. *Vancouver Sun*.

Hirsch, F. (2000). Trade and environment issues in the forest and forest products sector. *Geneva Timber and forest Discussion Papers* ECE/TIM/DP/19. Retrieved from http://www.unece.org/trade/timber.

Hoberg, G., and Harrison, K. 1994. It's not easy being green: the politics of Canada's green plan. *Can. Pub. Pol.* 10: 119-137.

Howlett, M. (2001). *Canadian forest policy: Adapting to change.* Toronto: University of Toronto Press.

Hummel, M. (1989). *Endangered spaces: The future for Canada's wilderness.* Toronto: Key Porter Books.

Humphreys, D. (2003). The United Nations forum on forests: Anatomy of a stalled international process. *Global Environmental Change*, 13, 319-323.

International Centre for Trade and Sustainable Development (ICTSD) (2003). *Trade and environment. Doha round briefing series: Cancun Update, August, V2, 9.*

Kiekens, J.P. (2000). *Forest certification.* Retrieved from http://www.forestweb.com/APAweb/ewj/2000_spring/f_forestcertification.html.

Jaakko Poyry Consulting. (2002). *New Brunswick crown forests: Assessment of stewardship and management.* Report prepared for the New Brunswick Department of Natural Resources and Energy and New Brunswick forest Products Association, November.

Jackson, J. (2001). The role and effectiveness of the WTO dispute settlement mechanism. In J. Jackson, E. Weiss (Eds.), *Reconciling environment and trade.* Ardsley, New York: Transnational Publishers, Inc.

Kim, Q.S., & Carlton, J. (2001, May 23). Timber industry goes to battle over rival seals for 'green' wood marketplace. *Wall Street Journal*.

Klein, D. (2002). In Spulber, D., (Ed.), *Famous fables of economics: Myths of market failures* pp. 49-69. Oxford: Blackwell Publishers.

Lackey, R. 1999. Radically contested assertions in ecosystem management. J. Sustain. For. 9(1/2): 21-34.

Levy, P. (2003). Non-tariff barriers as a test of political economy theories. Economic Growth Centre, *Center Discussion Paper No. 852*, Yale University, February.

McDermott, C., & Hoberg, G. (2003). From state to market: Forestry certification in the U.S. and Canada. In B. Shindler, T. Beckley, M. Finlay (Eds.), *Two paths towards sustainable forests: Public values in Canada and the United States.* Corvallis, Oregon: University of Oregon Press.

Maestad, O. (2001). Timber trade restrictions and tropical deforestation: A forest mining approach. *Resource and Energy Economics*, 23, 111-132.

Magee, S., Brock, W., & Young, L. (1989). *Black hole tariffs and endogenous policy theory.* Cambridge, UK: Cambridge University Press.

Meidinger, E. (1997). Look who's making the rules: International environmental standard setting by non-government organizations. *Human Ecology Review*, 4, 52-54.

Meidinger, E.. (2002). *Law Making by the Global Civil Society: The Forest Certification Prototype.* Retrieved from http://www.law.buffalo.edu/eemeid.

Montreal Process. (2003). *Montreal process first forest overview report.* Retrieved from http://www.mpci.org.

National Aboriginal Forestry Association (NAFA). (2003). *Aboriginal-held forest tenures in Canada 2002-2003.* October. Ottawa.

National Aboriginal Forestry Association and the Institute on Governance (NAFA-IOG). (2000). *Aboriginal forest sector partnerships: Lessons for future collaboration.* June. Ottawa.

Natural Resources Canada. (2003). *The state of Canada's forests 2002-2003.* Ottawa.

Nelson, H., & Vertinsky, I. (2004). The US Canada softwood lumber dispute. In A.M. Rugman (Ed.), *North American economic and financial integration.* Volume 10 of Research in Global Strategic Management. Oxford: Elsevier.

Nelson, H., & Vertinsky, I. (.2003). The Kyoto Protocol and climate change mitigation: implications for Canada's forest industry. Edmonton: *SFM Network Research Communication Series*. Retrieved from http://sfm-1.biology.ualberta.ca/english/pubs/en_rc.htm.

Neumayer, E. (2001). *Greening trade and investment*. London: Earthscan Press.

New Brunswick Department of Natural Resources. (Undated) *The Canadian biodiversity Strategy and New Brunswick Progress: Incentives and Legislation*. Retrieved May 15, from http://www.gnb.ca/0263/pdf/bio4-e.pdf.

New Zealand Forest Research Institute Limited. (1999). *Study of non-tariff measures in the forest products sector*. Rotorua, New Zealand.

New Zealand Institute of Economic Research. (2000). Real barriers to trade in forest products. *Working Paper 2000/4*, Wellington. Retrieved from http://www.nzier.org.

Nyborg, K. (2000). Homo economicus and Homo politicus: Interpretation and aggregation of environmental values. *Journal of Economic Behavior and Organization*, 42, 305-322.

Oberthuür, S., Buck, M., Müller, S., Pfahl, S., Tarasofsky, R., Werksman, J., & Palmer, A. (2002). *Participation of non-governmental organisations in international environmental governance: Legal basis and practical experience*. Prepared for the Umweltbundesamt. Ecologic. June. Retrieved from http://www.Ecologic.de.

OECD. (1999). *Voluntary approaches for environmental policy: An assessment*.

Ontario Ministry of Natural Resources (OMNR). (2004). *Ontario promotes forest certification*. April 1. Retrieved May 15 from http://www.mnr.gov.on.ca/MNR/csb/news/2004/apr01nr_04.html.

Oxley, A., & Osborne, K. (2002). A study of the trade and environment issue. *Australian APEC Study Centre*, Monash University. Retrieved from http://www.apec.org.au.

Pearson, C. (2000). *Economics and the global environment*. New York: Cambridge University Press.

Pedersen, L. 1996. Sustainability in the crucible of competing demands. For. Chron. 72: 609-614.

Pitts, G. (2004, June 3). CEOs heading to unique global summit. *Globe and Mail*, B3.

Porter, G, Brown, J.W., & Chasek, P.S. (2000). *Global environmental politics*. Boulder, Colorado. Westview Press.

Prestemon, J. (2000). Public open access and private timber harvests: Theory and application to the effects of trade liberalization in Mexico. *Environmental and Resource Economics*, 17, 311-334.

Raunetsal, J., Juslin, H., Hansen, E., & Forsyth, K. (2002). Forest certification update for the UNECE region, Summer 2002. *Geneva Timber and forest Discussion Papers ECE/TIM/DP/25*. Retrieved from http://www.unece.org/trade/timber.

Saunders, J. (2003, March 27). US lumber auction has critics. *Globe and Mail*, B6.

Schwab, O. (2002). *International trade regimes-implications for forest product trade flows*. Mimeograph. July. FEPA working paper. University of British Columbia, Vancouver, BC.

Scudder, G. 2003. *Biodiversity Conservation and Protected Areas in British Columbia*. March. University of British Columbia Biodiversity Research Centre, Vancouver, BC. Accessed March 2, 2004 at www.sierralegal.org.

Sedjo, R., & Simpson, R. (1999). Tariff liberalisation, wood trade flows, and global forests. *Discussion Paper 00-05*. Washington: Resources for the Future.

Sizer, N., Downes, D., & Kaimowitz, D. (1999). Tree trade liberalization of international commerce in forest products: Risks and opportunities. *World Resources Institute*. Accessed August 20, 2001 at http://www.wri.org/wri/.

Southgate, D., Salazar-Canelos, P., Camacho-Saa, C., & Stewart, R. (2000). Markets, institutions, and forestry: The consequences of timber trade liberalization in Ecuador. *World Development*, 28(11), 2005-2012.

Sprinz, D., & Vaahtoranta, T. (1994). The interest-based explanation of international environmental policy. *International Organization*, 48(1), 77-105.

Stavins, R., & Barrett, S. (2002). Increasing participation and compliance in international climate change agreements. *Kennedy School of Government Working Paper*, RWP02-031.

Taxpayers for Common Sense. (2001). *Lost in the forest: How the Forest Service's misdirection, mismanagement, and mischief squanders your tax dollars*. Washington, DC.

Tomberlin, D., Buongiorno, J., and D. Brooks. 1998. Trade, Forestry, and the Environment: A Review. Journal of Forest Economics 4(3): 177-206.

United States Trade Representative and White House Council on Environmental Quality. (1999). *Accelerated tariff liberalisation in the forest products sector: A study of economic and environmental effects*. Washington, DC.

United States Government. (2003*). Market access for non-agricultural products: Indicative list of key non-tariff barriers.* Submitted to the WTO. Dated February 13. Retrieved from http://www.ustr.gov/2003-2-13-ntb-submission-final.pdf.

US Newswire (2004). Retrieved May 11, 2004, from
 http://releases.usnewswire.com/GetRelease.asp?id=139-12012003).

van Kooten G.C. 2002. Economic Analysis of the Canada–United States Softwood Lumber Dispute: Playing the Quota Game. Forest Science, November, vol. 48, iss. 4, pp. 712-721(10).

VanGrasstek, C. (1992). The political economy of trade and the environment in the United States Senate. In P. Low (Ed.), Trade and the environment. Washington, DC: World Bank.

Vertinsky, I., & Zhou, D. (2000). Product and process certification: Systems, regulations and international marketing strategies. *International Marketing Review,* 17(2-3), 231-252.

Victor, D., Rautsiala, K., & Skolnikoff, A. (Eds). (1998). *The implementation and effectiveness of international environmental commitments.* Cambridge, MA: MIT Press.

Wear, D., & Lee, K. (1993). U.S. policy and Canadian lumber: Effects of the 1986 memorandum of understanding. *Forest Science,* 39(4), 799-815.

Weintraub, S. (2003). Lack of clarity in US trade policy. *Issues in International Political Economy,* 43. Washington, DC: Center for Strategic and International Studies.

Whiting, K. (2001). Impacts of multilateral environmental agreements on international trade and sustainable forest management. *Forest Economics and Policy Analysis Working Paper.* Vancouver, BC. October.

York, G. (2004, May 13). Myanmar mired in deforestation crisis. *Globe and Mail,* A14.

Young, O. (2002). *The institutional dimensions of environmental change.* Cambridge, MA: MIT Press.

Zhang, D. (2001). Welfare impacts of the 1996 US-Canada softwood lumber trade agreement. *Canadian Journal of Forest Research,* 31(11), 1958-1967.

Zhou, D., & Vertinsky, I. (2002). Can protectionist trade measures make a country better off? A study of VER's and minimum quality standards. *Journal of Business Research,* 55, 227-236.

# CHAPTER 13

# SUSTAINABLE FOREST MANAGEMENT: CIRIACY-WANTRUP'S DEFINITION OF CONSERVATION IN TODAY'S FOREST RESOURCE CONTEXT

WILLIAM R. BENTLEY

*Professor Emeritus of Forest Policy and Management, SUNY College of Environmental Science and Forestry, Syracuse NY 13210 USA
Email: billbentley@cox.net*

RICHARD W. GULDIN

*Director, Science Policy, Planning, Inventory, and Information, USDA Forest Service, Washington DC 20250 USA
Email: rguldin@fs.fed.us*

**Abstract**: Economic theory has limitations when guiding normative decisions over long periods of time, but empirical estimates suggest that forest owners regularly make decisions based on short-term assessments that lead to long-term sustainability. Ciriacy-Wantrup's 1952 definition of conservation – shifting resource use toward the future – is an interesting rule that leads from the short-term to the long-term. This rule is applied in a case study of America's Forest Inventory and Analysis program using the Montréal Process criteria and indicators. The results provide a basis for discussing how to deal with ambiguity in our understanding of the future and guiding policy analysis.

## 1. INTRODUCTION

Sustainable forest management is, in one sense, what the normative side of forest economics was about historically. What are the ecological, economic, and social conditions under which a forest will be managed in perpetuity for valued goods and services? Over much of the last century, the focus was on sustained yield of timber, but the question of sustainable forest management is about the sustainability of yields or flows for any good or service (e.g., wildlife, watershed protection) or multiple combinations of desired results. We also can examine the likelihood of land remaining in forests as opposed to a more developed use, such as crop agriculture, housing, or commercial expansion. Our focus is on the narrower picture, but if

*Kant and Berry (Eds.), Institutions, Sustainability, and Natural Resources: Institutions for Institutions for Sustainable Forest Management, 297-309.*

sustainable forest management does not make economic sense, the broader question of sustainable forested landscapes is mute.

Our instinct as economists and quantitative managers is to find optima. Both economists and ecologists often talk about sustainability as if there is a defined and measurable point where either sustainability begins or is the "best" of possible sustainable solutions. We begin our discussion by acknowledging the limitations of economic theory when seeking long-term optima. Risk, uncertainty, and ambiguity regarding the future make it difficult to be precise regarding the future, especially over the 30 to 100 plus years in forest production cycles. Optimal rotations are an idealized result that we hope guides us away from sub-optimal solutions.

However, we note several positive economic studies that suggest the future will flow from past and present economic, social, and ecological relationships. We are looking for workable solutions. In particular, we explore Ciriacy-Wantrup's (1952) definition of conservation and the contemporary concern with "criteria and indicators for the conservation and sustainable management of temperate and boreal forests," often referred to as the Montréal Process. In passing, we note several similar ideas suggested by other observers. We use an empirical case study of the USDA Forest Service's recent experience with preparing the *National Report on Sustainable Forests: 2003* (USDA Forest Service, 2003). The agency used the Montréal Process criteria and indicators and data from its Forests Inventory and Analysis (FIA) program and other sources to assess the nation's movement toward sustainable forest management.

## 2. RISK AND UNCERTAINTY RESTRAIN THEORY REGARDING SUSTAINABLE FOREST MANAGEMENT

Over the past 25 years or so, several quantitative models have been developed and tested in context of particular timber markets. Berck (1979), Sedjo and Lyons (1990), and Berck and Bentley (1997), for example, used variants of dynamic programming to develop models that tested the rationality of timber owners vis. a vis. markets and timber prices. In each case, the authors concluded that the markets and players exhibited behaviour remarkably close to economic rationality or at least behaved as if they held "rational expectations" about the future. Although less rigorously tested, many other papers report results from analyzing data that has similar strong patterns (e.g., Holmes, Bentley, Hobson, & Broderick, 1990).

Berck and Bentley's (1997) research on stumpage prices for old-growth redwood forests is an example of research that demonstrates forest owners and markets are reasonably good at making sound short-term economic decisions. Removals from total inventory were regular and steady, so supply shifted backward in a predictable fashion. Demand shifted in and out in response to a few cyclical factors, primarily housing starts. In this supply-demand dynamic, owners harvested timber at a rate that earned almost exactly 6% in real interest. The earnings were entirely from rises in real redwood timber prices because there was no physical growth or change in quality characteristics. Even more important, investment returns from old-growth redwood fit into a portfolio of investments with similar risk.

Old-growth redwood is like Hotelling's (1931) exhaustible coal mine; indeed Berck and Bentley's (1997) research supports his theory. Old growth redwood is not a sustainable resource, but young-growth redwood is. The price of young-growth redwood demonstrated a parallel pattern by the 1970s.[1] In this case, sustainability makes sense from an economic viewpoint when scarcity adds to biological growth and quality changes over time. In the redwood case, a group of large and small investors were managing a renewable forest resource in a way that leads to sustainable production of redwood timber by substituting a renewable resource – managed young growth –for a non-renewable one – non-sustainable old growth.

Despite these results, ex ante models are at best guides to help planners. The fundamental problem is that risk or probabilistic models compound with time. Over a few economic and ecological events, the models "explode," thus becoming useless for making predictions that can guide decisions. The old-growth redwood model, for example, was of little use in predicting real timber prices more than a year or two beyond the 1953-1980 data series.

The challenge is to think about what information forest managers and policy makers need in the short run. Financial maturity models are useful for normative purposes. Comparing what you know today with your short-term expectations for tomorrow allows a forest manager to make decisions that minimize risk and avoid serious sub-optimization over a year or two. The common example is comparing timber value growth rates with opportunity costs measured by internal rates of return or external interest rates set by various markets loanable funds. Timber value growth is produced by biological growth, changes in prices, and increased quality (usually defined by value of final products made from logs). The timber value growth rate can be 8% real or more when scarcity leads to higher real timber prices over time.

Capital budgeting for timber investments, in contrast, requires estimation of risks, but also uncertainty, where probabilities cannot be estimated. Ambiguity, where even the outcomes are unknown, presents even more difficult analytic issues.[2] Even with some actuarial data, the risk of fire, insect attack, or diseases cannot be converted into insurance premiums that are useful except for self-insurance purposes[3]. While some uncertainty problems can be analysed using game theory, most long-term forest investments have ambiguous outcomes. For example, certain markets of quality northern birch logs looked excellent a generation ago, prompting some well-considered advice to invest in silvicultural treatments to increase clear white birch log production for small turnings like spools and golf tees (Irland, 2004). With technological changes, however, the markets disappeared and with them the supposedly low risk investments in high quality birch peeler production.

### 3. PRACTICAL QUESTIONS ABOUT SUSTAINABLE FOREST MANAGEMENT

Forest managers have four questions that are pragmatic in character, but central to moving forward on sustainable forest management.

  i.   Will we know when we get there? How will we know that we are practicing sustainable forest management?

ii. Will we know if we are on the wrong path? What signals should tell us something is amiss?

iii. What should we measure? Each of the proposed and practicing systems has many criteria or indicators. Do we really need to measure all these things or are some (or many) superfluous?

iv. How should we interpret changes in the variables? So there is a change. So what?

Siegfried von Ciriacy-Wantrup (1952) spent his career thinking about such questions in various resource conservation contexts. His definition of conservation is especially relevant in context of sustainability. Conservation is the shift of use toward the future. For example, it is either reducing rate of use of timber or growing more. Depletion is the opposite. With his related definition of critical zones, Ciriacy-Wantrup anticipated the Endangered Species Act of 1972 and the emerging field of conservation biology. A critical zone for some flow resources is the zone beyond which you cannot save a population or other resource flow.

The most widely read approach that considers critical zones and boundaries of possibility is the Brundtland Report – *Our Common Future* (WCED, 1987). The Brundtland Report established sustainability as an important development policy issue, including giving impetus to our concern with sustainable forest management among many resource-specific problems.

Several other ideas follow the same general logic. The broad field of portfolio analysis, which uses empirical risk estimates as its base, is looking for turning points. Project SNAFOR (Simulated National Forests in early 1970s) was a cooperative project between the US Forest Service and The University of Michigan. The simulated reality created a framework for decision models based on projected futures. The exploding or compounding probabilities created a funnel of projected near-term reality and more distant and less reliable futures (Countryman and Bentley, 1973). This conception of uncertainty and ambiguity leads to a focus on near-term results with longer term outcomes more of a guide or constraint. This idea is akin to Ciriacy-Wantrup's definitions of conservation and critical zones, especially his notion that better choices leave open more choices in the future.

Fedkiw's (1998) pathway hypothesis is that the Forest Service is learning about ecosystem and social complexity over time; it is refining its understanding of the funnel of possible future realities. His pathway is another metaphor, in our opinion, for choices that retain or even open up more future choices.

While never a popular literature in the US or forestry, G.L.S. Shackle developed some ideas that have merit in the sustainability context (see his collected papers in Frowen, 1987). Shackle's basic question is, "Would you be surprised if this happened?" He distrusted subjective probabilities because he believed that they led to an artificial sense of knowledge and begged the real question. The notion of a range of outcomes, over which one would not be surprised, combines the definition of uncertainty from game theory with a psychological appreciation of ambiguity. While we have not explicitly coupled this with the sense of an exploding projected future in SNAFOR, one could envision a funnel within the funnel which bounds the range of possible outcomes that would not be a surprise.

Reinterpreting both the SNAFOR "funnel" model of information and knowledge about the future and Fedkiw's pathway model of learning through time in light of Shackle could lead us to a practical understanding of Ciriacy-Wantrup's definition applied to sustainable forest management. We return to that task after developing our case study.

## 4. THE GLOBAL PERSPECTIVE ON SUSTAINABLE FOREST MANAGEMENT

At the global level, the U.N. Food & Agricultural Organization (FAO, 2001) tracks and reports forest land and volume statistics every decade. The historic data is acreages, production, and consumption, but the FAO added spatial data during the past decade to provide more precise estimates of change in specific locations. The Global Forest Resources Assessment 2000 (FRA, 2000) summary provides an overall context in which to consider the US forest inventory and assessment. To summarize:

> Forests cover about 3,870 million ha, or 30 percent of the earth's land area. Tropical and subtropical forests comprise 56 percent of the world's forests, while temperate and boreal forests account for 44 percent. Forest plantations make up only about 5 percent of all forests; the rest is natural forest. FRA [The Global Forest Resources Assessment 2000] revealed that the estimated net annual change in forest area worldwide in the 1990s was -9.4 million ha, representing the difference between the estimated annual rate of deforestation of 14.6 million ha and the estimated annual rate of forest area increase of 5.2 million ha.

The world's forests are concentrated in the north temperate nations and in the tropics, with an important but smaller concentration in the south temperate zones of Latin American and Oceania. Russia, Brazil and Canada have the largest forest areas, with the US, China, and Indonesia forest areas relatively smaller but still large in the global context. Most forested nations are at a point of stability in area, but East Asia is adding area, primarily through plantations, and selected tropical nations are still suffering from deforestation.

North America is the largest consumer of wood and fiber. Europe collectively would be the second largest consumer, but only Germany stands out as being large. Japan and China follow. The US and Canada lead the world in industrial round wood production, with China, Brazil and Russia considerably lower, followed by Scandinavia, and several Southeast and South Asian nations. Fuel wood, in contrast, is focused in Asia and Africa, with India, China, and Indonesia being the dominant producers.

### 4.1 The Montréal Process

Following the Rio de Janeiro UNCED meeting in 1992, the United States and 11 other countries with temperate and boreal forests formed the Montréal Process. Together, they developed and voluntarily adopted 7 criteria and 67 indicators for reporting on the conservation and sustainable management of their forests

*Criterion*: a category of conditions or processes by which sustainable forest management may be assessed; the categories are characterized by a set of related indicators, which are monitored periodically to assess change.

*Indicator*: a quantitative or qualitative variable that can be measured or described.

| *Montréal Process Countries* | |
|---|---|
| ▪ Australia | ▪ Mexico |
| ▪ Canada | ▪ New Zealand |
| ▪ Chile | ▪ Russian Federation |
| ▪ China | ▪ USA |
| ▪ Japan | ▪ Uruguay |
| ▪ Republic of Korea | ▪ Argentina |

The Montréal Process includes the following Criteria and Indicators (indicator numbers in parenthesis):

i.  Conservation of biological diversity (9)

ii.  Maintenance of productive capacity of forest ecosystem (5)

iii.  Maintenance of forest ecosystem health (3)

iv.  Conservation and maintenance of soil and water resources (8)

v.  Maintenance of forest contribution to global carbon cycles (3)

vi.  Maintenance and enhancement of long-term multiple socio-economic benefits to meet the needs of society (19)

vii. Legal, institutional, and economic framework for forest conservation and sustainable management (20)

The philosophical approach of the Montréal Process has several premises. A basic assumption is that forests are essential to long-term well-being of local populations, national economies, and biosphere. Thus, the practical goal of sustainable forest management is meeting the needs of present and future generations. Evaluation or assessment looks for changes over time. We must monitor trends and then evaluate looking at the whole not the parts. Consequently, the set of "Criteria and Indicators" tells the story – not the individual measures. Finally, the process follows the lessons of social forestry movements all over the planet; that is, an informed, aware, and participatory public is indispensable to promoting sustainable management of forests. The basic premise is that improving quality of information will better inform policy debates. This approach to criteria and indicators flows nicely from Ciriacy-Wantrup's definition of conservation – sustainability is conservation in more contemporary terms.

## 5. US NATIONAL REPORT ON SUSTAINABLE FORESTS – A MACRO CASE STUDY

The Montréal Process countries set a goal. By 2003, each country would prepare a report on the status, condition, and trends in their forests using these criteria and indicators. In the United States, the Forest Service led the team of federal agencies to prepare and analyse the data for the U.S. report.

The Forest Service has a long history of collecting, analysing, and reporting information about the status and condition of America's forests—both public and private. This history began even before there was a Forest Service. In 1872, Congress asked the Department of Agriculture to prepare a report on the condition of forests. Dr. Franklin Hough, a physician with strong interests in America's forests, prepared that initial report.

Until creation of the Forest Service in 1905 as a separate Department of Agriculture agency, the Bureau of Forestry provided information and assistance to private landowners and state agencies to improve the practice of forestry. The McSweeny-McNary Act of 1928 created a permanent responsibility within the Forest Service to inventory and report on the status and condition of America's forests. The Forest Inventory and Analysis (FIA) program within the Research & Development area developed and implemented the sampling procedures and analytical approaches.

Initially, the FIA program was focused solely on the extent of forests and their use for timber production. By the early 1990s, responses to expanding public interest in forests for uses other than timber production led to a suite of measurements. This enlarged set included other information important to understanding the ecology and health conditions of the forests. For example, all the vegetation on the site was sampled, not just trees; soil samples were analysed; and protocols were developed to sample lichens as indicators of ozone damage.

Between 1950 and 1970, the Congress called several times for special national reports summarizing the nation's timber situation. These requests came at irregular intervals, in response to contemporary situations, such as the rapid acceleration of home construction after World War II. In 1974, the Congress recognized the need for a permanent, regular reporting cycle of national conditions for all forest resources—water, recreation, range forage, minerals, in addition to timber – and passed the Forest and Rangelands Renewable Resources Planning Act (RPA. Beginning in 1975, the Forest Service summarizes available data and reports on conditions and trends of the resources and values every 5 years. This RPA reporting cycle and the expertise developed at several of the research stations and with university collaborators made possible the preparation of a comprehensive 2003 national report on sustainable forest management. Much is known about some indicators—both in terms of solid data and trends—but less is known about other indicators and some data gaps exist.

## 5.1 Lessons Learned by the USDA Forest Service

It is not yet possible to make a firm, defensible statement about whether forests in the United States are being managed on a sustainable basis. This ambiguity is not just the result of data availability issues. Sustainable forest management requires integration of forest conditions and trends with the values that society places on the many different facets of the forest. The 67 indicators represent these values. Some people place a higher value on certain indicators than they do on others. Consequently, their view about whether forest management is sustainable may differ from persons who have a different set of values. Rather than endorsing any particular suite of values, the National Report presents indicator data in as robust a form as is available, which should encourage a values-based public dialogue about future outcomes desired and the actions needed to move the Nation towards the goal of sustainable forest management.

Dr. Guldin prepared the results illustrated in Table 13.1 for a series of oral/visual presentations in 2003. Indicators are shown for each of the seven criteria. These results in aggregate provide an overall picture of the United States forests moving toward sustainability. While this overall picture is positive, we will note some counter movements and instances where the micro-measures are sometimes moving in the opposite direction.

For those among us who are concerned with optimization in some rigorous sense, this information is at best hints of first and second order considerations of a dynamic system. In the more qualitative "Wantrupian" perspective of this paper, however, these indicators are reasonable evidence of direction of change in terms of conservation and sustainability.

The five biophysical indicators are generally quite positive, but do hint about local situations that are not so positive. For example, "Threatened and endangered species tend to be concentrated in 12 hotspots along the coastline, in the arid Southwest, and in the southern Appalachians" points to 12 micro-level situations where serious problems may exist. While soil and water indicators are quite positive, some local situations include threats to anadromous fisheries and public health of forest-based water supplies. While the nation's forested area is stable, major shifts in ownership from industry toward other kinds of institutional owners and from larger non-industrial owners to smaller, much fragmented holding will affect the supply of wood and fiber at various prices.

In contrast, the Socio-Economic and the Legal, Institutional, and Economic Framework criteria are less positive. These criteria and the forces they represent are major causes of the potential issues hinted at in the biophysical indicators. The forces affecting prices, in-country vs. overseas supplies, and continuing increases in consumption of commodity and service values are all increasing. These social forces cause biophysical changes, not all of which are positive. Over longer time periods, changes in the biophysical variables will cause supply shifts for timber and the many forest-based services. Increased scarcities will be reflected in higher prices, more political activity, or other social signals.

**Table 13.1.** *Summary of the U.S. Criteria and Indicator Results, 2003*

---

*1 – Biological Diversity*

■ Since 1900, total forest area in the United States has fluctuated less than 5 percent.

■ Forest types are changing.

■ The acreage in bigger trees is growing.

■ Fragmentation is more natural in western forests than in eastern forests.

■ Threatened and endangered species tend to be concentrated in 12 hotspots along the coastline, in the arid Southwest, and in the southern Appalachians.

■ The geographic ranges of most terrestrial animal species have not been reduced, when range is analyzed at the State level.

■ Parcelization of private ownerships is getting worse.

*2 -- Productive Capacity*

■ 504 million areas of forest (67% of all forest) are classified as timberland—forests available for timber harvesting.

■ The current volume of wood on timberland suitable for making products is 859 billion cubic feet, 39% greater than the volume that existed in 1953.

■ Net annual growth (23.6 billion cubic feet nationally) exceeds annual removals (16 billion cubic feet nationally) by 47%.

■ Privately owned land provides the majority of the nation's timber.

*3 – Ecosystem Health*

■ Pre-European settlement conditions will never again exist in the eastern U.S.

■ Since 1952, annual mortality has been constant at 0.75% of growing stock volume.

■ Since 1960, the acreage burned annually by wildfires has average 4.1 million acres. But in 3 of the past 7 years, the acreage burned has exceeded 150% of the 1960-2002 average.

■ Recent insect outbreaks and the damaging effects of exotic invasive species are considered beyond the range of historical variation.

■ Atmospheric emissions have not caused recent, widespread damage to America's forests.

*4 – Soil and Water Resources*

■ Water quality from forests is generally very good compared to non-forest land.

■ Soil erosion from undisturbed forests is generally very low.

---

*Table13.1 (cont.)*

*Table13.1 (cont.)*

## 5 – Global Carbon Cycles

- From 1953 to 1997, the amount of carbon stored in trees and other forest vegetation increased 46%, largely due to increases in the size and age of trees and of trees taking over abandoned croplands.
- Forests in the U.S. play an important role domestically and internationally in carbon sequestration.

## 6 – Socio-Economic Benefits

- The harvest-growth relationships suggest that timber volumes on national forests will continue to accumulate.
- The data are consistent with intensive forest management and high productivity of forest industry lands.
- The U.S. produced 203 million tons of wood and paper products and 2.7 quads of wood energy in 1999, compared to 83 million tons and 1.7 quads in 1950.
- Annual per-capita consumption of wood and paper products in the U.S. has been increasing while the annual per-capital harvest of wood has been stable at 65-70 cubic feet per person.
- International trade is an important component of the markets for wood and paper products.
- Forests are very important to many kinds of recreation, yet the acreage open to the public free of charge is declining while population growth is increasing the total demand for forest-based recreation.
- Outdoor recreation consumption is increasing.
- Eighty four percent of the U.S. population aged 16 or older reported that clean air and water is the value which management should emphasize on forests.

## 7 – Legal, Institutional, & Economic Framework

- Urban forest is large and growing; primarily due to suburbanization.
- Markets and legal frameworks are well established, yet continually changing. Markets for special property arrangements, such as conservation easements, are increasing in number and acceptability.
- Design of market institutions may become a very important t area of research.

Three broad kinds of responses to such information are possible. First, while we have little information on risk probabilities, we can apply our models for dealing with uncertainty and ambiguity. In particular, we can outline outcomes in the future that would not surprise us. Rather than projecting trends as such, we can pose some alternative futures that seem possible if not probable. A list of non-surprising but not necessarily high likelihood outcomes might include:

i. US industrial ownership declines to near zero for fiber-based companies and only a few middle-sized firms focused on solid wood products own and manage timberlands.

ii. With the shift in timberland ownership toward Real Estate Investment Trusts and Timber Investment Management Organizations, managerial intensity and productivity of historically industrial lands slowly declines.

iii. Most public forestlands are either in reserves or under very low intensity management for multiple values, most of which are not sold in the marketplace.

iv. The urban/rural fringe keeps moving outward and with it water quality and habitat quality decline, which prompt more regulation and other social control of private development.

v. Some creative new institutional arrangements emerge that allows the bundle of property rights associated with forestlands to be broken apart in ways that encourage working landscapes but preclude intense development for housing or commercial establishments.

While such "not surprised" lists would please G.L.S. Shackle, many of you in our audience will say, "So what do you do with such flimsy perhaps even whimsical outlooks on the future?" That prompts the second response.

The list of "Not surprising outcomes" is, with few exceptions, responses to market and political forces. If we want to reinforce the likelihood of a particular outcome, we can tilt the playing field in that direction. For example, we can adjust tax codes for income and property to favor moving timberlands into working landscape categories with the development rights stripped or conservation easements sold to land trusts or similar non-governmental entities. This would go along way in dealing with water quality, wildlife habitat, and other positive externalities society receives from larger (rather than smaller) blocks of private forestland. We can tilt the rules to favor shifts toward renewable resources like solid wood and tree fiber by our support of public forest research, reduced taxation of capital gains, and other institutional changes that shift the future toward desired rather than undesirable outcomes.

Nothing in this set of is new, but our sense is that institutional economics does not receive the attention it deserves among applied economists. We admire the rigor of many of you – indeed sometimes we are in awe of it. However, but we are reminded of a colleague in forest biology who chided one of us with, "Don't confuse rigor with usefulness!" Our reinterpretation of Professor Ciriacy-Wantrup and the US case study of the Montréal criteria and indicators have offered as a suggestion of what we can glean from such assessments to guide our thinking on important policy questions.

A third response is to reinforce measurement of all the indicators in 5 to 10 years and to add new indicators that help measure progress on the legal-institutional front that will positively affect biophysical and other socio-economic indicators. In this sense, the Montréal Process is a learning process that helps us, in Fedkiw's (1998) sense, determine if we are on the right path.

## 6. CONCLUSIONS

Sustainable forest management properly shifts the focus of management decisions from, "How can I use the forest today?" to "What is left behind after I use the forest today, and what will continue to be productive enough to meet the future needs of a growing population?" This is a shift in focus towards the future. This kind of shift follows Ciriacy-Wantrup's emphasis on conservation shifting resource use towards the future. Analyses at multiple geographic scales are useful in focusing dialogue at the community to national levels on the values citizens hold about forests and how they should be used today, while also conserving the ability to use them to meet future needs.

Focusing on Ciriacy-Wantrup's conservation criteria keeps the arguments concrete and practical in a subject that tends to be elusive. The sustainable forest management perspective allows us to find many working examples of movements toward sustainable forest management. These examples are useful in learning how to improve our managerial and policy decision models and process steps.

## NOTES

[1] This point is developed in Bentley, W. R., and P. Berck (1984). Price forecasting models for old-growth redwood stumpage. Third preliminary report to the US Department of Justice. 55 p.

[2] The definitions for risk and uncertainty are commonly used by economics, but ambiguity in this sense is more common in some of the practical management literature (e.g., Miller et al., 1994).

[3] In general, fire insurance is not available for private forest owners, but several corporate managers we have talked to over the years do describe fire risk as one where self-insurance is possible. This leads to strategies of not blocking up holding through trades or acquisitions. The same proposition seems applicable for reducing risk from insect attacks as well.

## REFERENCES

Berck, P., & Bentley, W.R. (1997). Hotelling's theory, enhancement, and the taking of the Redwood National Park. *American Journal of Agricultural Economics*, 79(2), 287-298.

Berck, P. (1979). The economics of timber: A renewable resource in the long run. *The Bell Journal of Economics*, 10(2), 447-462.

Ciriacy-Wantrup, S.V. (1952). *Resource conservation.* University of California, Division of Agriculture Science.

Countryman, D.W., & Bentley, W.R. (1973). *Project SNAFOR: A gaming simulation technique for training natural resource managers.* SNAFOR Report No. 1. Univ. Michigan. 86 p.

FAO (Food and Agriculture Organization of the United Nations). (2001). *Global forest resources assessment 2000*: Main report. Rome: FAO.

Fedkiw, J. (1998). *Managing multiple uses on National Forests, 1905-1995: A 90-year learning experience and it isn't finished yet.* Washington, D.C.: USDA Forest Service.

Frowen, S.F. (Ed.) (1987). *Business, time, and thought: Selected papers of G.L.S. Shackle.* New York: New York University Press.

Holmes, T.C., Bentley, W.R., Hobson, T.C., & Broderick, S. (1990). Hardwood stumpage price trends and characteristics in Connecticut. *N. Jour. Applied Forestry*, 7(1), 13-16.

Hotelling, H. (1931). The economics of exhaustible resources. *The Journal of Political Economy,* 39(2), 137-175.

Irland, L.C. (2004). Are we wasting money on intensive management? A Powerpoint presentation at the Northeastern Forest Economists meeting, New England Society of American Foresters annual meeting, Quebec City, March 25, 2004, plus unedited speaking notes (courtesy of Dr. Irland).

Sedjo, R.A., & Lyon, K.S. (1990). *Long-term adequacy of world timber supply.* Washington, D.C.: Resources for the Future.

USDA Forest Service. (2003). *National Report on Sustainable Forests:* 2003. U.S. Department of Agriculture, Forest Service.

WCED (World Commission on Environment and Development). (1987). *Our common future.* New York: Oxford University Press.

# CHAPTER 14

# STAKES, SUSPICIONS AND SYNERGIES IN SUSTAINABLE FOREST MANAGEMENT–THE ASIAN EXPERIENCE

## CHERUKAT CHANDRASEKHARAN

*TC 15/174, 'Prathibha', F-1, Althara Nagar, Vellayambalam*
*Thiruvananthapuram – 695 010, India*
*Email: cherukat@vsnl.com*

**Abstract:** In spite of the intensive global dialogue about the spirit and substance of Sustainable Forest Management (SFM) during recent years, there is little agreement on the scope and definition of SFM among the different stakeholders. Criteria and Indicators, and Forest Certification processes have provided primacy for stakeholders, and they substantially influence the way the forest resources are managed, through their claims for benefits and related tactics. Most often they are competitors and their interests are in conflict, as a result they tend to view each others with suspicion and get involved in power struggles. Synergies are, however, developed in situations where there are no 'better' alternatives to co-operative action or where clear policy incentives foster and nurture development of such synergies. The understanding, and appreciation of SFM by stakeholders are conditioned by the extent to which their claims are satisfied; this decides the success or otherwise of SFM implementation. Asia presents a wide spectrum of forestry situations and stakeholder roles. This paper discusses four cases drawn from four different countries, to illustrate stakeholder influences, to identify relevant issues and to draw indicative lessons in SFM.

The four countries – India, Indonesia, Malaysia and Papua New Guinea (PNG) represent various stages in the forestry "development" scale. Their forest resource endowment and problems of management are considerably different. Based on historical background, their institutional frameworks also vary. Similarly, the four cases reviewed illustrate different management needs and focus. The Out-grower Farms of Clonal Trees of ITC Paperboard and Specialty Paper Division in India is a case of partnership between a large private sector corporation and farmers to establish pulpwood plantations in private farmlands. PT. Sari Bumi Kusuma in Indonesia is the case of a well-managed private logging concession, which has been in operation since 1978 and has recently been renewed for a further period of 70 years. Matang Mangrove Forest of Perak State, Malaysia is a case of integrated mangrove management for wood and non-wood products, with a 100 year history of SFM. Vanimo Forest Products Ltd in PNG is a 20 year non-renewable concession in a customarily owned forest, due to expire in 2010. These cases and the situation in the respective countries highlight that major constraint to SFM is not lack of technology, but the institutional factors, which militate against the application of the best available technology. These institutional factors appear in the form of short-term perceptions and time preferences of the investors and other stakeholders. Due to the long time horizon involved, technology-based models of SFM, often, face implementation problems, and plans are vitiated by intervening developments in political, economic and social arenas. To address such situations, long-term, policy-based commitment of the stakeholder community is crucial.

*Kant and Berry (Eds.), Institutions, Sustainability, and Natural Resources: Institutions for Sustainable Forest Management, 311-339.*

"A first rate theory predicts; a second rate theory forbids; and a third rate theory explains after the event" - Aleksander Isaakovich Kitaigorodskii (from a lecture given at IUC, Amsterdam, August 1975).

How does SFM theory rate?

## 1. INTRODUCTION

Asia (including Oceania) has experienced rapid economic growth, accompanied by enormous social and political transformation, but its environmental performance has not been commensurate. In some cases unsustainable utilization of natural resources, with forest resources prominent among them, has caused skewed economic growth and permitted poverty to persist unnecessarily. In Asia forest conservation has been a live issue since the 4th century B.C. With unprecedented loss and degradation of forest resources, expressions of concern have become louder and global during the last 3 to 4 decades – for example, the Stockholm Conference of 1972, World Conservation Strategy of 1980, Brundtland Commission Report of 1987, Earth Summit of 1992, UNEP's report on Global Environmental Outlook (GEO) of 2000 and Johannesburg Summit of 2002.

Historically, the foremost objective of forest management has been the sustained-yield production of forest products. In recent years, a new emphasis on the environmental and conservation roles of forests has led to greater emphasis on overall sustainable forest management (SFM) with multiple objectives for meeting changing social needs. International Tropical Timber Organization (ITTO) (1998) defines SFM as the process of managing forests to achieve one or more clearly specified objectives of management with regard to the production of a continuous flow of desired forest products and services, without undue reduction in its inherent values and future productivity and without undue undesirable effects on the physical and social environment. SFM is the specific and practical action for translating the concept of sustainability into reality in forestry. With many different definitions, there appears to be no consensus regarding the precise context, scope and use of SFM. As in the case of the fabled elephant and the blind men, SFM means different things to different people. Special features of forestry such as its long gestation period and investment horizon (often not matching with the investors' time preference), difficulty to distinguish between forest capital and incremental growth (often leading to over-exploitation and capital consumption), and high level of externalities (often causing disinterest on the part of private investors) add to the complexity of SFM.

## 2. FORESTRY SITUATION IN ASIA

Asia and Oceania together account for 746 million ha of forests, comprising 626 million ha of natural forests and 120 million ha of forest plantations (Table 14.1). While accounting for more than half of the world's population, the Asia and Oceania region covers 30% of the world's land, and has less than 20% of the world's forests and the lowest ratio of forest per person. Asia and Oceania does account for 64% of the world's forest plantations. As of 2000 wood and biomass volumes were 63 cum/ha and 82t/ha for Asia and 55 cum/ha and 64 t/ha for Oceania, against the global average of 100 cum/ha and 109t/ha. The impact of land degradation is also evident in forest plantations due to poor site selection, lack of species-site matching, absence of maintenance, and inadequate protection from fire and grazing. For Asia and Oceania, the share of industrial wood in the total wood harvest in 2000 was only about 25%, influenced by the low level of forest-based industrialization, log exports, and dependence on fuelwood for meeting local energy needs. While industrial wood production in 2000 accounted only for 6.9% of the total for India, the corresponding figure was 26.0% for Indonesia, 87.3% for Malaysia, and 35.6% for Papua New Guinea (PNG) (FAO, 2002).

*Table 14.1. Regional Distribution of Forest Area, 2000*

| Region | Land Area (Million ha) | Total Forest (Natural Forests & Plantations) | | | Natural Forest (Million ha) | Plantation Forest (Million ha) | Forest Area Per capita (ha) |
|---|---|---|---|---|---|---|---|
| | | Area | % of Land Area | % of World's Forests | | | |
| Asia | 3,085 | 548 | 18 | 14 | 432 | 116 | 0.15 |
| Oceania | 849 | 198 | 23 | 5 | 194 | 4 | 6.58 |
| Asia-Oceania | 3,934 | 746 | 19 | 19 | 626 | 120 | 0.20 |
| World | 13,064 | 3,870 | 30 | 100 | 3,682 | 187 | 0.65 |

*Source:* FAO (2001)

Forest degradation is making Asia a grey continent (UNEP, 1999; ITTO, 2001; Chandrasekharan, 2003). Conversion of forest lands to other uses and improper management of surviving ones has led to a continuous fall in forest area; compensatory efforts to rehabilitate degraded forest land and to afforest bare and unproductive lands have been inadequate to offset this loss. The rate of change in net area under forest in Asia and Oceania has however shown considerable improvement, particularly due to increased afforestation activity through people's participation (Table 14.2). During the 1980s, the region lost 3.8 million ha of natural forest per year; with an afforestation rate of 1.5 million ha per year, the net forest loss annually amounted to 2.3 million ha. In the 1990s, even though deforestation of the natural forests remained at a relatively high rate, afforestation increased to over 3.5 million ha/year, thus reducing the net forest loss to 729,000 ha/yr. The extent of the forest resources has been falling, but consumption (including wasteful uses) of forest products have continued to increase. Sustainability requires that the formation of new capital equals the sum of resource depletion through extraction and environmental damages. For sustainable development, capital formation has to be even higher, requiring higher levels of investment (Panayotou, 1995). It has been estimated that in the early 1990s, the forestry sector in the developing world suffered a net negative investment (difference between disinvestment through forest loss and degradation and actual gross investment in forestry development) of US$ 24.6 billion per year. For the developing countries of Asia, the net negative investment was US$ 6 billion per year (UN-CSD, 1996; Chandrasekharan, 1996). Even though it is now several years since the international community began discussing new approaches to the financing of sustainable development, nothing tangible has happened to benefit forestry in Asia.

*Table 14.2. Forest Cover Change in Asia and Oceania, 1990-2000*

| Region | Total Forest 1990 (000 ha) | Total Forest 2000 (000 ha) | Annual Change (000 ha) | Annual Rate of Change (%) |
|---|---|---|---|---|
| Asia | 551,448 | 547,793 | -364 | -0.07 |
| Oceania | 201,271 | 197,623 | -365 | -0.18 |
| Asia-Oceania | 752,719 | 745,416 | -729 | -0.10 |

*Source:* FAO (2001)

Some of the institutional arrangements for forest management adopted, based on ownership and operational responsibilities, in the region are the following:

- Forest owned and managed privately, subject to government regulations.

- Private entrepreneurs manage customarily owned community/tribal forests, subject to legally valid arrangements.

- Small private forests (woodlands) managed by co-operatives.

- Forest owned and managed, and product processing and marketing fully or partly carried out, by government, either directly or through state-owned companies.

- Forests owned and silvicultural management carried out by government, with harvesting, processing, and marketing undertaken by the private sector.

- Forests owned by government and managed (except for protected areas) by the private sector on concession (lease) arrangements, including harvesting, processing and marketing.

- Forest privately or customarily owned by individuals or communities and managed directly by government or through state owned companies, for the benefit of the owners.

There are serious inadequacies in forest governance. The most serious shortcoming is the failure to control illegal activities and to protect the forest. While Indonesia and Malaysia, have established national level forest and timber certification bodies, overall, certification is still a marginal activity in Asia. Behind many of the sectoral deficiencies, there exists the influence of corruption, coming in many shades and at many levels. Other proximate contributing factors to the unsatisfactory forestry situation include weaknesses of the SFM framework, inadequate forest protection, unrecorded production and consumption (as much as 80% of total wood use in some countries), neglect of NWFP management, bio-piracy, lack of boundary demarcation and open access to forests, encroachments, incendiary fires, shifting cultivation, and the changing forest resource landscape. An overview of apparent and underlying causes for the present state of affairs, ranging from poverty and population pressure to inadequate governance, lack of transparency, accountability, discipline and commitment is presented in Figure 14.1.

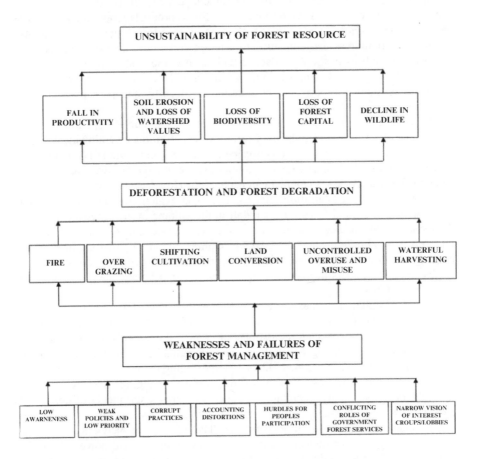

*Figure 14.1. Illustrative Flow Diagram – Problem Tree*

*Source:* Chandrasekharan (2003)

A number of countries in the region have followed a system of forest concessions of varying duration and with typically unsatisfactory results. Thailand terminated its forest concessions some years after they were introduced and later introduced a logging ban in natural forests; Cambodia has suspended the concessions recently for an in depth evaluation; Philippines is phasing out the remaining concessions to eliminate them completely by 2011. The concessionaires in Malaysia and Indonesia are increasingly being urged to have the forest management units (FMUs) certified by the certification bodies. In Papua New Guinea, the concession companies, particularly the expatriate companies, are coming under considerable scrutiny by international agencies, NGOs and the government agencies. In a number of countries a sizeable share of the total wood (and other forest product) requirements are now being supplied by non-forest sources such as homestead forests and farm wood lots.

In order to address the rampant deforestation and environmental degradation, a number of countries in the Asia-Pacific region, over the past 15 to 20 years, have completely or partially banned logging in natural forests. However, in some countries, logging is reported to be continuing illegally, and perhaps even more destructively than in the past.

In other cases, to address the weaknesses in the current situation, a new instrument of Criteria and Indicators for SFM was introduced following UNCED 1992. The countries in the region subscribe to the C&I under the ITTO and CIFOR processes, and some countries have introduced (or revived) their own C&I. The C&I and certification processes complement each other. However, in most cases C&I are yet to be implemented faithfully and rigidly, as is evidenced from the continuing deforestation of natural forests and the fall in the volume of growing stock. Neither the existence of an elegantly designed C&I nor a system of certification will by themselves be able to ensure SFM, which requires, above all, a strong institutional backing. Institutions (and institutional instruments) are generally designed, with stakes and stakeholders in mind; but, power plays can sabotage the due functioning of the institutions.

Policy formulation is an exercise in satisfying the needs of the different groups of stakeholders who together are to derive the benefits and bear the costs of forest management. The rules and regulations of policy implementation are meant to address the conflicts among stakeholders. The problems lie in distinguishing genuine stakeholders from spurious stakeholders and in deciding on the relative weights to be assigned to the interests of the various groups, ranging from government officials/foresters, the local community, concessionaires, consumers of forest goods and services, eco-tourists, environmentalists, animal lovers, tree lovers and others. Often stakeholders are competitors, since the stake of one adversely affects the stake of others; they involve themselves in "stakeholder politics" and power plays. The multiplicity of forest products and their uses, and the correspondingly large number of stakeholders, make forestry planning and management very complex. As a result, if the roles, rights, privileges and obligations of the stakeholders are not fully discussed, their respective positions clarified and spelt out, they tend to view each other with suspicion. They execute their private agenda either secretly or in collusion with other amenable groups. Synergies are, however, developed in desperate situations where there are no 'better' alternatives for co-operative action or where clear policy incentives encourage development of such synergies.

There appears to be a systematic pattern in the way the natural forests in a country are systematically and progressively exploited, destructively and unsustainably, through the combined force of the competing and colluding stakeholders. Under the pressure of the emerging crisis new synergies are then, sometimes, established to recreate forests. Situations where sustainability prevails through the harmonious efforts of all stakeholders are rare and exceptional. However, institutions can influence behavior through the vision, mission and values which they embody.

Asia presents a wide spectrum of situations in this regard. Four cases are discussed from four different countries representing the range of situations in Asian forestry, to illustrate how outcomes are influenced by the overall situation in the country, to identify the issues involved and to draw indicative lessons for sustainable forest management.

## 3. FORESTRY SITUATION IN THE FOUR SELECTED COUNTRIES

The four countries – India, Indonesia, Malaysia and Papua New Guinea (PNG) - represent four stages of the forestry development. India and PNG, with a population density of 336 and 10 per sq. km, a per-capita forest area of 0.1 ha and 6.5 ha and a per-capita round wood consumption of 0.29 and 1.4 cum respectively, represent the two extremes; Indonesia and Malaysia with population densities of 112 and 71 per sq. km., per-capita forest areas of 0.5 and 0.9 ha and per-capita round wood consumption of 0.54 and 0.52 cum respectively, represent positions in between. While Indonesia has the largest amount of forest in absolute terms, Malaysia leads in terms of crop condition represented by growing stock volume and quantity of above ground biomass. It may be noted that while the volume of growing stock is partly dependent on the type of forest and land involved, it is mainly influenced by the quality of forest management (see Table 14.3).

These four countries illustrate a sequence of stages in the march towards forest degradation and destruction – low population pressure with forest exploitation for financing development; expanded logging activities eventually going out of control, an increasing level of illegal activities and deforestation, forest governance becoming weak and succumbing to corruption; the shrinking forest base being declared (though often, ineffectively) out-of-bounds for further exploitation; and finally, people/farmers beginning to "cultivate" forests to meet market demand.

In country after country, this pattern is repeating itself. Though details of the unfolding scenario affect the actual paths and the speed of events, the general direction appears to be the same. India has about reached the final stage of the march; Indonesia is not far behind; Malaysia has taken the hesitant decision to stop the march or to go at a slow pace; PNG has just started on the path and seems to assume that it has inexhaustible resources, enough to last forever. Considering the facility to move logs easily and speedily across oceans, PNG's logging activities can also go out of control and catch up with others – unless it can call forth great determination to break the trend.

*Table 14.3.* Forestry Situation in the Four selected Countries in the Asia-Oceania Region, Year 2000

| Country | Land Area (000 ha) | Natural Forest (000 ha) | Plantation Forest (000 ha) | Total Forest (000 ha) | Forest as percentage of land area | Forest area per capita (ha) | Growing Stock Volume and above ground biomass | |
|---|---|---|---|---|---|---|---|---|
| | | | | | | | Cum/ha | t/ha |
| India | 297,319 | 31,535 | 32,578 | 64,113 | 21.6 | 0.1 | 43 | 73 |
| Indonesia | 181,157 | 95,116 | 9,871 | 104,986 | 58.0 | 0.5 | 79 | 136 |
| Malaysia | 32,855 | 17,543 | 1,750 | 19,292 | 58.7 | 0.9 | 119 | 205 |
| PNG | 45,239 | 30,511 | 90 | 30,601 | 67.6 | 6.5 | 34 | 58 |

\* Of the total area of plantation forest, rubber accounts for 5,534,000 ha (12.5%)

*Source:* FAO 2001, Forest Resources Assessment 2000.

*3.1 India*

India has a long tradition in forestry, having started scientific forest management in the first half of the 19th century. India has one of the lowest per-capita forest areas (0.1 ha) and the lowest per-capita wood consumption (0.29 cum) in the world. Over 90% of the wood produced is consumed as fuel wood. About 22% of the country is under forest cover, comprised of 31.5 million ha of natural forest and 32.6 million ha of planted forests. The Forest Resources Assessment 2000 (FAO, 2001) shows annual natural forest loss during the 1990s (1.90 million ha) to be slightly exceeded by area of annual plantations (1.93 Million ha). Whether there has been a net gain is however unclear, due to the inclusion of rubber plantations, farm wood lots and home gardens as forests, which hitherto were considered as outside the definition of forest. Other indicators point to continued loss. Average wood volume has fallen from a level of 47 cum/ha in 1990 to 43 cum/ha in 2000; the stock of above ground biomass has fallen from 93t/ha in 1990 to 73t/ha in 2000. In view of the need to conserve biological diversity, and to protect the remaining natural forests, logging in the natural forests has been banned in several parts of India.

Joint Forest Management (JFM), informally initiated in the early 1970s to enlist the participation of local people in forest rehabilitation efforts, has become the flagship programme of India in people's participation. JFM is a strategy in which the government (represented by the Forest Department) and a village community enter into partnership agreements to jointly protect and manage forestland-adjoining villages and to share responsibilities and benefits. JFM has spread throughout the country, bringing under its aegis around 14 million ha of forestland. There are around 63,000 JFM Committees in 27 States engaged in protection and regeneration of degraded forests, in return for certain usufructs and other benefits. A major complaint against JFM was that it covers only the protection and maintenance of degraded forests. In January 2000, the Government of India issued a new circular, which envisages extension of JFM to better-stocked forests also. Additionally, it provides for mandatory (50%) involvement of women in JFM activities, and women membership (33%) in Executive Committees. However, informed, active and organized participation is yet to become a normal feature of forest management.

Private sector plays an important, though different, role in Indian forestry. Traditionally, rural people in India have grown trees and useful plants on their farms, homesteads, and community lands, primarily to meet the household requirements for fuel, poles, timber and medicinal plants. Several different combinations of agro-silvo-pastoral systems have been practiced. With the advent of social forestry, a promotional drive was launched to plant trees in wastelands, institutional lands, and non-forest public and private lands. A large number of tree farming and agro-forestry enterprises have sprung up all over the country and they are performing an important role as suppliers of forest raw materials as well as of market products.

The private sector is dominant in the area of harvesting and processing. About 90% of forest-based products are manufactured in the private sector while about 97% of the forests are owned and managed by the Government (GOI/MOEF, 1999). The raw material requirements of the privately owned forest industries had been met through a variety of arrangements such as auctions, negotiated sales and allocation agreements. With logging restrictions, they are now in the grip of raw material shortages. Many industrial units are forced to rely on private non-forest sources.

With the advent of the Forest Conservation Act of 1980 and the National Forest Policy of 1988, the production role of the Government Forest was assigned a lower priority. Logging operations in natural forests were discouraged and in several cases locally banned. Imports of logs and wood products are now allowed. The wood scarcity situation has provided an impetus for development of farm forestry, homestead forestry, agro-forestry and trees outside forests. Some wood-industry units have made efforts to grow captive tree plantations. Several industrial units are also promoting out-grower tree farms. Currently about 50% of the wood supply in the country comes from non-forest sources. Of the rest, a considerable portion is accounted for by imports, with the balance obtained from public forests, mainly forest plantations. While logging is being legally banned in natural forests, there has not been adequate management and protection; this is leading to their further deterioration.

The 1988 National Forest Policy had directed that, as far as possible, forest industries should meet their raw material requirements from timber grown in collaboration with farmers and the local community. To improve the raw material situation, a number of companies took the initiative and experimented with various approaches to form company-community partnerships.

- Supply of free or subsidized planting stock with or without a buy back guarantee;

- Bank loan schemes under which the company helps the farmer to get the loan and provides planting stock, technical extension and a buy back guarantee;

- Leasing or sharecropping schemes under which the company raises and maintains plantations on farmers' lands, based on appropriate arrangements;

- Intensive research and development and commercial sale of clonal planting stock to farmers by companies with or without a buy back guarantee.

The various schemes started under the company-community (farmer) partnerships have two major achievements to their credit. First, they have generally popularized the concept of tree farming; second they have directly contributed to the cultivation of a large number of commercial trees on private lands. It has been widely accepted, however, that there is an urgent need to loosen the bureaucratic control and simplify procedures to allow the private sector to contribute more effectively (Saigal *et al.*, 2002). The timber transit rules and regulations have been relaxed by some states as a means to encourage private tree planting.

In summary, successful implementation of SFM in India faces many severe challenges. Increasing population pressure on a decreasing resource base makes forest management a difficult task. Conflicts exist in several of the interfaces between forestry and community – in watershed management, plantation development and biodiversity conservation. Also, several components of SFM are clearly missing – e.g. inventory of resources, functional and land capability classification, waste-free utilization, and sustained investment. There have been some positive changes in the forestry sector during the past decade, such as the acceptance and expansion of JFM, increasing involvement of farmers in growing trees, private sector participation in forestry, partnerships of forest product manufacturing companies with local farmers and so on. Converting this feeble trend into a strong march towards SFM will be a major challenge. One contributing factor is a serious under appreciation of the economic value of forests. Unrecorded (and/or misclassified) uses of forest products distort income accounting and reduce budget allocations for the forestry development. The value of forest-provided benefits (wood products, wood energy/fuel wood, fodder and forest grazing, thatching and construction materials, medicinal plants, edible products and other non wood products and so on) in India during the early years of 1990s was estimated as US $ 43.8 billion annually, against a reported GNP share of forestry of US $ 2.9 billion in the year 1993, representing 1.2% of India's GNP. Most of the forest benefits are left unrecorded. The value of forests reflected in System of National Accounts represents less than 10% of the real value (UNCSD, 1996).

## 3.2 Indonesia

Indonesia's forest resources have suffered serious depletion in recent years. Its net annual forest loss during the 1990s was 1.3 million ha (the average annual loss of natural forests was about 1.7 million ha but annual plantation was about 375,000 ha) accounting for an annual deforestation rate of 1.2% (FAO, 2001). The remaining natural forests suffer serious degradation due to a multitude of factors including excessive logging, illegal activities, and forest fires. Between 1990 and 2000 the stock of woody biomass fell drastically from 203t/ha to 136t/ha and wood volume from 179 cum/ha to 79 cum/ha. WRI (1999) estimates that by the turn of the century, Indonesia has lost 72% of its original forest cover. Overexploitation and poor management imperil Indonesia's forest resources and threaten the livelihoods of the communities whose existence is intimately linked to these resources.

In the 1960s Indonesia introduced a system of forest concessions for managing and utilizing its forest resources. The number of concessions steadily increased from 45 (covering about 5 million ha) in 1970 to 584 (covering about 68 million ha) in the 1990s. Development of wood processing industries closely followed the expansion of forest concessions. This uncontrolled expansion of logging and processing led to misuse and overuse of forests. Assessment by the Ministry of Forestry (MOF) indicated that less than 20% of the concessions were maintaining an acceptable standard (as defined below). Over the years, some 128 concessions are reported to have been cancelled. Forestry Studies carried out by FAO found that many of the shortcomings encountered in the management and utilization of concession forests may be attributed to the non-compliance with the prevailing regulations (relating to boundary demarcation, logging practice, post harvest silviculture and rehabilitation, fire protection, protection against illegal practices, etc.) rather than to any inherent weaknesses of the concession system (FAO/MOF, 1990). A recent ITTO Mission to Indonesia (ITTO, 2001), however, has surmised that the serious flaws of the logging concession system appear to be the source and strength of illegal logging in Indonesia. Irrespective of the validity of the dominant cause, the quantum of illegal logging has reached, nationally, a level of 30 to 50 million cum annually, compared to the legal logging of about 20 million cum. The main reasons for illegal logging include inadequate management, mismanagement, non-compliance of prevailing regulations, and weaknesses of the concession system. Examples of inadequate management are lack of concession and forest management unit (FMU) boundaries, lack of fire protection measures, and lack of maps and inventory information. Mismanagement can be seen in the removal of trees against silvicultural principles causing damage to the standing crops and cutting from outside the prescribed felling area.

In order to meet the mounting criticism from within and outside the country, the MOF recently arranged to have performance appraisal conducted on 415 logging concessions (comprising of 116 units applying for license renewal and 299 units having ongoing licenses) to help to decide whether to revoke the concessionaire's license or allow them to continue their operations. Initial results indicate that only about 40% of the concessions have been maintaining moderate to good standards.

Inappropriate structuring of forest industries and flaws in the forest plantation (timber estate) development programme are other areas suffering serious deficiencies. As of December 2000, there were 176 approved plantation concessions, with a land allocation of 7.76 million ha. A recent evaluation of 92 plantation concessions showed that only 31 (33.7%) could be considered as technically and financially feasible, with 51 (55.4%) both technically and financially unfeasible (Press Release of MOF dated 01 March 2003). Gaps and inadequacies in implementing the recently introduced decentralization and reform laws have further exacerbated the situation.

With the promulgation of decentralisation laws (UU 22/1999 and UU 25/1999), authority and responsibility for forest management have been decentralized to local level governments, mainly to the *Kabupaten* (District) level. All forestry activities with certain exceptions (e.g. natural resource conservation) now fall under local government management. The local governments welcome this new authority in the expectation of improving their revenue. Unfortunately, these governments are quite unprepared to manage forestry. They do not have the needed institutional set-up (local regulations, organization, skilled manpower) or experience in managing the forestry sector. It will take some years to develop the *Kabupaten* forest services into an able organization. Efforts in that direction are required urgently if they are to derive the potential benefits from the valuable forest resource and not to be the cause of further loss of the forest resources now under their authority. In addition, the scenario is made ambiguous by a series of inadequate and contradictory regulations, without clear indication as to which regulation supersedes other regulations. Some districts are already exercising their powers by giving permits to private entrepreneurs to run small forest concessions, and have established District Forestry Services by appointing personnel to manage forestry in the district. Some *Bupatis* (elected heads of *Kabupaten* or district) have even issued their own respective decrees, superseding those issued by centre. In general, however, many regions continue to follow the old system; or are doing nothing because of the prevailing confusion. Hence, lack of preparedness on the part of local governments and their acting beyond the authority granted by law, contradictions and lack of clarity in the legal provisions, and misperception and unrealistic expectations on the part of the people are major problems.

In summary, though Indonesia's timber production and processing output rose to about US $ 20 billion annually within a short period, disproportionately heavy environmental and social costs were incurred in the process. A daunting range of problems threatens the sustainability of forestry; many of them trace back to; corruption, collusion and nepotism; inadequate monitoring and evaluation; a lack of political commitment, and defects of governance, which now include contradictions and confusion relating to decentralization. Reversing this trend is a major challenge.

## 3.3 Malaysia

With the forests covering about 59% of Malaysia's land area, forestry is an important economic sector. Net annual forest loss during the 1990s was 237,000 ha (the average annual gross loss of natural forests - 272,000 ha and the annual rate of plantation - 35,000 ha) representing an annual deforestation rate of 1.2%. The remaining natural forests still carry a reasonable growing stock, even though there has been some qualitative degradation. The forestry growing stock, which stood at 214 cum/ in 1990, fell to 119 cum per ha in 2000, and the corresponding fall in the stock of above ground biomass was from 261 t/ha in 1990 to 205 t/ha in 2000 (FAO 2001).

In 2001, Malaysia had net export earnings of US $ 2.02 billion from primary wood products alone. Wood-based panel products accounted for 61% of it, followed by sawn wood (21%) and logs (14%). In recent years the average annual traded value (international and domestic) of all wood products has reached a level of US$ 4.5 billion (MPI 2002). A total of 226,000 people were directly employed by the forestry sector in 2000. The forestry sector's share of GDP fell from 5.3% in 1996 to 4.4% in 2000, reflecting the fast growth of the economy as a whole.

The National Forestry Policy (NFP) of Malaysia was approved in 1978, and then revised in 1992 to take cognizance of current global concerns for the conservation of biological diversity, sustainable utilization of genetic resources and participation of local communities in forestry. Malaysia has given adequate priority to the development of non-wood products and forest based recreation. The existence of consultative committees linked to sectoral programmes and activities at the village, *mukim,* district, state and federal levels helps to ensure good social relations. However, conflicts between timber operators and aboriginal people do arise, particularly with regard to their traditional rights and rights of occupancy. The existing mechanisms of conflict resolution need to be strengthened.

Implementation of two of the important policies concerned with the conservation and sustainable management (the National Forestry Policy 1978, as revised in 1992, and the National Policy on Biological Diversity 1998) has been made difficult by the dual responsibility between the Ministry of Natural Resources and Environment and the Ministry of Plantation Industries and Commodities. Overall, Malaysia's forest resources are reasonably well managed. Its forestry administration and profession are keenly vigilant to ensure that the orientation towards SFM is not lost. This, however, is not easy in the current global environment where multiple pressures, misinformation and conflicts can distort the long-term vision, in favor of short-term preferences. There is need for a combination of flexibility and tenacity to make SFM work. It is, therefore, necessary for Malaysia to continue its focus, without any let-up, on such important aspects of SFM as: demarcation of forest management units, detailed inventory and bio-prospecting for planning multiple use forestry, balancing of resource management and utilization, providing clarity in institutional roles and responsibilities, resolving conflicts among stakeholders, a clear policy and strategy regarding plantation forests, and research support for policy development and refinement.

### 3.4 Papua New Guinea

The net annual forest loss during the 1990s was 113,000 ha, an annual deforestation rate of 0.4%. Annual plantation development was just 4,000 ha. Degradation of the remaining natural forests in terms of fall in growing stock compared to the benchmark of 1991 was serious, caused by excessive and wasteful logging, *swidden* agriculture and forest fires. Between 1990 and 2000, the growing stock of wood fell dramatically from 168 cum/ha to 34 cum/ha and the above ground biomass from 191 t/ha to 58 t/ha (FAO 1995, FAO 2001). PNG is one of the major exporters of tropical round wood in the region, about 20% of its total wood harvest being exported in this form; 64% is used as fuel wood and only 16% undergoes any form of processing. Net export earnings of forest products are about US$ 200 million annually, over 90% coming from round logs. However, logs accounted only for 5% of total exports of PNG during 1998-2000. Butterflies, live birds, eagle wood, sandalwood and rattan products are important sources of local income. Maintenance of traditional cultural values and forest-based recreation are some of the other important aspects of SFM in PNG.

People in PNG follow a traditional life style and grow, gather or hunt most of their food. Land ownership is vested with customary landowners who comprise a large share of the rural population; and virtually all forestland is owned by clans or tribal groups under customary law (PNG-FA/NFS, 2002). Traditionally, customary owners never considered their land as property but as a domain for survival of land-group members, past, present, and future. This constitutionally guaranteed customary land ownership is the key institutional element influencing forest use. Accordingly, to develop forest resources, the government must first acquire timber rights from customary owners on the basis of Forest Management Agreements (FMAs).

Under the FMA, the PNG Forest Authority is designed to secure the commitment of resource owners to follow recommended forest management practices, while simultaneously offering investors access to the forest for a minimum period of 35 years. Implementation involves the State issuing a Timber Permit (TP) under which it establishes the returns due to the land owner (which include a package of social, economic and infrastructure benefits) and takes responsibility for the management of the forest on behalf of the customary owners; this management role can be implemented through a developer, including harvest and construction of infrastructure. To date, about 10 million ha of forests (one third of the total) have been acquired for commercial logging through FMA, of which 6.9 million are considered suitable for sustained yield management. The acquired areas have normally been allocated to foreign developers with financial capabilities. Currently there are 32 logging concession projects covering 195 acquired areas, with an extent of about 5.6 million ha distributed over 15 provinces of the country.

Prior to the promulgation of the 1991 National Forest Policy and Legislation, timber rights were acquired by a process referred to as Timber Rights Purchase (TRP). The rights acquired under this system were only for the harvesting of merchantable timber and did not transfer to the State or concessionaires the rights to manage the forest. The New Forest Policy aims, among other things, to promote management and protection of the forest resources as a renewable natural asset. The silvicultural system prescribed is selective logging, involving removal of mature and over-mature trees, facilitating the remaining crop to grow naturally to maturity. Even though the pre-FMA (prior to 1991) system was also qualified as selective harvesting, the cutting of all trees above the prescribed limit over the area was completed within a period of 10 to 20 years, thus consuming the resource at a faster and unsustainable rate. From 1991/92 onwards, all new forestry operations have a cutting cycle of 35 years. But concession logging is in fact undertaken as a one-time activity, without any legal provision in the FMA system for continued production and post-logging silviculture. The World Bank and donor agencies have expressed concerns at the manner in which timber projects are awarded and controlled in PNG.

The landowner company (LOC) concept was developed as part of the 1979 National Forest Policy in order to increase national participation in the forestry sector. Many of the LOCs have been issued with TPs, to develop their own resources. Whilst the concept is good, the practical reality has been different. Most of the LOCs have been plagued by mismanagement, corruption, and in-fighting between different landowner factions.

In the existing landowner situation in PNG, a rational forest revenue system is important to ensure that local land owners receive their due share of 'rent'. In 1993, the National Forestry Development Guidelines had proposed sharing of rent (surplus of product price after adjusting logging cost, operators minimum profit and minimum stumpage) between PNG Government (85%) and the operator (15%). The Government's share, in turn, is to be apportioned with landowners after adjusting for development and administration expenses. The proposal has not been adequately implemented. There have been inequities in the sharing the resource rent (Filer and Sekhran, 1998) as the benefits from the forestry operations have generally not filtered to the genuine landowners, nor has income been saved or invested to ensure long-term development (PNGFA/NFS, 2002).

PNG has established a number of regulatory instruments to support SFM, even though there are gaps in implementation, particularly in relation to long term development of forest resources. Key Standards for Selection Logging in PNG provides for monitoring of timber operations at every stage, and independent surveillance of log movements and inspection of log shipments to control malpractices and transfer pricing. This function has been carried out for about the last 10 years, on contract, by Societe Generale de Surveillance of Switzerland, facilitated through an EU-supported project, and with beneficial impact.

Available analysis/studies suggest that there are gaps between policies and public interest, as well as between policies and practice. Two basic deficiencies to be addressed on a high priority basis are: (i) lack of adequate education and human resource capability of the land-owning communities and (ii) lack of clear definition and delineation of rights/ownerships over the natural resources, including lack of land use planning, land survey and land settlement. In addition, many factors conspire against "development" in PNG: high operating costs; dearth of skills; small domestic market; law and order problems; cumbersome government procedures; inter-tribal conflicts and rivalries; greed and corruption among community leaders.

## 4. THE FOUR CASES

The four cases chosen for analysis here cover a range of situations and activities, and different facets, needs and foci of SFM. They reflect the different impacts of stakeholder suspicions, stakeholder politics, as well as development of synergies. The Out-grower Farms of Clonal Trees program of Indian Tobacco Company (ITC) Paperboard and Specialty Paper Division (ITC-PSPD) in India is a case of partnership between a corporation and farmers to establish plantations of genetically improved and high yielding varieties of pulpwood species. PT. Sari Bumi Kusuma in Indonesia is a well-managed private logging concession, which has been in operation since 1978 and has recently been renewed for a further period of 70 years. Matang Mangrove Forest of Perak State, Malaysia is the case of integrated mangrove management for wood and non-wood products, with a 100-year history of SFM. Vanimo Forest Products Ltd in PNG is a 20-year non-renewable concession in a customarily owned forest, due to expire in 2010. These cases and the surrounding conditions in the respective countries bring out a range of interesting issues and valuable lessons regarding SFM.

*4.1 The Out-grower Farms of Clonal Trees of ITC – PSPD, India*

ITC is one of India's largest multi-business private corporations with business segments covering fast moving consumer goods, hotels, agri-business, information technology and paperboards, specialty papers and packaging. ITC-PSPD operates an integrated pulp and paper mill located at Sarapaka, near Bhadrachalam, in the Khammam district of Andhra Pradesh state. The mill, established in 1979, currently has an installed capacity of 65,000 metric tons of pulp and 182,500 metric tons of paper and paperboard per year. The present requirement of cellulose raw material of ITC-PSPD of about 400,000 tonnes per day (tpa) will grow to 800,000 tpa as production capacity and product range increase. An interesting feature of ITC-PSPD is the company's sponsorship and support of out-grower (small farmer) production of pulpwood.

When the company established the paperboard mill in Sarapaka, it had the commitment of the Government of Andhra Pradesh (GOAP) to supply the major part of the required raw material (bamboo and hardwoods) from government forests. With continuing deforestation and forest degradation, and consequent forest policy changes restricting wood removal from natural forest, the raw material source on which the company has relied was closed off and the Andhra Pradesh Forest Department (APFD) could not maintain its commitment beyond 1986. After 1988, with a view to the conservation of natural forests and in keeping with the National Forest Policy, "clear felling" of forests was terminated. By the 1990s, pulp and paper industrial units in the State obtained 78 to 82% of their raw material requirements from non-government sources. The dire raw material supply situation had been building for some time. To save their investment, the company needed alternative sustainable raw material sources. It realized that raising captive plantations would be difficult due to land ceiling laws and restrictions on leasing forest lands. One interesting option was to promote tree planting by farmers, using genetically improved and high yielding varieties of pulpwood species, mainly clonal plantations of *Eucalypts, which* can be managed on a 3-4 year cutting cycle and can stand 4 coppice cuttings before being replanted. The company started distributing free *Eucalyptus* seedlings in 1982 but decided to discontinue it in 1986; farmer response was below expectations and they did not take adequate care of the seedlings supplied free of cost to them (Kulkarni, 2002).

Between 1987 and 1995 the company ran a bank loan scheme to promote farm forestry, supported by the National Bank for Agriculture and Rural Development (NABARD). During this period 7,441 ha of tree plantations were raised on the holdings of 6,185 farmers in 1,138 villages, on the basis of a comprehensive package consisting of quality planting stock, technical extension services and a buy-back guarantee at a minimum support price or the market price, whichever was higher. Again, however, the achievement was far less than the target set. Productivity of these plantations was too low (6 to 10 cum/ha/yr) to be acceptable to farmers as a land use option. There were logistical problems, especially in getting the farmer's loans sanctioned. Worst of all, the program failed to achieve its primary objective of getting raw material for the mill, because after availing themselves of the loan, most farmers sold their produce elsewhere, often harvesting their plantations earlier than the rotation period stipulated. Thus, ITC-PSPD decided to discontinue the scheme after 1995 (Saigal *et al,* 2002).

The need for research to improve the quality of pulpwood plantations and their productivity had been recognized by the company, which in 1989 launched an R&D and tree improvement programme. Based on performance of individual clones in the field trials, promising, fast growing and disease-resistant clones of *Eucalyptus tereticornis* and *Eucalyptus camaldulensis* were identified. Planting stock of the most promising "Bhadrachalam clones" was released to farmers from 1992 onwards. The company is currently supplying (selling) eleven different *Eucalyptus* clones (called Bhadrachalam clones) on a commercial basis to farmers along with continuing extension services and offers a buy-back guarantee at an agreed price. Up to 1999-2000, the company had sold over 7.2 million clonal seedlings of *Eucalyptus.* The trees are disease resistant and self-pruning, with survival rate is as high as 95%. An analysis of costs and returns of a number of tree farms of varying quality indicated that net annual gains ranged from Rs. 18,000 to Rs. 49,000 (equivalent to US$ 400 to 1,100) per ha, giving an IRR of 14 to 35% depending on site quality and management inputs, during the first cutting cycle 3 of years. This return far exceeds that of alternative land uses. Profits increase in the subsequent coppice cuttings, since the cost involved in maintaining a coppice crop is lower. Further, since the tree farms are raised under a system of agro-forestry, additional income will be earned from the harvest of the associated agricultural crop (Rao, 2004). The clonal tree farm programme is understandably popular with the farmers. The company is working on genetic improvement of *Casuarina,* and plans to produce and sell improved clonal seedlings of this variety in the near future. ITC-PSPD also realizes the need for continuing research to enhance the various aspects of plantation management; it is carrying out trials on silvicultural/agro-forestry practices (e.g. spacing in planting rows, type of inter-crop). It is also in process of implementing a core area development programme, to intensively promote tree farms in areas falling with in a radius of 150 km surrounding the mill, offering additional incentives to farmers. This will help to reduce the cost of transportation of the bulky raw material (Rao, 2004).

At the end of 2002, the number of farmers participating in the clonal plantation program of ITC-PSPD was 6,372 (involving an area of 10,200 ha) and is steadily increasing. Currently, about 40% of the pulpwood requirement of the company is being met from clonal tree farms. The company expects that its entire pulpwood requirement can be supplied from this source by 2007. The average size of clonal "tree farms" (i.e. plots) is about 1.6 ha; 50% of the plots fall under the size of 1 to 5 ha; 42% have less than one ha under clonal pulpwood plantation. Only the remaining 8% have clonal tree plots of over 5 ha. According to information available, small farmers participating in the clonal tree farm programme are utilizing 25 to 50% of their total farmlands for raising pulpwood crops. The situation is typical of the raw material catchment of ITC-PSPD.

The tree farmers of Andhra Pradesh have formed an Association to address common problems and concerns. Based on "pressure" from the Association, the GOAP, in June 1999, exempted *Eucalyptus spp, Leucaena leucocephala, Casuarina* and some others from the purview of the Andhra Pradesh Forest Produce Transit Rules 1970, to facilitate procurement and transport of pulpwood, as an encouragement to the tree growers. Other than this policy incentive, the farmers are not provided with any assistance for their tree farming activities.

The clonal plantations so far established have the potential to sequester 0.5 million tons of carbon, thus helping in the reduction of green house gases. This 0.5 million tons of carbon is worth approximately US$ 1.5 million, at the rate of US$ 3 per metric ton, in terms of carbon credit. This benefit, currently an externality, amounts to about 10% of the direct benefits obtainable from pulpwood production. The clonal tree farm programme is being proposed for support under the clean development mechanism, which will help the farming community to grow more plantations and thereby further carbon sequestration (Rao, 2004). The mutually beneficial productive linkages between PSPD and the farming community are expected to result in vital multiplier impacts on the larger economy of the region (Chandrasekharan, 2003).

*4.2 PT. Sari Bumi Kusuma, Indonesia*

PT. Sari Bumi Kusuma (PT. SBK) is a private logging concession belonging to the Alas Kusuma Group, located somewhat remotely in Central Kalimantan province of Indonesia, and essentially comprised of tropical rain forest covering an area of 208,300 ha.   Logs produced are processed into sawn timber, mouldings and commercial plywood at the company mills in Pontianak, West Kalimantan. Company operations are marked by good quality of planning and implementation.

PT. SBK started its operation under the original concession agreement in 1978. The original 20-year period expired in 1998 and the concession was renewed for a period of 70 years (1998-2068) under a new agreement. As per the new 70-year concession, PT. SBK has a mandate from the MOF to manage the forest under the Tebang Pilih Tanam Jalur or selective cutting and strip planting (TPTJ) system. This system, which had become mandatory for renewed concessions by the time the new license was issued, requires the company to practice enrichment planting in strips. The 70 year period of the concession will involve a first cycle of felling and planting for 35 years; PT. SBK is then granted an additional 35 years to harvest those trees planted during the first cycle.

To ensure sustainability, concession operations are based on an elaborate system of management plans – 35 year Perspective Plan; Environmental Management Plan; 5 year Working Plan; and one year Operations Plan. Forest management planning is supported by a rural development diagnostic study. The 35-year perspective plan provides broad goals and targets and essentially serves as a general guideline for development. The one-year plan or Rencana Kerja Tahunan (RKT) is the basis for action. The contents of RKT include, among other things: accomplishments during the previous one year plan, compared to targets set and detailed operational plans for the current year. Cutting area and yield are prescribed, based on growing stock inventory and assessment of crop condition in terms of distribution of diameter classes and species. The volume of felling is determined by the forest potential (based on pre-harvesting inventory), which could change from year to year.

During the initial concession period, the company followed Tebang Pilih Tanam Indonesia (TPTI) or Indonesian Selective Cutting and Planting System with a cutting diameter limit of 60 cm for limited production and 50 cm for normal production forest. Under TPTI, if there is sufficient regeneration in logged over area, there is no need for planting. Studies by PT SBK have shown that under TPTI the space available for planting is limited and only 2.3% of the planted seedlings receive conditions suitable for good growth, due to congestion. The idea of clearing strips for planting was originated to overcome this deficiency of TPTI, so that planted seedlings can be nurtured properly (Suparna, 2004). Since 1998, PT. SBK has been implementing TPTI in its virgin forests (103,262 ha with an effective area of 81,606 ha) and TPTJ in its logged-over areas (76,235 ha, with an effective area of 67,333 ha) as mandated by government. TPTJ is a modification of TPTI, where nursery raised seedlings of valuable species are planted in line; it enables progressive improvement in the productivity of the forest involved. It is expected that log yield per ha can potentially be raised to 300 cum (compared to the current yield of about 50 cum), because of the additional yield available from the trees planted in strips. TPTJ as a system and technique integrates the beneficial aspects of selection system (i.e., conservation) and the system of clear felling and planting (i.e., high productivity and yield). PT. SBK has made a substantial contribution to the development of TPTJ through experimentation within the concession area (Suprana, 2004).

In both the cases (TPTI and TPTJ), forest and timber inventories are undertaken at intensities of 1% at the time of the preparation of the long-term plan, 5% for the five year plan and 100% for the one year plan. As a pre-harvest operation, all harvestable trees are marked and measured to estimate the yield. Nucleus trees which will serve as seed sources for future regeneration are also marked to ensure that such trees are not damaged during felling. All logging is carried out mechanically. Directional felling is insisted on to reduce damage. The company is introducing reduced impact logging by stages. Post-harvest activities prescribed (both for TPTI and TPTJ), among others, include: protecting the forest from fire, illegal activities, pests and diseases; measures to conserve bio-diversity;

decommissioning of skid trails, log landings, temporary roads and camps, and planting up such areas to avoid soil erosion; and boundary maintenance. In respect of TPTJ, trees planted along strips are regularly tended and nurtured. Since TPTJ involves planting and tending of trees in strips it adds to the management cost per unit area. However, the initial estimates of the company indicate that the cost per unit volume of wood produced will be lower, giving a profitability (competitive) advantage (Chandrasekharan, 2003).

As a step towards obtaining timber (and forest management) certification, the company has signed an agreement with the Smartwood Program of the Rain Forest Alliance, to conduct a preliminary scoping. While appreciating the intrinsic merits of certification, the company is of the view that some of the C&I will be very difficult to implement in view of the present state of affairs in forestry – such as boundary conflicts, illegal activities, availability of illegal logs at low price, and misinterpretations of law. In its view there is need to further rationalize and simplify the principles and steps involved in certification.

The socio-economic contributions of PT. SBK are substantial. Its annual average contribution towards forest-related tax revenue over 1995-2000 was Rps 35.7 billion (US$ 4.5 million). The company's contribution to local employment and income is equally substantial. Some 64% of all company workers are locally recruited. While the concession agreement does not permit management of non-wood products by the company, it supports the non-wood activities of the local community for subsistence and family income. The company has a long tradition of supporting local communities to improve their livelihood and living conditions. The long-term vision of the company of a harmonious forestry enterprise, and its commitment to support local community development makes it a unique operation. With the support and collaboration of the local community, the company has been able to keep the concession area safe from illegal logging. On average the company spends Rps 2.97 billion (US$ 375,000) per year on social welfare and development, thus sharing a portion of its profits with the local community. The successful social development practice of PT.SBK underlines the need and feasibility of effectively correcting this lapse wherever it exists. The company also undertakes environmental conservation and research activities and collaborates with universities and national research organizations. The staff development programme of the company has paid good dividends through improved performance.

A stickler to the conditions set forth for SFM, and to all the rules and regulations in that regard, PT. SBK plans its concession operations meticulously, and implements them faithfully. The company has been able to practice SFM in spite of a number of difficult hurdles and has received consistently good reviews and ratings for its good work (Chandrasekharan, 2003).

### 4.3 Perak State Forestry Department Operations in Matang Mangrove Forest

Perak State Forestry Department's operation in Matang mangrove forest, covering a total area of 40,151ha and productive area of 32,746 ha, is an integrated activity to produce wood and non-wood forest products on a sustainable basis. Its main product is charcoal. Matang mangroves came under forest "reservation" in 1902. During the early years, these forests were the main source of fuel wood for the Malayan Railways and for the tin mining industry. As the demand for mangrove fuel wood dwindled due to competition from cheaper inland supply, by 1930 some local entrepreneurs had taken to making charcoal in kilns. The need to manage the mangrove forest was realized immediately after the area came under reservation. Management practices from then onward underwent continual change (Noakes 1952). Trials were made with many systems to find the one that would require minimal planting, or no planting at all, to regenerate the felled areas. Since 1950, the Matang mangroves have been managed under prescriptions of a comprehensive working plan, which is revised regularly every 10 years. The primary objective is the sustainable production of quality greenwood of *Rhizophoraceae*, for charcoal manufacturing and poles, on a sustained yield basis, with provision for conservation and protection of the environment.

The silvicultural system currently followed in Matang mangrove management is clear felling in periodic blocks, followed by natural regeneration, supplemented by artificial planting as necessary. The standard rotation has been fixed at 30 years. Estimation of yield has been carried out on a 10 year periodic basis, and conducted only for the periodic block which will come up for final felling in the 10 year period of the working plan. The estimated average yield per hectare of mature crop is about 175 metric tons. All clear felled areas are systematically rehabilitated by planting and occasional weeding. There are two thinnings, at the ages of 15 years and 20 years, before the new crop undergoes final felling at the age 30 years.

The charcoal and pole industries based on the Matang mangroves are very strictly regulated. The number of charcoal kilns permissible is decided every 10 years, coinciding with the 10-yearly revision of the working plan, and made proportionate to the amount of forest resource available for that period. Allocation of areas (for charcoal burning) is, accordingly, made for a period of 10 years. Allocated areas are released for harvesting to the contractors on a year-by-year basis. During the period 1990-1999, the productive area of 7,980 ha (annual average of 798 ha) was allocated to 75 charcoal contractors who operated 336 kilns (approximately 2.3 ha to feed a kiln). Similarly, there were 70 pole contractors approved for the period 1990-1999, for an annual thinning area of 2,136 ha, half of which corresponds to first thinning (15 years) and half to second thinning (20 years). Additionally, non-wood products, such as fronds of *Nypa fruticans* and *Acrostichum aureum* are harvested, on a limited scale. The potential of collecting raw distillate from the charcoal kilns, for refining into pyroligneous acid is currently being tested by a few kiln owners. Matang mangroves support a large fishery industry, including capture fisheries and aquaculture. Within Matang mangroves there are 34 village settlements. In addition, there are several traditional fishing villages along the upstream banks of mangrove tidal rivers. These communities are engaged in various mangrove-related economic activities. Eco-tourism is also emerging as an economic activity in the Matang mangroves.

Sustainable utilization of Matang mangroves has been a profitable operation for both private and public participants. Government revenue collected from timber extraction alone has recently averaged RM 1.56 million (US 425,300) per annum. As against this, government expenditure on administration, forest development, forest operations, conservation and protection activities over the whole of Matang mangroves was RM 0.93 million per annum. Leaving an average net public sector net revenue of RM 0.63 million (US$ 171,500) per annum.

The direct tangible economic value of annual harvest of produce has been estimated at RM 155,474,500 (US $42 million) (RM 25,350,100 from wood production RM 130,124,400 from fishery and aquaculture). The fishery and aquaculture value, to a large extent, is attributable to the presence of the mangroves; and the sustainable fishery in the area is a function of sustainable mangrove management.

Matang mangroves support a viable population of a variety of fauna. The diversity of fauna and flora was never directly threatened by the intensive management of the mangrove resources for the sustainable production of fuel wood and poles. Other than the permanent stand of non-productive forest which has hardly been exploited, the productive forest is being managed with due consideration for the conservation of the environment. By managing the mangroves on a 30 year rotation, in any one year approximately 1,100 ha or 2.7 % of the total area of Matang are clear felled. Clear felling of the annual coupe (harvest area) is not carried out in one single cut, but progressively over a period of 12 months, at approximately 100 ha a month. Subsequently, the clear felled areas are ensured total regeneration either through natural or artificial means or a combination of both. As an additional precautionary measure, the annual coupes are further divided into smaller sub-coupes, which are spread out over the whole Matang mangroves. In so doing, disturbance to any particular habitat is localised and the wildlife has ample opportunities to find safety in the adjoining forested areas.

In the past, a constraint, which reduced the capacity of the Forestry Department to fully restock the clear felled areas, was the damage to the plants caused by crabs, monkeys and deep flooding. These problems have been overcome through the introduction of a locally developed potted seedling technique, in 1986. The major constraints to be addressed now are in the area of product development and utilization, including the lack of research and skilled personnel in the area of non-wood products.

With a history of about 100-years of sustainable management of the fragile mangrove ecosystem (which has disappeared in several parts of the world due to encroachments of prawn culture and illegal logging), the Matang experience is worthy of emulation. The contributing factors in this regard include: a strong policy and legal framework; high quality of planning, implementation, supervision and monitoring; strong and continuing political commitment; regular ten yearly revision of working plans; clear objectives of management; long-term security of concession tenure; support for local educational and research organizations; adequate staffing; and above all the goodwill of the local community. The Matang experience particularly underlines the oft-repeated lesson: SFM is achieved not by *ex situ* discussions and elegance of concepts, but by *in situ* actions (Chandrasekharan, 2003).

*4.4 Vanimo Forest Products Ltd., Papua New Guinea*

Vanimo Forest Products Ltd. (VFP) in the Sandaun Province (a subsidiary of WTK Realty Pty Ltd of Sarawak, Malaysia) is a private concession operation undertaken on contract with the Government, within the Timber Rights Purchase (TRP) System. The tropical rain forest covered in the contract, having an area of 287,240 ha and commercially productive area of 190,160 ha, is customarily owned by the tribal communities. The timber license of VFP covers a period of 20 years, 1990-2010.

Logs produced by VFP are partly used as input into sawmill in Vanimo, and partly for export as raw logs. During 2001, the company's sawmill processed 60,369 cum of logs against 134,149 cum of logs exported. Logging and log/timber export activities are monitored by the Societie Generale de Surveillance of Switzerland, on contract with the Government of PNG.

The PNG Forest Authority provides guidelines and regulations regarding standards and procedures for timber management. As stipulated in the guidelines, the company follows a sophisticated system of forest management and harvest planning. Yield is regulated by a combination of area and volume specified in the annual allowable cut (AAC). Only mature trees of commercial species are harvested leaving a healthy residual stock for the future. Except for major roads, which will add to the rural infrastructure and benefit the communities, all temporary developments such as skid trails and log landings are decommissioned, such that after harvest the forest will soon be returned to its natural state (Tiong, 2004). An Environmental Plan is required under the Environmental Protection Law, which stipulates that logging is not to be done in areas of over 30% slope; buffers of natural vegetation should be maintained in 100m width around villages, 50m width on river margins, 10m width on stream and creek margins. Post-logging operations include those required for reporting completion of logging activities in order to obtain clearance from PNG-FA and authorization to start work on the next set-up, and silvicultural operations to support SFM, although this second type of operations currently are not a requirement.

VFP activities generate considerable government revenues and income/welfare benefits to the land-owning local community. It is the prime revenue source of the Sandaun provincial government. The annual average of charges and levies remitted by VFP during 1999-2000 amounted to K 13.1 million (US$ 4.5 million). The company regularly employs 946 persons, with peak season employment going over 1,000. The number of nationals on the regular pay roll is 889. The company makes all efforts to employ as many local community members as possible (even though they often lack the self-discipline required for work under strenuous conditions), in the logging and milling operations. Additionally, the company invests in developing social infrastructure such as schools, playgrounds, health centers, water supply, and communication systems, as well as in land and agricultural development. The company claims to accommodate, to the extent possible, the demands (including fabricated compensation claims) of the community, which go beyond the agreed conditions, in the interest of maintaining social harmony and friendly relationships.

The customary ownership of VTA is shared by 1976 households, located in 55 villages and clustered into 26 clan groups. The landowners are encouraged to

334    CHERUKAT CHANDRASEKHARAN

participate in the preparation of annual logging/working plans. But poverty and malnutrition are major problems among the villagers of VTA, surprising in view of the wealth of natural resources at their disposal (24 ha per person or about 145 ha per family). Flaws in the policies, weak implementation by the government, and the community attitude towards the cash economy and related work ethics are cited as contributing factors. The part of the timber revenues which is supposed to be utilized for "development" by the government, often does not benefit the landowners. The "representatives" may not share the timber royalty accruing to the landowners equitably among the members; and the money so distributed is not usually invested in land development but rather wastefully and quickly spent on "consumption".

The company has been strictly following all regulations relating to timber operations and trade, paying all the levies and premiums due, and meeting other obligations such as infrastructure development, establishment of processing unit(s), provision of social amenities etc. VFP had voluntarily established 300 ha of natural regeneration area, 5 permanent sample plots and several mother trees for promoting natural regeneration, but later discontinued them as they are not required under current TP regulations which stipulate only logging plans but not silvicultural plans for inducing natural regeneration, protecting the crop from fire and other injurious agents, scientifically assessing growth and yield, etc. The company management is of the view that it will be possible to obtain an assured harvest of 25-30 cum/ha in the next cut through appropriate silvicultural management of the residual stock.

Overall the company has performed creditably in terms of its strict adherence to the conditions of the TP; its capacity to address the constraints in implementing a sophisticated harvesting operation, in a customarily–owned forest, in a remote location; its support for social capital formation and HRD, to the benefit of the local community; its contribution to government revenue, local employment and income; and finally, its respect for local culture and sensitivities. However, in the existing land tenure situation (for more on this see the earlier section on PNG), and considering the continuous changes in government policies and regulations (relating to species, sizes, AAC, charges, technical prescriptions, clearances, continuation of TPs, and so on), the primary concern of the entrepreneurs will be to safeguard their investment. They will find it difficult to make increased investment, if they cannot conduct their legitimate business and forestry operations in a peaceful and unhindered manner. Inadequate knowledge on the part of some NGOs about the nature of logging under selection system, and the special conditions existing in PNG, has led to the VFP being blamed for violations of TP conditions. There is also very little awareness or comprehension, at the field and landowner's level about international initiatives such as C&I for SFM, eco-labelling and certification. The government is not properly playing its role of supporting the development of the customary landowners and the village community (through extension, education, infrastructure development and so on, appropriately utilizing the revenue from timber resources) in a transparent and accountable manner (Filer & Sekhran, 1998).

## 5. SCENARIOS

Both the positive synergies and the difficulties found in the four cases appear to be influenced by differences in the behavioral environment, which are related to the relative scarcity of forest resource, effectiveness of institutions, ownership pattern and investor attitudes. In the case of the Out-grower Farms of Clonal Trees of ITC-PSPD, India, the synergy between the company and the farmers of the locality was driven by the dire situation of raw material scarcity affecting the company, which provided an avenue leading to higher income for the farmers. In respect of PT. Sari Bumi Kusuma of Indonesia, it is the long standing social commitment of Alas Kusuma Group which has about 50 years of beneficial relationship with the Indonesian forestry sector, that facilitated synergy with the community living in and around the concession area. The driving force of synergy in the case of the management operations in Matang mangrove forest in Perak State, Malaysia has been the excellent public forest administration and governance, which involved all stake holders (charcoal operators, pole contractors, local fishermen and villagers) in the resource management process. In respect of Vanimo Forest Products Ltd. of PNG, it is the investor's principled attitude and flexibility in its social relationships with resource owners, which has helped to promote some form of synergy. These and similar potential synergies are suggestive of possible future scenarios.

Scenarios can be influenced by trends and vision. While the current situation and trends often provide a depressing prospect, an achievable vision calls for vital imagination. The global vision of the World Bank/WWF Alliance was that by 2050, about 80% of the global forests will be managed for preservation of bio-diversity, environmental services and non-commercial uses; commercial scale production forestry will be concentrated in about 20% of the global forest areas accounted for by intensively managed secondary forests and plantations. Other positive scenarios envisioned (and partly reflected in the cases discussed) include: a shift of forestry emphasis to non-wood forest products and services; intensive natural forest management through continuous/intensive enrichment; industrial tree plantation development, as captive to processing units; and community and farm based sustainable forest management. The luxury of one-time logging followed by a long period of recovery is now a highly restricted scenario. In the absence of a stable framework guiding the course of events, these visions are unfortunately likely to be vitiated by a large number of problems affecting the sector.

### 5.1 Issues and Challenges to SFM

Issues are root causes of problems, about which there will often be differences in perception. Symptoms of the presence of such issues include inconsistencies, contradictions, unrealized objectives, lack of focus, and the prominence given to processes and peripheral aspects. Many of the issues, which have been highlighted in the foregoing discussion fall into one of the following categories.

*Governance related issues* include a lack of political/policy commitment to the cause of SFM; prevalence of illegal logging and smuggling; lack of urgency to address deforestation and forest degradation; inadequate accountability and transparency; lack of enforcement capability; lack of forest demarcation; inadequate emphasis on productivity; inadequate civil society participation; and spurious stake-holding and the need to identify genuine stakeholders to be involved in SFM.

*Institutional issues* often underlie problems of governance. They include: inadequate planning capability; inadequate policies and strategies; weaknesses of enforcement agencies; corruption in public administration; gaps and flaws in rules and regulation; lack of reliable information; lack of adequate and independence monitoring; differences in stakeholder perceptions and preferences; the absence of relevant research; and the lack of a forest resource accounting system.

*Science and technology related issues* are often exacerbated by institutional weaknesses. These include: inadequate funds, facilities and human resources for research and development; inadequate dissemination of technological packages; a greater professional preoccupation with conceptual aspects and with "re-inventing the wheel than with practical problems"; lack of studies on forest resource use and ecosystem response; inadequate research on non-wood products; lack of client participation in research planning.

*Issues related to forest utilization and trade* directly and indirectly impact on development, and include, among others: wastages in harvesting and processing; frequency and volume of low value uses (e.g. fuel wood, unprocessed non-wood products); lack of emphasis on competitive advantage (efficiency), while reaping the comparative (nature provided) advantage of natural resource endowment; inadequacy of studies on controlling of timber (forest) sales as a means of stabilizing price and consumption; need for a proper mix of market forces and pricing policies; continuing inability to use certification as a market-based instrument and the contradiction implicit in promoting RIL and certification.

*Environmental issues* are linked to questions of wise use vs. non-use or misuse and cover, *inter alia*: the poor condition of protected areas; wildlife and human conflicts; endangered species and ecosystems; loss of bio-diversity; eco-piracy; degraded ecosystems; poor quality of forests and plantations; watershed deterioration; dead rivers and sterile soils; wise use and non-use on paper while abuse occurs in practice; a need for waste re-cycling; appropriate valuing of environmental services, carbon sequestration, emissions trading, clean development mechanisms and the use of carbon sinks.

*Socio-economic issues* often assume serious proportions as a result of inadequate stakeholder participation and inadequate commitment on the part of the leadership. Prominent among them are: poverty of resource owners; lack of adequate social concern on the part of individuals and government institutions; changes in resource landscapes; lack of rent-capture capability; lack of benefit/cost balancing; need for policy and behavioral research; need to balance ecological and economic concerns; the role of forestry in poverty alleviation; lack of downstream processing of forest products; seasonal and irregular nature of forestry employment; net disinvestment resulting from inadequate investment and excessive deforestation; and customary land ownership.

*General and over-arching issues* include: global and local pressures on forest management from various stakeholders and interest groups; changes in behavioral norms and creative dissatisfaction; poverty and macro-economic instability; inadequate understanding and lack of conviction about the real and fundamental role of forestry on the part of policy makers and decision takers; lack of long range vision and vital imagination about forestry; need for enhancing forestry science and profession; and need to define genuine stake-holding.

The importance of forestry to the national economy has significantly waned in several countries, particularly in those where depletion and degradation of forests have been serious. Most such countries have lacked powerful movements pushing for rehabilitation of forests and forestry. In view of this situation, the question has been raised in several quarters of whether forestry science and forestry as a profession are fated to be of transitory historical significance.

*5.2 Lessons Learned*

The country experiences, analyses of cases and the issues arising there from highlight some following lessons.

i.   Technology or theory-based plans by themselves, in the absence of good governance and a modicum of socio-economic equity, cannot guarantee SFM.

ii.  Stakeholder synergies for SFM often have to pass through painful stages due to conflicts and suspicions; once synergies can be developed, however, SFM will in most cases be possible.

iii. In developing synergies for SFM, it is necessary to adequately/appropriately balance the multiple functions of forest and to address the issues linked to the "people factor".

iv.  In the absence of SFM with adequate emphasis on its capability to provide externalities, forestry as a sector will be relegated to the status of a residual land use; SFM may easily remain an illusion, rather than a productive vision.

v.   The rare cases of well functioning FMUs, favoured by special conditions or circumstances, cannot alone do much to enhance the socio-economic situation of local communities, since the situation is the result of much deeper problems which would call for well thought out political interventions at the national level.

vi.  In the ultimate analysis SFM is not achieved by ex-situ declarations but by in-situ actions.

## 6. CONCLUSIONS

The major constraint for SFM is not lack of technology, but the institutional factors, which militate against the application of the best available technology. These institutional factors appear in the form of short-term perceptions and time preferences of the stakeholders, inadequate concern on the part of policy-makers to adopt and follow an appropriate forest management system, and inability of the public forest administration to enforce policy and regulations.

The problems of today in the forestry sector are mostly the result of the shortsighted solutions applied to yesterday's problems. To get out of this vicious cycle, reforms need to be based on a long-range vision. For this to happen, the prime factor is an adequate system of governance, to properly guide forestry development and to create competitive advantages in the sector. Sustainable growth in the forestry sector will ultimately depend on developing competitive advantages through increasing efficiency (and improved technology), rather than extracting effortless benefits on the basis of nature-provided comparative advantages, which can soon be exhausted.

Conceptual development or the existence of C&I for SFM or elegantly prepared plans will not by themselves ensure improved and sustained forest management. Focusing on such "advances" belies the fact that effective SFM involves an enormous task requiring money, manpower, materials and management. What it requires is absolute commitment, supported by appropriate institutions, science and technology, human resources, co-ordination mechanism and targeted funding.

## REFERENCES

Chandrasekharan, C. (1996). Cost incentives and impediments for implementing sustainable forest management. UNDP/Republic of South Africa Workshop on Financial Mechanisms and Sources of Finance for Forestry. Pretoria, South Africa. 4-7, June 1996.

Chandrasekharan, C. (2003). *Sustainable Forest Management – A mirage? Issues arising from private sector experiences.* Regional Report for the Asia-Pacific Region. Joint Malaysia – ITTO Project PD 48/99 (Rev 1).

FAO. (1995). Forest Resources Assessment 1990 – Global Synthesis. FAO Forestry Paper, 124, Rome.

FAO. (2001). Global Forest Resources Assessment 2000. Main Report FAO Forestry Paper 140. Rome.

FAO. (2002). *Yearbook of forest products 2000.* Rome.

FAO/MoF. (1990). *Situation and outlook of the forestry sector in Indonesia.* UTF/INS/065/INS Forestry Studies. Technical Report No. 1. (Indonesia) 4 Vols.

Filer, C., & Nikhil, S. (1998). *Loggers, donors and resource owners: Papua New Guinea country study.* Policy that Works for Forests and People Series No. 2.

GOI/MOEF. (1999). *National Forestry Action Programme – India, 2 vols.* New Delhi.

ITTO. (1992). *ITTO guidelines for the sustainable management of natural Tropical forests.* Yokohama.

ITTO. (1998). *Criteria and indicators for sustainable management of natural Tropical forests.* ITTO Policy Development Series No. 7, Yokohama.

ITTO. (2001). *Achieving Sustainable Forest Management in Indonesia.* Report of the ITTO Technical Mission to Indonesia. Yokohama.

Kishore, N.M., & Constantino, L.F. (1994). Sustainable forestry. Can it compete? *Finance and Development*, 31(4), 36-39.

Kulkarni, H.D. (2002). "Bhadrachalam" clone of Eucalyptus – An achievement of ITC. Paper presented at the International Workshop on Forest Science and Forest Policy in the Asia Pacific Region. Building Bridges to a Sustainable Future. IUFRO/USDA-FS/MSSRF, Chennai, 16-19, July 2002.

MPI. (2002). Malaysian country report on National Forest Policy review. Presented at the 19th Session of the Asia-Pacific Forestry commission, 28-30 August 2002. Ulaanbaatar, Mongola.

Noakes, D.S.P. (1952). *A working plan for the Matang Mangrove Forest Reserves, Perak.* Forestry Department, Federation of Malaya, 172 pp.

Panayotou, T. (1995). Matrix of financial instruments and policy options: A new approach to financing sustainable development. Paper for the Second Group Meeting on Financial Issues of Agenda 21. 15-17 February 1995, New York.

PNG-FA /NFS. (2002). National Forest Policy review: Country report presented at the 19th session of the Asia Pacific Forestry Commission, 26-30 August 2002, Ulaanbaatar, Mongolia.

Power, A.P. (2002). Land mobilization program in PNG. Paper for the Eco-forestry Policy Workshop, Lae-Rabaul-Port Moresby, Nov-Dec.2002.

Rao, S.N. (2004). Clonal pulpwood tree farms change the rural landscape in Andhra Pradesh – A case study. Presented at the International Conference on Sustainable Management of Tropical Forests – Private Sector Experiences. 13-15 April 2004, Kuala Lumpur, Malaysia.

Saigal, S., Arora, H., & Rizvi, S.S. (2002). *The new foresters: The role of private sector enterprise in the Indian forestry sector.* A research report prepared by Eco Tech Services (India) Pvt Ltd., New Delhi in collaboration with IIED, London.

Suparna. N. (2004). Local community support and technological refinements are crucial for sustainable forest management – A case study. Presented at the International Conference on Sustainable Management of Tropical Forests – Private Sector Experiences. 13-15 April 2004. Kuala Lumpur, Malaysia.

Tiong, P. (2004). Timber production promotes regional and community development in remote regions of Papua New Guinea – A case study. Presented at the International Conference on Sustainable Management of Tropical Forests – Private Sector Experiences. 13-15 April 2004. Kuala Lumpur, Malaysia.

UN– CSD. (1996). Report of the Secretary General. E/CN17/IPF/1996/5 to the Second Session of CSD Ad-Hoc Inter-Governmental Panel on Forests (IPF), Geneva, 11-22, March 1996.

UNEP. (1999). GEO 2000. Global Environmental Outlook. *UNEP's Millennium Report on Environment*, Nairobi.

WRI. (1999). *World Resources.* 1998-1999. Washington, DC.

# CHAPTER 15

# INSTITUTIONS, SUSTAINABLE FOREST MANAGEMENT, AND POST-NEWTONIAN ECONOMICS

## SHASHI KANT

*Faculty of Forestry, University of Toronto*
*33Willcocks Street, Toronto, Canada M5S 3B*
*Email: shashi.kant@utoronto.ca*

**Abstract.**   This Chapter synthesizes the contents of this volume, and provides an overview of a new paradigm of economics, to which I assign the term Post-Newtonian Economics. To put the synthesis in perspective, first some causes of the current status of Newtonian or neo-classical economics are discussed. Second, direct and indirect correspondences between the different concepts discussed in the thirteen Chapters, Chapters 2 to 14, of this volume and Kant's basic principles of the economics of sustainable forest management are established. Finally, the basic differences between Newtonian and Post-Newtonian economics are discussed.

## 1. INTRODUCTION

In the last Chapter (Kant, 2005) of the companion volume (Kant & Berry, 2005), I argue that mainstream economics—which has been termed Walrasian or neo-classical economics, and which I call Newtonian economics[1]—is a good example of "positive feedbacks" and "increasing returns due to information contagion,"[2] and path-dependence, "lock-in" due to small historical events, and inefficiencies that are some of the main characteristics of this stream of economics. I also argue, in that Chapter, that the inefficiencies are due to its "locked-in" position in *Chicago man*, which is convenient, successful, unnecessarily strong, but false (McFadden, 1999), and a single *(General) Equilibrium*, which is conceptually simple, analytically strong, but difficult, if not impossible, to exist. Here, I add some other causes, which are prevalent in Newtonian economics, for these inefficiencies—a narrow focus, over-dependence on markets, and the negligence of the role of institutions, other than market, in economically optimal resource allocation and decision-making mechanisms.[3]

*Kant and Berry (Eds.), Institutions, Sustainability, and Natural Resources: Institutions for Sustainable Forest Management, 341-356.*
© 2005 *Springer. Printed in Netherlands.*

Some differences between neo-classical and institutional economics are discussed in Chapter 1. However, the focus of neo-classical economics can be very well understood using the definitions of economics given in graduate and undergraduate text books, for example Eaton and Eaton's (1991) definition: "Economics is the study of the allocation of scarce resources to the production of alternative goods." Irrespective of the diverse terminology used by various authors, the focus of neoclassical economics is the allocation of scarce resource, which is too narrow compared to the complexity of economic systems, as discussed in the previous Chapters of this volume and in the companion volume. In addition, in neoclassical economics, resource scarcity is indicated or measured by market prices, but market signals are just one category of scarcity signals, relevant only for those resources that are traded in the market. History provides strong support for the claim that scarcity signals of environmental goods, such as biodiversity and clean air, come through environmental movements and not through the market. Similarly, scarcity signals of Aboriginal goods (values) are coming from courts or through demonstrations by Aboriginal groups (Kant & Lee, 2004). Hence, market signals of scarcity alone are insufficient for achieving the objective of the socially desirable distribution of scarce resources.[4] In fact, the emergence of the SFM paradigm itself is a proof of the limitations of the market and market signals, and a total dependence of SFM economics on markets will be self-defeating for SFM.

In addition, market-based approaches to SFM have several other fundamental problems (Kant & Lee, 2004). *First*, in the economic theory of revealed preference techniques, with a foundation based on market mechanisms, a preference is simply defined as the binary relation underlying consistent choice which may have some rationale for analysing the preferences of competitive consumers but not necessarily for the choices made by government agencies, consumers in an imperfect market (Sen 1982, p. 5), Aboriginal people, or environmental groups. *Second*, market-pricing mechanisms are based on preferences or choices restricted to alternatives belonging to a class of convex polyhedra, (i.e., budget triangles, in the two-commodity case) (Sen 1971) while there are no such restrictions on preferences or choices for merit goods, environmental goods, and/or goods not traded in the market. *Third*, many features of SFM are grounded in collectivism and generational equity. The selfish individual as an operator in a market is counter-intuitive to collectivism,[5] and an alternate view of the individual as a citizen—an agent who judges the alternatives from a social perspective which includes his own well-being but also quite possibly, many other considerations (Arrow 1951, pp. 17-18; Sen 1996)—would facilitate them pursuing the objectives and features of SFM. These limitations clearly demonstrate that market-based mechanisms are not the answers for many issues related to SFM, and the economics of SFM has to incorporate all possible institutional mechanisms.

The main objectives of this volume are to add an institutional perspective, theoretical as well as applied, to the economics of sustainable forest management and to provide a holistic perspective of the emerging concepts of the new economics. Direct and indirect correspondences between the different concepts discussed in the thirteen Chapters of this volume and Kant's (2003) basic principles of the economics of sustainable forest management are established in the next section. The basic differences between Newtonian and Post-Newtonian economics are discussed, and the similarities between Kant's basic principles and the main features of Post-Newtonian economics are highlighted in section 3.

## 2. THE BASIC PRINCIPLES OF THE ECONOMICS OF SFM

Kant (2003) argues that the basic idea behind SFM is to manage forests in such a way that the needs of the present are met without compromising the ability of future generations to meet their own needs, and economic models of SFM should be able to capture both orientations—individualistic as well as altruistic and/or commitment—of an individual's behavior. The incorporation of such behavior will be possible in economic models that are based on a "both-and" principle rather than an "either-or" principle. Under the umbrella of the "both-and" principle, Kant (2003) proposes four sub-principles of the economics of SFM: existence, relativity, uncertainty, and complementarity. Kant (2003) concludes that the two dominant requirements of the economics of SFM are a consumer choice theory different than the theory of *Chicago man* and the economics of multiple equilibria different than the economics of *General Equilibrium*. In this volume, the focus is on institutional aspects, hence the issues related to consumer choice theory are not addressed directly and in detail (please refer to the companion volume for details), but some chapters in this volume raise and discuss some fundamental issues related to consumer choice theory. In addition, many chapters of this volume provide strong evidence for the need for an economic theory of multiple equilibria resulting from market as well as non-market institutions, and many chapters confirm, directly or indirectly, the relevance of Kant's four sub-principles of the economics of sustainable forest management and emerging economic thinking.

### 2.1 Consumer Choice Theory

The second chapter of this volume, by Martin Luckert, even though focused on institutions for SFM, is similar, at least in its approach, to the second chapter, by Colander (2005), in the companion volume. Both chapters focus on complexity. Colander focuses on the complexity of economic systems in general while Luckert focuses on the complexity of institutions. However, institutions being the foundations of economic systems, the outcomes of the two chapters are similar. Luckert raises multiple questions about traditional consumer choice theory. He questions one-way, cause-and-effect relationships and proposes co-dependence between institutions and economic behavior:

> Connecting institutions to economic behavior in pursuit of social objectives may require further refinements in our understanding and characterization of institutions, our understanding of non-institutional determinants of behavior (...), a wider recognition of potential co-dependence (as opposed to cause and effect relationships) between institutions and economic behavior, more explicit recognition of transactions costs and belief systems.(Chapter 2, p.21)

> The complexity of institutions (...) is one key reason why economists have had a hard time predicting behavior under alternative institutional structures... Problems arise

because it is not clear how the complex incentives created by combination of rules, taken together, influence behavior. (Chapter 2, p.26)

Luckert attacks the foundation of the traditional consumer choice theory—optimality conditions—and by doing so he is indicating, at least indirectly, the need to move from the principle of maximisation to the principle of satisfaction:

> As economists, we frequently think about issues in terms of optimality, optimal levels of production, optimal time paths, optimal rotation etc. Given the levels of complexity that we witness, I question whether we are ready to apply the concept of optimality to issues regarding institutions for SFM. (Chapter 2, p.22)

Luckert also recognises the role of belief systems and social structures in economic behavior and the context-specificity and diversity of human behavior:

> Economists undertaking social choice analysis typically rely heavily on influences of assigned values on economic behavior. ...if we were to better understand institutions and organizations, we must seek to better understand the belief systems that underlay these social structures. Assigned values are only the tip of the iceberg in understanding a broader view of economic behavior implied in combining social choice and public choice perspectives. (Chapter 2, p. 33)

> We also find that these subsistence economies tend to be much more tied to the land than developed economies. Thus, the link between household production and the environment may be direct and visible. The closeness of economic activities to the land, and the localized nature of these economies may cause economic behavior to approach ecological behavior. (Chapter 2, p. 28)

Diaw, in Chapter 3, and Kant and Berry, in Chapter 4, also attack neo-classical economics approach to institutional change, which suggests that the bundles of private property rights will increase with the economic scarcity of the property, and that private property rights are essential for economic efficiency. The analyses and observations, in these two chapters, contribute directly to the theory of multiple equilibria, rather than consumer choice theory. However, both the chapters, at least implicitly, are arguing for social rationality or rationality of a social agent rather than the rationality of a traditional economic agent, termed a rational fool or social moron by Sen (1977).

Vatn, in Chapter 5, questions another cornerstone of neo-classical consumer choice theory—gross substitution—specifically with respect to environmental goods. He suggests that biodiversity, an environmental good, is a systems good, a common good, and an ethical good, and that the market is the wrong metaphor for issues where such goods are involved. He writes:

> In this sense, ecosystems are functionally opaque (). The exact and full contribution of a function or species in an ecosystem is not known, indeed is probably unknowable, until it ceases to function. Furthermore, it will then be very difficult to establish what has really happened. This is the essence of the perspective of complex systems and challenges the idea of substitution at its fundamentals. (Chapter 5, p. 118)

Similarly, Vatn examines the ethical issues associated with environmental goods, and questions the use of trade-off calculations, a tool of neo-classical consumer theory:

> This way the rights of nature, as each individual perceives it though, becomes part of the calculus. ... this reveals a serious misunderstanding of the character of moral

claims. Such claims have to go beyond *individual* evaluations, since ethics and morality are social phenomena. They belong to another category from those to which ordinary trade-off calculations are appropriate. Commodity preferences and norms are incommensurable entities (Chapter 5, p. 121).

In addition to these four chapters in the first section, Chichilnisky, in Chapter 7, introduces the concept of privately produced public goods, and suggests that efficiency in carbon emission trading market will require equity in the distribution of carbon emission rights. This may not seem a direct challenge to conventional consumer choice theory, but it definitely adds a new dimension to the efficiency of markets which was missing from neo-classical economics. Similarly, the observations of Binkley, in Chapter 6, about the establishment of markets for environmental services as "the great tragedy of science" indicate the weaknesses of a frictionless world of markets as conceptualized by neo-classical economists. Bentley and Guldin, in Chapter 12, also highlight the limitations of neo-classical economic theory when guiding normative decisions over long periods of time.

### 2.2 Economics of Multiple Equilibria

The convexities of production and utility (or consumption) functions, perfect markets, frictionless functioning of markets, absence of increasing returns and externalities, and no market failure due to uncertainties are essential ingredients of the economics of *Chicago man* and *General Equilibrium*. However, in real life, most of these essential ingredients are not available. In the companion volume (Kant & Berry, 2005), the authors provide multi-dimensional evidence for the absence of these ingredients and the presence of *Multiple Equilibria*. For example, Colander (2005) argues that sustainability literature fits into models with multiple equilibria. Similarly, post-Keynesian consumer choice theory for SFM by Lavoie (2005), behavioral economics and SFM by Knetsch (2005), and other chapters in the companion volume reject the notion of general equilibrium, and support the concept of multiple equilibria. In the companion volume, the sources of multiple equilibria are non-linearities and non-convexities in production, consumption, and management functions, lexicographic and context-dependent preferences, increasing returns, and externalities. In this volume, the main source of multiple equilibria is the incorporation of institutions, other than the market, in economic analysis and a different nature of markets for environmental goods, which are not private goods.

The concept of general equilibrium is an outcome of the Walrasian model, in which market adjustments are frictionless, market prices alone always suffice for all allocation problems, and there is no role for institutions other than the market. Hence, the recognition and incorporation of institutions, other than the market, will naturally lead to friction in market adjustment, and resource allocation will be a joint outcome of the market, other institutions, and interactions between the market and other institutions. In this scenario, the concept of general equilibrium becomes futile. Similarly, for non-market economies, where prices are eschewed, suppressed, or non-existent (Barzel 1989, p.99), the concept of general equilibrium is useless. All the chapters in this volume, either directly or indirectly, discuss the role of institutions, other than the market, in resource allocation for sustainable forest management. Some chapters also highlight the irrelevance of markets for the valuation of different attributes of forest resources, such as biodiversity. Hence, in summary, all the chapters in this volume are contributing towards the futility of the concept of general equilibrium, and the need for an economic theory of multiple equilibria. The essence of the role of non-market institutions in SFM and the need for an economics of multiple equilibria can be summarised in Diaw's words:

> There is massive evidence across Africa that rural lands, including forests, continue to be predominantly governed by indigenous tenure principles, mingled with state law and occasional private titling (). Since at least the 1960s, numerous researchers have reviewed the systems of rights that govern African land and forest tenure to find that, far from disappearing, these systems, already complex in pre-colonial times, had further evolved into multidimensional constructs of eco-niches and overlapping rights. This is consistent with observations made in other parts of the world (); it also comforts Shashi Kant's finding that property rights in tropical forest systems have evolved toward Pareto efficient pluralism rather than the singular private property optimum that economic theory had predicted. (Chapter 3, p. 44)

> This economic and legal pluralism and the underlying resilience of indigenous tenure institutions () are a formidable challenge to the theory of non-Western economic institutions....... African land and forest policies have been rooted in epistemologies of modern transformations that considered indigenous tenures and other forms of non-Western economic 'otherness' as doomed to be replaced by higher forms of modernity. That this did not happen as predicted should be a powerful incentive for revisiting the theoretical parameters under which the debate on Western and non-Western forms of economic organization was originally framed. (Chapter 3, p. 44)

With this discussion of the two broad features, I move to a discussion of the four sub-principles of the economics of SFM. Khan (2005), in the companion volume, put these principles in a broader and interdisciplinary context. Khan called these principles together an "ethics of theorizing" rightly observing "that these four sub-principles draw attention to the broader interdisciplinary framing that the subject demands, and emphasize, rather than a particular theory, the theoretical principles that go into its theorizing." In this volume, some authors recognise the importance of the concepts behind these principles to the economics of SFM explicitly, but without referring to the principles, and some authors implicitly use or refer to these principles to make their arguments.

## 2.3 The Principle of Existence

In Kant (2003), I emphasized the existing situations under the principle of existence, and that the word "situations" would require a broad interpretation including practices, models in operation, basins (in Colander's (2005) terminology), and norms. When I proposed this principle, I had a face-to-face communication between a forest manager and a forest economist in mind,[6] and my idea was to make a call for self re-examination by economists themselves of the so-called economically efficient models suggested by neo-classical economists[7]. Mitra (2005), in the companion volume and in his previous work, has proved that forest rotation based on maximum sustained yield, which is also known as a forester's rotation, is economically efficient from the perspective of inter-generational equity. In forestry literature, it is a common observation that for a regulated forest, Faustmann's rotation reduces to the rotation of maximum sustained yield for a zero rate of time preference. I was trying, therefore, to point out that there may be very good economic factors in existing situations and/or practices which do not fit in economic models based on *Chicago man* and *General equilibrium*, and that neo-classical economists might have ignored those economic factors for the sake of the mathematical convenience and elegance of their models. In a way, I was hinting at recent developments in emerging streams of economics, as confirmed by Colander (2005), Lavoie (2005), and Knetsch (2005) in the companion volume.

In this volume, Diaw (Chapter 3) and Kant and Berry (Chapter 4) clearly illustrate the importance of the principle of existence in terms of existing institutions, practices, and incremental dynamics. Diaw also provides some examples such as share-cropping and non-wage systems which may seem economically inefficient, similar to the forester's rotation, from the perspective of a narrow neo-classical economic vision, but they may not be inefficient from the perspective of a broader economic vision, which includes all (market and non-market) costs (including transactions) and returns. As Diaw observes:

> The "productive inefficiency" of sharecropping was a direct result of the theoretical postulates of marginalism. Assuming mutual equality and equal shares, neither owner nor tenant would invest its resources beyond the point where the marginal product equals half (and not all) of the product. This was a theoretical impossibility. But this result could not stand by itself in light of the need to explain the continued existence of this system. (Chapter 3, p. 48)

Kant and Berry, in Chapter 4, elaborate the critical role of institutional and organizational inertia in the dynamics of forest regimes which means that there are very high transaction costs associated with institutional change from their existing positions. In many situations, these transaction costs will make the proposed change economically inefficient, if transaction costs are included in the economic analysis. On other hand, if the proposed changes are pursued on the basis of narrowly defined economic efficiency, neglecting transactions and other non-market costs, the proposed changes may prove too costly to society.

The words of two Nobel Laureates, Stigler and Becker (1977), quoted by Luckert in Chapter 2, reaffirm the importance of the principle of existence: "de Gustibus non est Disputandum". Loosely translated, tastes are indisputable. Tastes are not right or wrong, they just are. Luckert's observation, "If institutions are reflections of held values, as discussed above, is it possible to think of them in terms of optimality?", furthers strengthens the role of the principle of existence in the economics of sustainable forest management.

## 2.4 The Principle of Relativity

The principle of relativity, as per Kant (2003), suggests that an optimal solution is not an absolute but rather a relative concept. Khan (2005), in the companion volume, draws parallels between the principle of relativity and Wittgenstein's binary of absolute and relative, and the appropriation of Wittgenstein's binary by Keynes to distinguish between absolute and relative needs. Khan rightly observes that "there is an important overlap, a common orientation if one prefers, between Kant's principle of existence and his principle of relativity." These two principles require a simultaneous reading. However, the principle of relativity should be read not only for making a distinction between the absolute and relative, but the broadest interpretation of "relative" will also be a part of this rule. In this sense, the principle of relativity, as per my reading of emerging streams of economics, seems embedded in Colander's (2005) new holy trinity of purposeful behavior, enlightened self-interest, and sustainability. Colander's discussion of different basins of attractions in complex systems and sustainability as a means of keeping within the existing basin of attraction, and not going to another basin that is considered less desirable, is also, at least implicitly, an indication of the principle of existence and the principle of relativity. Similarly, many principles of Post-Keynesian consumer choice theory discussed by Lavoie (2005), and various features of behavioral economics discussed by Knetsch (2005), confirm the relevance and importance of the principle of relativity to the economics of sustainable forest management.

In the economic framework that recognises and integrates the role of other institutions with the market, the relevance and importance of the principle of relativity seem critical, and this view is supported, either directly or indirectly, by almost every Chapter in this volume. In Chapter 2, Luckert discusses the differences between developing and developed countries, and highlights the possibility of different institutional arrangements achieving the desired goals of sustainable forest management. In Chapter 3, Diaw emphasises the critical role of different institutions, including indigenous tenure systems, state law, and private titling, and the positive outcomes of these institutions in different situations and contexts. In a way, legal and economic pluralism is a direct indication of the principle of relativity. Vatn's suggestion, in Chapter 5, to use value-articulating institutions, instead of the market, for valuation of biodiversity also, at least indirectly, hints at the principle of relativity. Similarly, the recognition of political transaction costs associated with the potential markets for environmental services by Binkley, in Chapter 6, and the requirement of equity for efficiency of privately produced public goods discussed by Chichilnisky, in Chapter 7, provide direct support to the principle of relativity. Hartwick's contrast, in Chapter 8, between countries that are geographically isolated and have a small resource stock, such as Easter Island and countries like Europe and cities in China, is an example of the principle of relativity. Similarly, Hyde's observation (Chapter 9) about the simultaneous existence of three forest situations in different parts of the same country and Sedjo's proposal (Chapter 10) for the three models for sustainable forest management are other examples in support of the principle of relativity.

## 2.5 The Principle of Uncertainty

The principle of uncertainty suggests that due to uncertainties in social and natural systems, an individual may never be able to maximize his outcomes, and will always search for positive outcomes. The complexity story, discussed by Colander (2005) in the companion volume, supports this principle. The focus of Colander's discussion is on the complexity of systems, and I believe, based on my reading of his text, that he has assumed uncertainty as an inherent property of complex systems. Similar to Colander's complexity story, the complexity of institutions, discussed by Luckert in Chapter 2, supports this principle, and I believe that Luckert has also assumed uncertainty as an inherent property of the complexity of institutions. The following text from Chapter 2 indicates the importance of positive outcomes and irrelevance of the idea of maximization:

> To speak of specifying an optimum institution may be at best arrogant, and at worst foolish. Instead, I would suggest that we portray our accomplishments and contributions as trying to understand processes and results that may lead to socially desirable SFM policies. However, this observation should not keep us from continuing our search for better forestry institutions. ... However, the "optimality" of these policies is not for us to determine. Rather, the policies and their impacts are for us to describe, while politicians and society will determine what is optimal by deciding whether they are improvements over what we currently have. (Chapter 2, p. 36)

Similarly, the following observations of Luckert about adaptive management and adaptive efficiency indicate the importance of the principle of uncertainty for the economics of SFM:

> As economists, we might also seek to mimic strategies that many other disciplines are adopting in pursuit of Sustainable Forest Management. That is, we may consider using concepts of adaptive management, commonly used in addressing natural science complexities of SFM ( ), for addressing complexities of institutions. Although politicians and publics may be hesitant to accept institutional experiments, they have also been historically hesitant of landscape level natural science experiments that ecologists are beginning to conduct. Just as attitudes have begun to change regarding large natural experiments, they may also be changing regarding the need for institutional experiments. North (1993) sums it up well: "there is no greater challenge facing today's social scientist than the development of a dynamic theory of social change that will give us an understanding of adaptive efficiency. (Chapter 2, pp. 37)

In the companion volume, the principle of uncertainty has received explicit recognition in the chapter on post-Keynesian consumer choice theory by Lavoie (2005). In addition, behavioral economics, evolutionary economics, and ecological economics have also recognised this principle. In this volume, Vatn (Chapter 5) and Bentley and Guldin (Chapter 13) recognise this principle explicitly. Vatn writes:

> The challenge is thus that it is impossible to predict what will happen if changes which appear are either too large or too frequent – i.e., beyond levels not earlier repetitively observed. They may change system performance in an essential way – i.e., an attractor shift is observed. We are, however, unable to predict where and when such a shift may happen. The system is characterized by radical uncertainty (). (Chapter 5, p. 118)

Similarly, to Vatn, Bentley and Guldin observe:

> The fundamental problem is that risk or probabilistic models compound with time. Over a few economic and ecological events, the models "explode," thus becoming useless for making predictions that can guide decisions. The old-growth redwood model, for example, was of little use in predicting real timber prices more than a year or two beyond the 1953-1980 data series. (Chapter 13, p. 299)

One other dimension of the principle of uncertainty is human behavior, which I did not mention in Kant (2003), and none of the Chapters either in this volume or in the companion volume has considered this. As it is discussed in the principle of complementarity, every individual is selfish as well as altruistic; the same individual may behave selfishly or altruistically in the same circumstances but at different periods. For example, an individual's behavior with respect to his kids and spouse may vary from one end of selfishness to the other end of altruism at different periods of time, holding all other things constant. Hence, the incorporation of uncertainty in human behavior in economic models is another challenge to future economists, and may require use of some of the tools of quantum physics.

## 2.6 The Principle of Complementarity

The principle of complementarity, as per Kant (2003), suggests that human behavior may be selfish as well as altruistic, that people can have economic values as well as moral values, and that people need forests to satisfy their lower level needs as well as higher level needs. Khan (2005) locates these binaries—economic/moral, lower/higher, and selfish/altruistic—in the work of Wittgenstein and Keynes, and adds two additional aspects. The principle of complementarity is fundamental to many emerging streams of economics: agent-based modelling, complexity theory, post-Keynesian consumer choice theory, and behavioral economics. For example, in agent-based modelling, every agent is not a *Chicago man*, and agents can be selfish as well as altruistic. Many principles of post-Keynesian consumer theory confirm the principle of complementarity. The recognition of the existence and the importance of lexicographic preferences require the acceptance of the principle of complementarity. Similarly, the recognition of the simultaneous roles of the market and other institutions, which is the main theme of this volume, provides the strongest support to the principle of complementarity. The simultaneous existence of formal and informal institutions, the need of complementarity between these two main categories of institutions, the embedded nature of tenures, and many other similar features of institutional structure also support the principle of complementarity.

I would close this section with Luckert's conclusion about a cross-disciplinary approach:

> As economists have grappled with concepts of optimal institutions, our tradition in theory and mathematics has caused us to abstract away many of the complexities that we face. These tools have allowed us to make unique and effective contributions in terms of understanding incentives and resulting behavior that institutions create. However, as we proceed, we must constantly weigh the benefits and costs of reductionist modeling relative to the problem we are trying to address. I would suggest, that in addition to pursuing our reductionist approaches, we expand our efforts cross-disciplinarily. In essence I am advocating that we expand our precise reductionism to include more holistic thinking. While as a group, economists tend to abhor the waffling that may be associated with such imprecision, without this balance, we may suffer from being precisely wrong, or precisely irrelevant. (Chapter 2, p. 38)

There is no doubt that the subject of economics, and for that matter all the social sciences, demands an integrative and holistic approach. All social sciences deal with humans, and the compartmentalized approach to social sciences, which tries to divide a living human being into different components that have no connections and interactions with each other, is an approach which may be possible only with a dead body and not with a living being. A cross-disciplinary approach is not only necessary, it is essential. However, in the economics profession, the first step may be to take an inter-stream (or intra-subject) approach because, in many situations, economists from one stream do not know or recognise what is going on in other streams of economics. In addition, neo-classical or Newtonian economists do not even want to acknowledge the developments which are going on in the emerging streams of economics.

## 3. POST-NEWTONIAN ECONOMICS

*Chicago man*, as McFadden (1999) observed, has become an endangered species; behavioral economics has severely restricted his maximum range, and he is not safe even in markets for concrete goods which was his prime habitat. McFadden (1999) issued a call to evolve *Chicago man* in the direction of *Kahneman-Tversky (K-T) man* by adopting those features needed to correct the most glaring deficiencies of *Chicago man*, and to modify economic analysis accordingly. Thaler (2000) predicted that *homo economicus* will evolve into *homo sapiens* who will have characteristics of less IQ, slow learning, heterogeneity, human cognition, and more emotions. Colander (2000a) declared the death of the term neo-classical economics and the birth of the new millennium economics.[8] Ormerod (2000) expected the re-birth of economics in the 21st century to give us a much better understanding of the world. Brian Arthur realised the need for a new paradigm of economics long ago, when he started working on the economics of increasing returns, and identified many differences between the standard approach and the complexity approach to economics (Colander, 2000b). In addition to these specific calls, all the emerging streams of economics, such as behavioral economics, complexity theory, evolutionary economics, evolutionary game theory, experimental economics, and the economics of increasing returns, have been contributing to the emergence of a new paradigm of economics which I have termed Post-Newtonian economics. In these calls and contributions to the new stream of economics, institutional aspects have not been able to attract the same level of attention as the *Chicago man*. I do not intend, by this statement, to decline the developments in institutional economics and their recognition in recent years, but I want to place a special focus on the integration of institutional aspects in the emerging paradigm of economics—Post-Newtonian economics.

The new paradigm will be fundamentally different from the Newtonian economics, as summarized in Table 15.1. The relevance of Kant's four sub-principles to the main features of Post-Newtonian economics is quite clear, but I leave a specific discussion on this issue for some future paper.

*Table 15.1.* Main Differences between Newtonian and Post-Newtonian Economics

| Feature | Newtonian Economics | Post-Newtonian Economics |
|---|---|---|
| **Holy Trinity** | Rationality, greed, and equilibrium | Purposeful behavior, enlightened self interest, and sustainability (non-equilibrium and multiple equilibrium) |
| **Agent** | *Chicago man* and rational fool (homo economicus), and homogeneous agents | *K—T man* and social agent (homo-sapiens), and agents are heterogeneous as well as versatile |
| **Rationality and information** | Mathematical or constructivist rationality and full information | Procedural and/or ecological rationality, and incomplete information |
| **Preferences** | Exogenous (as imposed by economists), self-regarding, and fixed preferences | Endogenous, reference-dependent, self as well as other-regarding and/or social preferences |
| **Needs and wants** | No difference between needs and wants | Difference between needs and wants, satiable needs, hierarchy of needs, and growth of needs |
| **Learning and emotions** | No learning and no emotions | Learning from others, frequency-dependent learning, and emotions may produce a behavioral response |
| **Actions of others and social interactions** | Market clearing prices and contractual exchanges | Agents interactions through market and non-market mechanisms, non-contractual social obligations |
| **Utility** | Scalar utility, expected utility theory | Vector utility, prospect theory and libertarian paternalism |
| **Uncertainty** | Risk | True or Keynesian uncertainty |
| **Elements** | Quantity and Prices | Patterns and Possibilities |
| **Principle** | Maximizing | Satisfying |
| **Modeling** | Modeling of Decision Outcome | Modeling of Decision (cognitive) Process |
| **Returns to Scale** | Constant and decreasing returns to scale | Constant, decreasing, and increasing returns to scale as well generalized increasing returns |
| **Feedbacks** | Negative | Negative as well as positive feedback, lock-in, path dependence, inefficiencies |

*Table 15.1 (cont.)*

*Table 15.1 (cont.)*

| Feature | Newtonian Economics | Post-Newtonian Economics |
|---|---|---|
| **Institutions** | Dominantly market, and no friction in market mechanisms Agent's behavior is independent of institutions Either no institutions, or formal institutions, are represented by a budget constraint , no role of informal institutions, institutions are treated as contractual interactions, institutions do not change | Market as well non-market institutions, and market mechanisms full of friction Two-way interactions between institutions and agents Outcomes are dependent on institutional setting, institutions are not freely available; role of formal as well as informal institutions, institutions are contractual as well as non-contractual social interactions, and institutions evolve over time |
| **Time, age, and generations** | Positive Discounting, No role of age and generations | Zero, positive, and negative discounting, individuals can age, generational turnover becomes central, age structure of population change, and generations carry their experiences |
| **Equilibrium** | General equilibrium | Multiple equilibria and non-equilibrium |
| **Society** | Aggregation of homogenous agents | Heterogeneous agents, similar populations may have different norms, tastes, and customs, resulting in local homogeneity and global heterogeneity |
| **Solutions** | Closed form solutions | Simple closed form solutions are not necessary; indeed, any solutions that are susceptible to simple interpretations may not exist |
| **Subject** | Structurally simple, deterministic, stable | Structurally complex, structures are constantly coalescing, decaying, and evolving. All this is due to externalities leading to jerky motions, increasing returns, transaction costs, and structural exclusions |

*Table 15.1 (cont.)*

*Table 15.1 (cont.)*

| Feature | Newtonian Economics | Post-Newtonian Economics |
|---|---|---|
| Approach | Tool driven | Problem and issue driven |
| Foundation | Non-cooperation | Cooperation |
| Basis | Newtonian Physics | Quantum Physics and Evolutionary Biology |
| Nirvana | Possible if there are no externalities and all had equal abilities | Not possible, externalities and inequalities are driving forces, systems constantly unfolding |
| Sustainability | Sustainability of neo-classical economics and neo-classical economists | Sustainability of society |

*Note:* This table is similar to a table by Kant (2005) in the companion volume.

## 4. CONCLUSIONS

The sustainability of global systems (social as well as natural) is a prerequisite for the existence of economics as well as economists. Hence, the goal of a new paradigm of economics should be "the sustainability of global systems" as opposed to the goal of "sustainability of Newtonian (neo-classical) economics and economists." The main elements of the emerging paradigm of economics, post-Newtonian economics, seem focused on the sustainability of global systems. However, a higher-level integration of all the emerging streams of economics and collective action by economists associated with these streams are necessary ingredients for the structural specifications, establishment, and growth of Post-Newtonian economics. In the companion volume, we have tried a partial integration of some emerging streams of economics, specifically behavioral economics, complexity theory, post-Keynesian economics, and social choice theory, with respect to the economics of SFM. In this volume, we have tried a partial integration of some themes of old and new institutional economics with respect to the economics of SFM. A comprehensive development of post-Newtonian economics will require many such efforts at different levels, and the recognition of such efforts. I believe that the establishment of an International Association of Post-Newtonian Economics, supported by regional chapters, may be a step in the right direction.

## NOTES

[1] Economists may debate the appropriateness or non-appropriateness of these terms, but I am sure that every economist understands which stream of economics is being addressed here. Similar to Hamilton (1970), I prefer the term Newtonian Economics because I believe that the concept of equilibrium came from Newtonian physics.

[2] Generally, increasing returns are associated with economies of scale in production, but the term refers more broadly to any situation in which the payoff for taking an action is increasing with the number of people taking the same action. Bowles (2004, p.12) termed it "generalized increasing returns". Arthur and

Lane (1993) also identified "information feedbacks" or what they called "information contagion" as a source of positive feedback and increasing returns.

[3] At a higher scale, all three causes can be attributed to either the *Chicago man* or the *General Equilibrium.*

[4] The sufficiency of market prices as an indicator of scarcity is doubtful even for market goods such as timber.

[5] The assumption of an individual welfare maximizing agent is a rationality condition that may not be unrealistic for some choices, but there is enough evidence that all choices in economic matters do not fall in this category. In addition, the interdependence between different people's welfare may make the pursuit of individual interests produce inferior results for all, in terms of those very interests (Sen, 1973). In fact, every member of the group might be better off with a norm involving systematic deviation from individual-welfare maximization (Sen, 1974).

[6] In one international conference, after a well-established forest economist finished his presentation about the optimal forest rotation, one practicing forester asked him about the optimal rotation age at which he should harvest a particular type of forest. The forest economist's answer was that it will depend upon so and so. The forester again asked, tell me the age at which I should harvest, and the forest economist did not have any answer. This incident forced me to think about the economic optimality of Faustmann's rotation.

[7] In this regard, Smith's (1985) observation is very useful: "The early polling of economists on Allais, Ellsberg, Second price, and other such 'paradoxes' makes it clear that economists will get it 'wrong' about as often as the sophomore subject until he or she has had considerable time to think and analyze. Incidentally, this observation provides an answer for that somewhat mythical business-man who asks, 'If you' re so smart why ain't you rich?'. My classmate, Otto Eckstein, didn't get rich by equating price to marginal cost."

[8] Colander (2000) also discussed the non-appropriateness of many other terms such as "new Classical", "mathematical economics", and "the era of modeling" to describe the recent development in economics.

## REFERENCES

Arrow, K.J. (1951). *Social choice and individual values.* New York: Wiley.

Arthur, W.B. (1994). *Increasing returns and path dependence in the economy.* Ann Arbor: University of Michigan Press.

Arthur, W.B., & Lane, D.A. (1994). Information contagion. In W.B. Arthur. *Increasing returns and path dependence in the economy,* (pp. 69-97). Ann Arbor: University of Michigan Press.

Auyang, S.Y. (1998). Foundations of complex systems theories in economics, evolutionary biology, and statistical physics. Cambridge: Cambridge University Press.

Barzel, Y. (1989). *Economic analysis of property rights.* Cambridge, New York: Cambridge University Press.

Bowles, S. (2004). *Microeconomics: Behavior, institutions, and evolution.* New York: Russel Sage Foundation and Princeton: Princeton University Press.

Colander, D. (2000a). The death of neoclassical economics. *Journal of the History of Economic Thought,* 22(2), 127-143.

Colander, D. (2000b). Introduction. In D. Colander (Ed.), *The complexity vision and the teaching of economics* (pp.1-28). Cheltenham: Edward Elgar.

Colander, D. (2005). Complexity, muddling through, and sustainable forest management. In S. Kant, & R.A. Berry. (eds.). *Economics, Sustainability, and Natural Resources: Economics of Sustainable Forest Management (pp.23-37 ).* Amsterdam: Springer.

Camerer, C.F., Loewenstein, G., & Rabin, M. (2004). *Advances in behavioral economics.* New York: Russell Sage Foundation , and Princeton: Princeton University Press.

Eaton, B.C., & Eaton, D.E. (1991). *Microeconomics* (Second Edition). New York: W. H. Freeman and Company.

Friedman, D. (2004). *Economics lab: an intensive course in experimental economics.* London: Routledge.

Hamilton, D. (1970). *Evolutionary economics.* Albuquerque: University of New Mexico Press.

Kant, S. (2003). Extending the boundaries of forest economics. *Journal of Forest Policy and Economics,* 5, 39-58.

Kant, S. (2005). Post-Newtonian economics and sustainable forest management. In S. Kant, & R.A. Berry. (eds.) *Economics, Sustainability, and Natural Resources: Economics of Sustainable Forest Management (pp.253-268 )*. Amsterdam: Springer.

Kant, S., & Lee, S. (2004). A social choice approach to sustainable forest management: an analysis of multiple forest values in Northwestern Ontario. *Forest Policy and Economics*, 6(3-4), 215-228.

Kant, S., & Berry, R.A. (eds.) (2005). Economics, Sustainability, and Natural Resources: Economics of Sustainable Forest Management. Amsterdam: Springer.

Lavoie, M. (2005). Post-Keynesian consumer choice theory for the economics of sustainable forest management. In S. Kant, & R. A. Berry. (eds.) *Economics, Sustainability, and Natural Resources: Economics of Sustainable Forest Management (pp.67-90)*. Amsterdam: Springer.

Kahneman, D. (2000). *Choices, values, and frames*. Cambridge: Cambridge University Press.

Khan, M.A. (2005). Inter-temporal ethics, modern capital theory and the economics of sustainable forest management. In S. Kant, & R.A. Berry. (eds.) Economics, Sustainability, and Natural Resources: Economics of Sustainable Forest Management (pp.39-66 ). Amsterdam: Springer.

Knetsch, J. (2005). Behavioral economics and sustainable forest management. In S. Kant, & R. A. Berry. (eds.) *Economics, Sustainability, and Natural Resources: Economics of Sustainable Forest Management (pp.91-103)*. Amsterdam: Springer.

McFadden, D. (1999). Rationality for economists? Journal of Risk and Uncertainty, 19(1-3), 73-105.

Ormerod, P. 2000. *Death of economics revisited.* Keynote address given to the Association of Heterodox Economists, June 29, 2000. Retrieved September 2, 2004, from www.volterra.co.uk/ Docs/dofer.pdf.

Schmid, A. A. (2004). Conflict and cooperation: institutional and behavioral economics. Malden: Blackwell Pub.

Sen, A.K.(1971). Choice functions and revealed preferences. *Review of Economic Studies,* 38, 307-317.

Sen, A.K. (1973). Behavior and concept of preferences. *Economica,* 40, 241-59.

Sen, A.K. (1974). Choice, ordering, and morality. In S. Korner (ed.). *Practical Reason.* Blackwell, Oxford.

Sen, A.K. (1977). Rational fools: A critique of the behavioral foundations of economic theory. *Philosophy and Public Affairs*, 6, 317-344.

Sen, A.K (1982). *Choice, welfare, and measurement.* Harvard University Press, Cambridge.

Sen, A.K. (1996). Environmental evaluation and social choice. *Japanese Economic Review,* 46(1): 23-37.

Smith, V.L. (1985). Experimental economics: reply. *The American Economic Review*, 75(1), 265-272.

Smith, V.L. (2000). *Bargaining and market behavior: essays in experimental economics.* Cambridge: Cambridge University Press.

Stigler, G.J. & Becker, G.S. (1977). De Gustibus non est Disputandem. *American Economic Review*, 67(2), 76-90.

Thaler, R. (2000). From homo economicus to homo sapiens. *The Journal of Economic Perspective,* 14(1), 133-141.

# INDEX